KU-370-893

About Island Press

Island Press is the only nonprofit organization in the United States whose principal purpose is the publication of books on environmental issues and natural resource management. We provide solutions-oriented information to professionals, public officials, business and community leaders, and concerned citizens who are shaping responses to environmental problems.

In 2001, Island Press celebrates its seventeenth anniversary as the leading provider of timely and practical books that take a multidisciplinary approach to critical environmental concerns. Our growing list of titles reflects our commitment to bringing the best of an expanding body of literature to the environmental community throughout North America and the world.

Support for Island Press is provided by The Bullitt Foundation, The Mary Flagler Cary Charitable Trust, The Nathan Cummings Foundation, Geraldine R. Dodge Foundation, Doris Duke Charitable Foundation, The Charles Engelhard Foundation, The Ford Foundation, The George Gund Foundation, The Vira I. Heinz Endowment, The William and Flora Hewlett Foundation, W. Alton Jones Foundation, The John D. and Catherine T. MacArthur Foundation, The Andrew W. Mellon Foundation, The Charles Stewart Mott Foundation, The Curtis and Edith Munson Foundation, National Fish and Wildlife Foundation, The New-Land Foundation, Oak Foundation, The Overbrook Foundation, The David and Lucile Packard Foundation, The Pew Charitable Trusts, Rockefeller Brothers Fund, The Winslow Foundation, and other generous donors.

About The Shire

The Shire: John Yeon Preserve for Landscape Studies is a center of the School of Architecture and Allied Arts at the University of Oregon. It comprises seventy-five acres on the Columbia River within the Columbia River Gorge National Scenic Area. The John Yeon Preserve for Landscape Studies was founded in 1995 by the John Yeon Trust with the gift of The Shire and its endowment to the University of Oregon. The property is both a natural and a designed landscape preserved and envisioned by John Yeon (1910–1994), a Pacific Northwest architect, designer, and conservationist. Activities at The Shire support research and education that advance an understanding and respect for the land and for design inspired by place. The first Shire Conference was held in 1998. The Shire Conference provides an academic and professional meeting ground for professional designers, educators, and scholars to explore critical issues of landscape architecture, preservation, education, and design plus the companion issues surrounding our relationship to landscape, land use, and environmental planning.

Ecology and Design

Ecology and Design
Frameworks for Learning

EDITED BY
Bart R. Johnson and
Kristina Hill

ISLAND PRESS
Washington • Covelo • London

Copyright © 2002 Island Press

All rights reserved under International and Pan-American Copyright Conventions. No part of this book may be reproduced in any form or by any means without permission in writing from the publisher: Island Press, 1718 Connecticut Avenue, N.W., Suite 300, Washington, DC 20009.

ISLAND PRESS is a trademark of The Center for Resource Economics.

Library of Congress Cataloging-in-Publication Data
    Ecology and design : frameworks for learning / edited by Bart R. Johnson, Kristina Hill.
        p.   cm.
    Includes bibliographical references.
    ISBN 1-55963-813-3 (pbk. : alk. paper)
    1. Ecological landscape design. I. Johnson, Bart. II. Hill, Kristina.
    SB472.45 .E39   2001
    712'.2—dc21   2001006880

British Cataloguing-in-Publication Data available.

Printed on recycled, acid-free paper

Manufactured in the United States of America
10 9 8 7 6 5 4 3 2 1

This book is the first volume of The Shire Papers, developed from The Shire Conference of 1998, in cooperation between Island Press and the University of Oregon.

We dedicate this book to She-Who-Watches.
May her example help us learn patience and
wisdom in how we treat the land.

Tsagaglalal, "She-Who-Watches,"
is a petroglyph design that "is only found on the
lower Columbia river area and has been securely dated to the
Historic period between A.D. 1700 and A.D. 1840. Legend
tells how Coyote placed Tsagaglalal on the rock to watch over
her people" (Rock Art Research Education 1989).

Contents

Foreword

The editors of this volume propose a "deep reconceptualization" of landscape design, planning, and land management oriented around standards of "ecosystem health, biotic integrity, and cultural well-being." Their purpose is to explore the implications of that change for the education of landscape designers and the design professions in general. Because of resistance from one established interest or another, intellectual transformations of this kind are very rare. The changes proposed in this volume, accordingly, are likely to meet resistance within the design and planning fields not because they don't make sense, but because they run against the grain of academic fragmentation and our cultural proclivity to conquer by dividing. An education that equips a new generation of practitioners to see landscapes realistically is one that will equip them to see new and better possibilities for places and people. The challenge and excitement revealed in the pages that follow lie in the possibility that we might tap the professional skills and moral energy of a new breed of ecologically based landscape designers to heal, restore, and make whole. What does this mean for professional education and for the design professions?

It means, first, that education must equip such professionals with the capacity to work effectively both within the academy and beyond. Despite the rapid increase in numbers of journals, books, conferences, and advances in the technology of geographic information systems, society has continued to sprawl, pave, and pollute. Few educators have taken seriously the need to equip a new generation for the kind of thoroughgoing transformation proposed here. We will need a corps of ecologically grounded professionals with wide-angle vision willing to risk a great deal, including their professional standing. Some of these, hopefully soon, must emerge as "public intellectuals" able to convey complex ideas of land to a wider audience now largely removed physically and emotionally from the landscapes in which they live. In other words, to be effective, the "landscape realism" envisioned in this volume (Chapter 1, Chapter 18) must be embedded in a larger realism that enables a new generation of designers to deal creatively with the many interests indifferent to or threatened by the ideas of ecosystem health, biotic integrity, and cultural health. But here, too, are new and creative possibilities for rethinking the economic and political factors that affect landscapes.

It is unlikely, however, that the educational changes implied by landscape realism can flourish without wider changes in the priorities of educational institutions. In other words, there will be no revolution of the sort imagined in these pages without a larger idea of education and the obligations that attend professional training. For the transformation proposed here to take root and flourish, the academy itself must be transformed into a more effective agent of ecological and cultural health, and no amount of "tweaking around the margins" here will do either. But this is not the direction in which education is heading. Instead, the academy is being reshaped to fit corporate interests that have little understanding of land beyond its cash value. Not surprisingly, a majority of students faithfully emulating the larger society now have more interest in making money than in developing a moral worldview. Educators must come to grips with the fact that they have been complicit in the larger problems of land and land management. A good place to begin the institutional transformation is to harness the talents and energies of faculty and students to redesign their own campuses so that one day they are climatically neutral, discharge no waste, enhance biological diversity, and support the emergence of locally sustainable economies. This means converting the university from just a place where education happens to one that educates ecologically. The campus, in its entirety, is a means to a larger end of improving how we think about land. No student ought to leave twelve or sixteen years of education oblivious and unfeeling toward the land community. Can design educators take a lead in such collaborative efforts to make their proposals come alive in the settings in which they work?

Second, we will need a larger idea of the land and our place in it—something like the "science of land" that Aldo Leopold once proposed—an idea big enough to embrace farmland, wilderness, urban areas, and that everywhere and nowhere zone called suburbia. That larger vision of land must include the entire biotic community, it must protect evolutionary processes, and it must work for people as the foundation for a fair and durable prosperity. We need as well a larger idea of design in which human intentions are informed by ecological realism and disciplined by a competent affection for particular places. How will these ideas be manifested in practice? The best example I can offer is that given by Jaquetta Hawkes, who once described the evolution of human life on the land in preindustrial Britain as a "creative, patient and increasingly skillful love-making" (1951, p. 202).

Third, ideas alone will find no fertile ground unless accompanied by a revolution in public attitudes, but few people are paying attention to fundamental things implied by landscape realism. That larger transformation will require educating the public to understand land issues and to respect and care deeply about the land—a marriage of competence and affection. This would be considerably easier if many or most adults had as children "soaked in" a place, as Paul Shepard once put it. All too often, however, the young, in Thomas Berry's words, are initiated into "an economic order based on exploitation of the natural life systems of the planet" (1999, p. 15). The result is a kind of "soul deprivation that diminishes the deepest of their human experiences" (p. 82). Deprivation of this sort also means that there is no emotional peg in the mind on which the adult can later hang the important facts about the land and land health. People are not inclined to care much about what they have not first come to love, and the romance with land must begin early in the small and safe places of childhood. For landscape professionals the challenge of creating a new public aesthetic is, in part, a practical one of helping to design and create the kinds of places and communities that instruct by exhibiting the elements of landscape realism.

Finally, after all of the reconceptualizing and educating is done, it remains to transform the larger political and economic forces that will impede the advance of landscape realism. The people who strip-mine, clear-cut, pave, and poison the land are hard at work and they don't give a damn about deep reconceptualizations or any other ideas beyond those accruing to their pecuniary advantage. They will lobby, advertise, and spend millions to preserve their ability to do what they've done all along. They are supported by a judicial system that perversely defines the corporation as a person while failing to grant legal standing to land, ecosystems, or future generations. It would be difficult to imagine a more mischievous way to organize human affairs. The upshot is that landscape realists must reckon with issues having to do with who owns how much land and by what terms and tenure, and must help define and elevate a public dialogue surrounding issues at the confluence of ecology, politics, ethics, and economics. Too often that dialogue has bogged down in the dreary defense of exaggerated ownership rights stripped of obligations. Someday we may arrive at a more sane and sustainable relation with the land in which land ownership is widely distributed and regarded as a sacred trust (Freyfogle 1998).

Challenges and obstacles notwithstanding, a revolution in human attitudes toward land and nature has already begun to gather considerable momentum. Looking back from a century or more hence, that revolution will appear as a kind of ecological enlightenment beginning with the likes of Henry David Thoreau, George Perkins Marsh, Frederick Law Olmsted, Jens Jensen, Rachel Carson, Aldo Leopold, John Lyle, and Ian McHarg. It is now a global movement. And like that of the eighteenth century, it will be incompletely realized and full of surprises and paradox. But it will fundamentally change how humans engage the land. Landscape realism—taking nature on its terms, not ours—is the bedrock upon which that enlightenment must proceed. It will require seeing nature and ourselves as co-evolving partners, not as servant and master. It is not too much to say that the fate of humankind hangs on this simple but profound change in perception and behavior. This collection of essays and the work of the scholars represented here are important as a landmark in that transformation.

DAVID W. ORR
Chair, Environmental Studies Program, Oberlin College

Citations

Berry, T. 1999. The great work. Bell Tower, New York, New York, USA.

Freyfogle, E. 1998. Bounded people, boundless lands. Island Press, Washington, D.C., USA.

Hawkes, J. 1951. A land. Random House, New York, New York, USA.

Preface

Along the grand Columbia River, which joins together Washington and Oregon, there is a place of remarkable beauty within the majestic landscape of the Columbia Gorge and the Columbia River Basin. This place—The Shire—75 acres of designed and wild landscape, was a gift in 1995 to the School of Architecture and Allied Arts, University of Oregon, by the trustees of the estate of John Yeon. John Yeon (1910–1994) was an architect, designer, environmentalist, visionary, and fierce advocate for beauty in our everyday world. Throughout his life, he held fast to his beliefs of what was right for architecture, the decorative arts, and protection of our landscape heritage. These ideals lead him to save, protect, and enhance The Shire, through enlightened stewardship and elegant design.

As a member of a long-standing Oregon family, Yeon understood well both the natural and cultural values of the Gorge and its importance to both Washington and Oregon and to the people who live there. He understood that this powerful, often unforgiving, landscape was of national importance. He knew that the Columbia Gorge was a landscape to be savored, revered, treasured—and vigilantly protected and guarded.

The breadth of John Yeon's vision cut across disciplines. His love for the land, and for design attuned to the spirit of places, has left a legacy that reaches from the protection of the Columbia Gorge and the establishment of Olympic National Park to the development of a regional architecture. Yeon himself designed a number of critically acclaimed and style-setting houses in Portland and elsewhere in the West. The Watzek House (1937), his best-known design, was recognized in an exhibit of important architecture by the Museum of Modern Art in 1939, and again in 1944 as part of the *Built in the USA, 1932–1944* publication. The Watzek House, too, was gifted to the University of Oregon by Richard L. Brown, one of the trustees of the Yeon Trust.

John Yeon purchased The Shire in 1965. His intent was to conserve the land and to create a personal landscape preserve. From then until his death in 1994, Yeon created a landscape of both designed and natural features. In an area of burgeoning growth, land development, and economic pressure, The Shire became an oasis along the Columbia. Sitting on the river's edge in Washington, The Shire directly faces Multnomah Falls in Oregon, the second-highest waterfall in the continental United States. As a designed landscape with one mile of land along the river, The Shire is formed around a series of open meadows, connected by meandering turf paths with edges of tall grass. The effect is one of constant discovery, reinforced by the numerous framed views across the Columbia to Multnomah Falls. This is a landscape of peace and solitude, of reinvigoration and wonderment. It is a landscape meant to soothe and to inspire.

As The Shire changed from a private landscape preserve to a publicly held landscape entrusted to the University of Oregon, it was clear that while the role of The Shire would evolve, the essential purpose of celebrating our Pacific Northwest landscape legacy would remain at the center of its mission. This landscape is now known as "The Shire: John Yeon Preserve for Landscape Studies." The University of Oregon is responsible for this landscape, and it has become an integral component in the academic life of the School of Architecture and Allied Arts.

It is also, however, a landscape in which the school—and its students and faculty—puts into action what it teaches. The Shire has become a laboratory for the study of ecological systems, a source of inspiration for painters and photographers, a site for architectural design exercises, a case study for planning students, and a passionate cause for those concerned about historic landscape preservation.

In addition to taking care of this landscape and opening it for use by various programs, faculty, and students, the school supports "Shire Conferences," organized by faculty. The purpose of these conferences is to engage a small group of academic and professional leaders in consideration of critical issues in a field. The topics and ideas are proposed by faculty. This volume is the record of the first Shire Conference, on the teaching of ecology in landscape architecture programs.

This volume is much more than a record of a conference, however. It is a reflection of creative ideas and an agenda for future scholarship and education. Inspired by the beauty and complexity of The Shire and the Gorge region, these papers, responses, and records of groups' deliberations serve as the collective wisdom of this group. As the first set of "Shire Papers," it sets the standards for those that follow in this series.

Professors Bart Johnson, University of Oregon, and Kristina Hill, University of Washington, have brought together a distinguished group of scholars and practitioners to—first and foremost—think about what it means to teach landscape ecology and landscape understandings in professional academic programs in landscape architecture. This conference was based upon not only the presentation of past research and experience but the belief that knowledge in the field would be expanded by bringing together a group of academics and practitioners who hoped to learn from each other. The discussions at the Shire Conference were as important as the prepared papers. The exchange of ideas was as important as the individual presentation of knowledge. Attendees were expected to participate in and add to critical discussions that led to group essays whose work extended beyond the conference—so that all who came to The Shire might leave knowing more, thinking differently about their work, and having renewed appreciation for the essential nature of intellectual exchange.

To their credit, Professors Johnson and Hill have achieved this worthy goal. They are to be credited not only with this volume, but with the conception and execution of a larger idea: to focus attention on an emerging and developing area of teaching, research, and practice.

I am especially pleased that Island Press has committed its energies to the publication of this series. The agreement between Island Press and the School of Architecture and Allied Arts is based on our mutual desire to provide the broadest possible distribution for exciting new ideas in areas of environmental concern. The Shire Conferences will always be about challenging our ways of thinking, presenting new ideas, and bringing together scholars and practitioners who are eager to share what they know and learn from each other. A very special thanks goes to Heather Boyer of Island Press for her vision in support of these ideals and goals. Her work on this book, as well as her broader support for the idea of an ongoing series, was fundamental to the publication of this volume.

As the first in an anticipated series, this volume represents not only the work of the authors and the editors. It also reflects the vision first articulated by the trustees of the John Yeon Charitable Trust to support, in as many ways as possible, our ongoing commitment to both the product and the process of John Yeon's vision for our human-made and natural environment. These ideals have guided the first Shire Conference, and will serve to guide others in the future.

ROBERT Z. MELNICK, FASLA
Dean, School of Architecture and Allied Arts, University of Oregon

The Shire: John Yeon Center for Landscape Studies was founded in 1995 by the John Yeon Trust with the gift of The Shire and its endowment to the University of Oregon.

Acknowledgments

The work in this book developed through a four-year collaboration among faculty who care deeply about the quality of teaching in the design professions, and whose ideas contributed enormously to both the process and the product. Kristina Hill first approached Robert Melnick to see if the University of Oregon would consider hosting a conference on ecology in design and planning. After discussions with Melnick and David Hulse, Bart Johnson agreed to take on the conference for the University of Oregon. Bart and Kristina worked collaboratively as organizers and chairs for the conference, which became known as the 1998 Shire Conference, held July 16–19, 1998, in the Columbia River Gorge of Oregon and Washington.

We (Bart and Kristina) soon agreed on education as the focus for our efforts. There had been substantial development of ideas for ecological design practice, but it seemed that education itself had not been examined strategically or comprehensively at a national level. A core group of junior faculty from around the country began to coalesce around this issue, including Bart Johnson, Sharon Collinge, Shishir Raval, Kathy Poole, Kristina Hill, and Ken Tamminga. This group began discussions with their colleagues on this subject by holding workshop and panel sessions at annual meetings of both the International Association of Landscape Ecology (U.S. Chapter, March 1997) and the Council of Educators in Landscape Architecture (October 1997). We'd like to thank all of the people who participated in those workshops, either as panelists or by contributing to the discussions. Joan Nassauer, Bob Coulson, Frederick Steiner, Sally Schauman, and Jack Ahern participated as panelists in addition to the core group listed above and helped to generate conceptual direction for a subsequent meeting.

We would like to recognize the support of all the people and organizations who made that conference possible. Principal funding was provided by The Shire: John Yeon Preserve for Landscape Studies, which was founded in 1995 by the John Yeon Trust with the gift of The Shire and its endowment to the University of Oregon, School of Architecture and Allied Arts. Co-sponsors included the University of Washington, College of Architecture and Planning; the U.S. Environmental Protection Agency, Corvallis, Oregon; the USDA Forest Service Columbia Gorge National Scenic Area; and the Society for Ecological Restoration, Northwest Chapter.

The Shire Conference was coordinated through the exceptional talents of Peg Butler, who at the time was a graduate student at the University of Oregon. She worked with dedication for two years to make the conference run smoothly from start to finish. Prior to and during the conference, she had help from a number of other students at the Universities of Oregon and Washington, many of whom also participated in writing this book following the conference. The students from the University of Oregon were Sarah Birger, Jeff Lanza, Pam Mayes, David Richey, Philip Richardson, and René Senos. They were joined by Megan Atkinson, Miranda Maupin, April Mills, Stephan Schmidt, Rebecca Taylor, Wood Turner, and Missy Wyatt from the University of Washington. The students helped to make the conference a success by sharing their enthusiasm and intellectual talents.

Robert Melnick, Dean of the School of Architecture and Allied Arts at the University of Oregon, was instrumental in generating support and organizational ideas for the Shire Conference as a small working symposium. Karen Johnson of the University of Oregon provided essential logistical and budgeting support. David Hulse, Carl Steinitz, and Kenneth Helphand deserve special thanks for their thoughtful critique of the conference plans developed by Bart and Kristina. We received comments on potential panelists and participants from twelve respected colleagues that helped ensure we approached some of the best people for this endeavor. Thanks are also due to Rob Ribe, Jurgen Hess, Laurie Aunan, and Nancy Russell for coordinating and leading the field trip in the Columbia Gorge that introduced conference participants to the region.

Thanks to an interest expressed by Island Press, the conference materials became a book proposal in late 1998. Bart and Kristina took on the job of editing a set of papers prepared as talks for the conference, and also a new set of synthetic papers produced by the working groups that developed their initial ideas at the Shire Conference. We thank the people who helped to make this book possible, since there were many talented minds involved. Our authors and co-authors demonstrated sustained commitment to finishing this project by reviewing draft after draft of editorial comments from us and from outside reviewers. We gratefully acknowledge the contributions of colleagues who participated in conference plenary sessions and working groups but who were unable to continue on as coauthors for the book: Julia Badenhope, Ann Bettman, Robert Coulson, Nancy Diaz, Mike Faha, Keith Hay, Kenneth Helphand, Neil Korostoff, James Palmer, Rob Ribe, Richard Toth, and Richard Westmacott. René Senos did an admirable job as our editorial assistant over two years of writing and production, and Craig Carnagey helped us over the finish line as our editorial assistant during the final months of production.

We thank our two anonymous reviewers and also thank the hard-working reviewers who commented on drafts of chapters in this volume: Ann Bettman, Catherin Bull, Richard Forman, Frank Golley, Kenneth Helphand, David Hulse, Peter Jacobs, Doug Johnston, James Karr, Elizabeth Meyer, Steve Moddemeyer, Louise Mozingo, Joan Nassauer, Robert Ribe, Iain Robertson, Sally Schauman, Frederick Steiner, Carl Steinitz, Simon Swaffield, Sam Bass Warner Jr., and Joan Woodward.

Finally, we acknowledge the patience of our students, colleagues, friends, and families over the last two years. Their support made this effort possible for us and encouraged us to keep our standards high, while reminding us to hurry up and finish.

Ecology and Design

CHAPTER 1

Introduction: Toward Landscape Realism

Bart R. Johnson and Kristina Hill

Ecology, the study of interactions between organisms and their environments, has long been a compelling theme for faculty, practitioners, and students of landscape design and planning.[1] Frederick Law Olmsted's visionary public designs, Jens Jensen's native plantings, May Watt's observations of vernacular landscapes, and Ian McHarg's book, *Design with Nature*, are all milestones of ecological thinking in landscape design and planning. Many contemporary designers and planners identify an understanding of ecology as crucial to their work (Spirn 1984; Lyle 1985; Hough 1995; Thompson and Steiner 1997; Nassauer 1997). Yet ecology is a rapidly evolving field that has undergone major paradigm shifts in the past two decades. It no longer presupposes a "balance of nature," but instead describes the natural world in terms of flux and change. Moreover, new fields of applied ecology such as conservation biology and restoration ecology have emerged in response to the global biodiversity crisis, a crisis inherently linked to the need to provide burgeoning human populations a reasonable quality of life. How can designers and planners respond to these increasingly global challenges, which require an integrated understanding of human societies and ecosystems? How will new theories of nature affect the theory and practice of landscape design and the collaborations that take place between scientists and designers?

Designers and planners are not alone in grappling with the interdependence of humans and natural systems. Many ecologists have come to recognize humans as keystone species in most, if not all, ecosystems. In a key development, some are beginning to emphasize urban ecosystems as critical landscape features (Pickett et al. 1997; Parlange 1998; Collins et al. 2000). Other ecologists have identified training in human dimensions of landscape change as an emerging need in conservation science education (e.g., Jacobson and McDuff 1998). Scientific practitioners are looking for new ways to collaborate as well.

Landscape designers, planners, and applied ecologists belong to a diverse group of disciplines that face common needs to integrate cultural and ecological understanding toward prescriptions for land protection and change. The extent to which they succeed in this endeavor depends not only on how scholars and professionals rethink their research and practice, but also on the priorities they establish for the next generation of scholars and practitioners through education.

That education, particularly its foundations and methods, is the subject of this book. We intend to stimulate faculty to think broadly and creatively about how they incorporate ecological knowledge in design and physical planning curricula, and to offer specific approaches to teaching. We ask what conceptual foundation and practical skills are needed for practitioners to develop ecologically responsible practices, and how they can adapt to a regulatory environment that is increasingly shaped by technical debates about environmental trends and impacts. How can we initiate collaborations with colleagues from the natural sciences to stimulate mutual learning and improved design and planning? How can we bring ecological accountability to design education while supporting our traditions of innovation and inspiration through art?

Common Ground for Dialogue

Within the last two decades, new ecological subdisciplines that seek to use ecological science as a foundation for solving environmental problems have gained prominence. Each has its own focus and approaches, and each is rapidly evolving (Boxes 1-1 to 1-4). New ideas for design subdisciplines have also emerged, including ecological engineering and ecorevelatory design. Meanwhile the established design professions have increasingly recognized the need for ecological awareness and responsibility, and have begun to adopt ecological guidelines for professional practice (Boxes 1-5 to 1-8). Whether these fields choose to learn from one another at this critical time, and whether they build collaborative approaches to land development and conservation, could have impacts that resonate throughout this century. One thing is abundantly clear—no single discipline possesses sufficient knowledge or skills to address the combined complexities of cultural and ecological issues across the diverse set of contexts and scales in which they occur.

This book project began with the idea that educational restructuring can be a means to plant the seeds of future professional and research collaborations among many fields, including landscape architecture, urban

BOX 1-1. Conservation Biology

> Conservation biology is a multidisciplinary science that has developed in response to the biodiversity crisis.
> —Michael Soulé, What is conservation biology?

> Conservation biology is the field of biology that studies the dynamics of diversity, scarcity and extinction.
> —Reed F. Noss and Allen Y. Cooperrider, Saving nature's legacy: Protecting and restoring biodiversity

ORGANIZATION NAME: Society for Conservation Biology (SCB)

ESTABLISHED: 1985

PRINCIPAL PUBLICATION: *Conservation Biology*

MEMBERSHIP: 5,200 members worldwide include resource managers, educators, government and private conservation workers, and students.

PURPOSE: SCB is "dedicated to promoting the scientific study of the phenomena that affect the maintenance, loss, and restoration of biological diversity . . . the Society was formed to help develop the scientific and technical means for the protection, maintenance, and restoration of life on this planet—its species, its ecological and evolutionary processes, and its particular and total environment." To this end, members "encourage communication and collaboration between conservation biology and other disciplines (including other biological and physical sciences, the behavioral and social sciences, economics, law, and philosophy) that study and advise on conservation and natural resources issues." <http://conbio.rice.edu/scb/info/>

Meffe and Carroll (1997, p. 22–25) suggest that conservation biology has a number of key characteristics that differentiate it from many other sciences. They arise from its goal of preserving "the evolutionary potential and ecological viability of a vast array of biodiversity," which itself is necessitated by human predilections to attempt to "control, simplify and conquer" inherently complex and dynamic native ecological systems. In particular, it is a crisis discipline, based in science that is *multidisciplinary*, is necessarily *inexact*, is explicitly *based in values*, and requires both *an evolutionary time scale* and *eternal vigilance*.

BOX 1-2. Landscape Ecology

Landscape ecology is the study of spatial variation in landscapes at a variety of scales. It includes the biophysical and societal causes and consequences of landscape heterogeneity. Above all, it is broadly interdisciplinary.

—International Association for Landscape Ecology Web site <http://www.crle.uoguelph.ca/iale/>

ORGANIZATION NAME: International Association for Landscape Ecology (IALE)

ESTABLISHED: 1982

PRINCIPAL PUBLICATION: *Landscape Ecology*

MEMBERSHIP: 1,500 members worldwide include landscape architects, ecologists, land/nature managers, conservation biologists, land-use planners, biogeographers, GIS specialists, spatial statisticians, wildlife biologists, and ecosystem modelers.

PURPOSE: The mission of IALE is to "develop landscape ecology as a scientific basis for analysis, planning and management of the landscapes of the world. IALE advances international co-operation and interdisciplinary synergism within the field, through scientific, scholarly, educational and communication activities." IALE "encourages landscape ecologists to transcend boundaries and to work together building theory and developing knowledge of landscape pattern and process, developing integrative tools, and making them applicable to real landscape situations and applying them to solve problems." Its core themes include the spatial pattern or structure of landscapes ranging from wilderness to cities, the relationship between pattern and process in landscapes, the relationship of human activity to landscape pattern, process, and change, and the effect of scale and disturbance on the landscape. <http://www.crle.uoguelph.ca/iale/>

BOX 1-3. Restoration Ecology

Ecological restoration is the process of assisting the recovery and management of ecological integrity. Ecological integrity includes a critical range of variability in biodiversity, ecological processes and structures, regional and historical context, and sustainable cultural practices.

—Society for Ecological Restoration Web site <http://ser.org/>

ORGANIZATION NAME: Society for Ecological Restoration (SER)

ESTABLISHED: 1988

PRINCIPAL PUBLICATIONS: *Restoration Ecology* and *Ecological Restoration*

MEMBERSHIP: 2,300 members worldwide include scientists, planners, administrators, ecological consultants, first peoples, landscape architects, philosophers, teachers, engineers, natural areas managers, writers, growers, community activists, and volunteers, among others.

PURPOSE: The mission of SER is "to promote ecological restoration as a means of sustaining the diversity of life on Earth and reestablishing an ecologically healthy relationship between nature and culture." To this end, SER encourages "the development of restoration, including restorative management, as a scientific and technical discipline, as a strategy for environmental conservation, as a technique for ecological research, and as a means of developing a mutually beneficial relationship between human beings and the rest of nature." The society has endorsed nine *Environmental Policies* and seven *Project Policies* that offer specific guidelines for restoration efforts and their evaluation. <http://www.ser.org/>

Debates about the scope and nature of restoration and its sometimes imprecise or divergent usage have led to distinctions of the five Rs of restoration ecology: restoration, rehabilitation, reclamation, re-creation, and recovery (MacMahon 1997). When considering restorative approaches, it is important to recognize differences among these and to see the entire set as a toolbox of approaches, with "restoration" to some previous state as one among a continuum of possibilities.

BOX 1-4. Ecosystem Management

Ecosystem management integrates scientific knowledge of ecological relationships within a complex sociopolitical and values framework toward the general goal of protecting native ecosystem integrity over the long term.

—R. Edward Grumbine, What is ecosystem management?

Ecosystem management is management driven by explicit goals, executed by policies, protocols, and practices, and made adaptable by monitoring and research based on our best understanding of the ecological interactions and processes necessary to sustain ecosystem structure and function.

—Norman L. Christensen et al., The report of the Ecological Society of America committee on the scientific basis for ecosystem management

Despite (or perhaps because of) the fact that there is no organization dedicated to its development, ecosystem management has become the primary paradigm of federal land management agencies. At the same time, it is clear that different agencies, organizations, and individuals use the term in very different ways. Most definitions rely on some concept of sustainability that includes ideas of ecological health or integrity as well as the delivery of goods and services for humans, but the relative emphasis on those qualities varies, as does the level of confidence that they can be jointly optimized. The Ecological Society of America report emphasizes that ecosystem management focuses primarily on the sustainability of ecosystem structures and processes necessary to deliver goods and services, rather than on the "deliverables." To do so, it must incorporate eight key factors: long-term sustainability; clear operational goals; sound ecological models and understanding; complexity and interconnectedness; the dynamic character of ecosystems; attention to context and scale; humans as ecosystem components; and adaptability and accountability (Christensen et al. 1996).

design, planning, architecture, civil and environmental engineering, landscape ecology, conservation biology, and restoration ecology. There are significant opportunities for these fields to learn from each other and, in so doing, to increase their relevance to contemporary issues. We feel that the core of such collaborations is twofold: first, to develop deep and meaningful understandings of places, including how each place is imbued with interdependent cultural and ecological attributes; and second, to assist individuals, organizations, communities, and regions to envision new courses of action and select from among alternatives. The essays contained in this book focus on identifying practical strategies for teaching these concepts and skills.

In addition to a desire to encourage collaboration, our motivation for this book was to explore and debate the idea that all designs should be held accountable for their ecological impacts. We wanted to address the philosophical divide between designers who want to inspire through art and designers who want to sustain natural processes by asking both groups to pursue a higher standard. That standard would call for designs that are aesthetically challenging, in the best sense of fine art, and ecologically sustainable, enabling humans to coexist with the other species that have evolved on this planet.

Defining the "goodness" of design is clearly an ethical and philosophical question, and we wish our position as editors to be explicit. We do not believe there is an inherent trade-off between beauty and ecological integrity in landscapes that are built or managed by humans. We believe that design excellence must be judged by both aesthetic and ecological criteria. Indeed, we find it ethically unacceptable for our students and for practitioners in the design fields to decide to concern themselves with art but not ecology. The decision to pursue ecological sustainability without art also is flawed, because art—whether fine or folk—may be the key to touching human hearts and minds in new ways. Artful design can be a means to affirm that being human is a profoundly beautiful expression of nature, an expression of the fundamental bonds that we share with all other forms of life. Our art can offer inspiration and hope in the face of negative environmental trends that are linked to human behavior. Joining art with a scientific basis for design can be a profound means to anchor and firm our art in the realities of these same relationships. As designers we are interested in new visions for the future, not the paralysis that results from recrimination and blame. This book is about taking practical steps to achieve a future in which artists and scientists collaborate and understand each other.

Frameworks for Learning and Collaboration

Human cultures and ecosystems exist in a reciprocal relationship. In essence, all landscape design is ecological, whether by intent or default, because every

BOX 1-5. Landscape Architecture

> Landscape architecture is the art and science of analysis, planning, design, management, preservation and rehabilitation of the land.
>
> —American Society of Landscape Architects Web site
> <http://www.asla.org/nonmembers/hq.html>

ORGANIZATION NAME: American Society of Landscape Architects (ASLA)

ESTABLISHED: 1899

PRINCIPAL PUBLICATIONS: *Landscape Journal* (not affiliated with ASLA) and *Landscape Architecture Magazine*

MEMBERSHIP: 12,000 members. ASLA is a national professional society that represents landscape architects in the United States. It includes private, public, and academic practitioners.

PURPOSE: The mission of ASLA is "to lead, to educate and to participate in the careful stewardship, wise planning and artful design of our cultural and natural environments." <http://www.asla.org/>.

ASLA Declaration on Environment and Development

ASLA has attempted to secure a place for landscape architecture as a leading land stewardship profession, as reflected in the ASLA "Declaration on Environment and Development" <http://www.asla.org/nonmembers/declarn_env_dev.html>. The declaration offers a set of principles that it states represent fundamental and long-established values of ASLA. These include how the "health and well-being of people, their cultures and settlements [and] of other species and of global ecosystems are interconnected, vulnerable, and dependent on each other"; the rights of future generations to landscapes with environmental assets at least comparable to those of today; the interdependence of environmental and cultural integrity, human well-being, and long-term economic development; sustainable development through integrating environmental protection and ecological function in development processes; and the responsibilities of developed countries to pursue internal and international sustainability.

The declaration concludes that because landscapes are "living complexes" that encompass "the basic processes that support life, meeting human needs requires healthy landscapes." Thus, "nurturing the processes of regeneration and self-renewal in the world's healthy landscapes and reestablishing these in the vast areas of the world's degraded landscapes are fundamental purposes" of landscape architecture. The declaration follows with a conceptual framework of five objectives for the ethics, education, and practice of landscape architects, accompanied by specific strategies to achieve each objective.

BOX 1-6. Civil Engineering

ORGANIZATION NAME: American Society of Civil Engineers (ASCE)

ESTABLISHED: 1852

PRINCIPAL PUBLICATIONS: *Civil Engineering,* as well as 28 technical and professional journals, and a variety of books, manuals of practice, standards, and monographs

MEMBERSHIP: 123,000 members worldwide include civil engineers and those in related disciplines. The Environmental and Water Resources Institute (EWRI) was created in 1999 as a semiautonomous institute to attract allied professionals. EWRI has 20,000 members.

PURPOSE: The mission of ASCE is to advance professional knowledge and improve the practice of civil engineering. To this end, it attempts to provide a focal point for development and transfer of research results and technical, policy, and managerial information, and serve as the catalyst for effective and efficient service through cooperation with other engineering and related organizations. <http://www.asce.org>

Role of the Engineer in Sustainable Development

ASCE Policy Statement 418. <http://www.asce.org/govnpub/policy/pol418_sustain.html>

The *Code of Ethics* of ASCE requires civil engineers to strive to comply with the principles of sustainable development. Further, the *Vision of the ASCE Strategic Plan* calls for global leadership in the "promotion of responsible, economically sound, and environmentally sustainable solutions that enhance the quality of life, [and] protect and efficiently use natural resources." To this end, ASCE "will work to develop and encourage the use of evolving technologies required to achieve an ecologically sustainable world for future generations . . . in this role, engineers must participate in interdisciplinary teams with ecologists, economists, sociologists and professionals from other disciplines."

The key issue is that the "demand on natural resources is fast outstripping supply in the developed and developing world. Likewise, the ability of natural systems to assimilate wastes is taxed, almost to exhaustion. Environmental, technological, economic and social development must, therefore, be seen as interdependent concepts in which industrial competitiveness and ecological sustainability will be addressed together as complementary aspects of a common goal." Furthermore, "sustainable development requires broadening the education of engineers and finding new ways to do business: i.e., doing more with less—less resources, less energy consumption and less waste generation. It requires focus on upstream prevention in preference to 'end of pipe' treatment, new manufacturing processes and equipment, expanded use of recyclable materials and the development of regenerative/recyclable products and packaging. Sustainable development requires approaches that imitate natural or biological processes."

BOX 1-7. Planning

> Planning involves many tools, including economic and demographic analysis, natural and cultural resource evaluation, goal-setting, and strategic planning . . . [planners] offer options—so that communities and their citizens can achieve their vision of the future.
>
> —American Planning Association Web site
> <http://www.planning.org/info/whatis.htm>

ORGANIZATION NAME: American Planning Association (APA)

ESTABLISHED: Created by merging the American Institute of Planners (est. 1917) and the American Society of Planning Officials (est. 1934)

PRINCIPAL PUBLICATIONS: *Journal of the American Planning Association* and *Planning*

MEMBERSHIP: 30,000 members, including practicing planners, officials, and citizens involved with urban and rural planning issues.

PURPOSE: APA is "organized to advance the art and science of planning and to foster the activity of planning—physical, economic, and social—at the local, regional, state, and national levels." Its objective is to "contribute to public well-being by developing communities and environments that meet the needs of people and society more effectively." Planners' ethical responsibilities include fostering meaningful citizen participation in planning decisions and protecting the integrity of the natural environment and the heritage of the built environment. <http://www. planning.org/>

Endangered Species, Habitat Protection, and Sustainability

The APA *Policy Guide on Endangered Species and Habitat Protection* <http://www.planning.org/govt/endanger.htm> states that "protecting natural system functions . . . is critical to the support of human, animal and plant populations" and recommends "a proactive approach to protect natural communities." It states further that the "preservation and enhancement of wildlife and its habitat cannot be distinguished from preservation of human habitat and so is a core function of government." The policy guide lists ten specific policies endorsed by APA toward these ends.

The APA *Policy Guide on Planning for Sustainability* <http://www.planning.org/govt/sustdvpg.htm> relates physical, social, and economic patterns of human development to a definition of sustainability that includes communities as good places to live; societal values of individual liberty and democracy; diversity of the natural environment; and the ability of natural systems to provide life-supporting services. The guide lists social and biophysical factors that limit sustainability and provides a strategy for participatory planning that integrates environmental, economic, and social goals and actions.

BOX 1-8. Architecture

ORGANIZATION NAME: American Institute of Architects (AIA)

ESTABLISHED: 1857

PRINCIPAL PUBLICATIONS: *Architectural Record* (affiliated with AIA) and *Architecture,* among many other journals.

MEMBERSHIP: 66,500 members.

PURPOSE: The mission of AIA is to serve its members, advance their value, and improve the quality of the built environment. To this end, AIA serves architects and their clients by promoting ethical, educational, and practice standards for the profession and by advocating excellence in design, defined here but not restricted to aesthetics, functionality, constructability, and cost effectiveness (Aligning the Institute for the Millennium, <http://www.e-architect.com>).

Sustainable Design

The 1993 "Declaration of Interdependence for a Sustainable Future" of AIA and UIA (International Union of Architects) called on architects to "place environmental and social sustainability at the core of their practices and professional responsibilities." Toward this end, AIA has taken the position that "the sustainable redevelopment of the built environment must begin as rapidly as possible." This will include designing buildings to be minimal consumers, and even generators, of energy and other resources; using building materials that have a benign impact on the environment throughout their life cycle; constructing buildings with internal environments that are health-giving and inspiring; arranging buildings so they foster community; developing urban areas and regions so they have natural environments within walking distance of every residence; and developing the infrastructure of transportation, utilities, and communications to enhance human scale community, and so that the automobile is optional for most people, most of the time. <http://www.e-architect.com/pia/cote/AIA-COTE/main/hlth_bld.asp>

landscape place, no matter how large or small, includes multiple species and biophysical processes that will be affected by human actions. In similar fashion, every ecological conservation or restoration plan is cultural, involving and affecting people. In particular, such plans are likely to distribute costs and benefits differentially among people with different socioeconomic and cultural status. Moreover, a plan's success ultimately depends on satisfying human needs and values. To ignore this reciprocal relationship of human culture and ecosystems is to turn away from a fundamental reality of the landscapes we share with other people and other species. As a basic principle for collaboration among the design disciplines and the new fields of applied ecology, we propose that all landscape design, planning, and management should be evaluated through a thorough accounting of its consequences for ecological health, biotic integrity, and cultural well-being (human, social, and economic).

One important arena for building common ground in design and ecology, for example, would be developing an understanding of health that integrates ecological health and human health. In ecology, health has most often been thought of from the standpoint of biodiversity and sustainability (Chapter 13), whereas landscape design in its early formulations included human health as one of its core concerns through civic design. A more unified concept of health than either discipline has embraced to date might conceive of human health in ecological terms, and in ways that dispel the illusion that we and our bodies are somehow separate from ecological realities. To this end, Steingraber and Hill (Chapter 8) propose that health encompasses (1) relationships among living bodies, (2) relationships among processes and organisms within living bodies, and (3) relationships between living bodies and the physical earth.

We realize that many designers and planners want to believe that their work already has positive affects on both cultural well-being and ecological health. But are we doing enough? The authors we worked with in this book generally shared the belief that the design professions, and their educational programs, are not sufficiently committed to understanding and sustaining ecological health. To "get real," design and planning education must embrace ecological knowledge as deeply as it does cultural knowledge. This is not to imply that designers must become ecologists or vice versa. Rather, the education of each must be grounded in a framework that unifies ecological and cultural ways of knowing. As Joan Nassauer (Chapter 9) points out, design is cultural action that structures ecosystems. We believe that design and planning education must be steeped in knowledge of how humans, as biological, social, and spiritual beings, inhabit a world filled with myriad other species, a world that is maintained and changing through crucial biophysical processes often invisible to the immediate senses.

In David Orr's (2001) words, "ecological designers should aim to cause no ugliness, human or ecological, somewhere else or at some later time." As every designer knows, what is "ugly" is a matter of opinion. But we know Orr's work well enough to believe he did not mean it in a superficial or arbitrary manner but rather with deep respect for the beauty of life. He calls for responsible design in space and time, and in human and nonhuman terms that have deep implications for how we act as members of what Aldo Leopold named the "land community." We argue that to meet this charge we must move beyond separate visions for humans and nature, guided by the recognition that humans are a key species in contemporary earth ecosystems and the premise that cultural well-being and ecological integrity are intimately linked. Further, we advocate a vision of community that is founded in diversity and that explicitly includes consideration for the intrinsic values of all species and for cultural, social, racial, gender, and intergenerational equity. Social and human justice should not be divorced from biological justice.

To reach these higher standards for design, new approaches are needed in both education and practice. We must collaborate more deeply with applied ecologists. We must find ways to interpret and apply new understandings from ecological science in physical planning and landscape design. We must understand the implications of our work for both social equity and ecological sustainability. And we must heal the historical schisms that have developed between practitioners who base their design work on artistic principles and those who look for a basis in scientific principles.

A working knowledge of design and planning processes is also essential to the emerging fields of applied ecology, if they are to translate the essentially descriptive and predictive knowledge of science into normative prescriptions for land conservation and development. Collaboration is also essential to their success. We should work together to build shared knowledge among people who are learning to restore ecological systems, people who are trained in the rigorous rules of testing and evidence that have shaped science, and people skilled in articulating and responding to human needs and aspirations.

Designers, however, have a longer record of considering the relevance of ecology to their work than ecologists have for considering the relevance of human cultural action to theirs. It is only recently that the practice of prescribing spatially explicit landscape futures has assumed a prominent role in ecology. While the fields of restoration ecology and landscape ecology have been jointly developed by designers, planners, and ecologists, among others, this collaborative relationship has not been prominent in other applied ecologies. Notably, those writing about partnerships needed for conservation biology and ecosystem management rarely mention a role for designers and physical planners, even when they offer wide-ranging lists of needed disciplines (e.g., Grumbine 1994; Meffe and Carroll 1997; Jacobson and McDuff 1998;

Kohm et al. 2000). In part, this omission may be due to what Howett (1998) characterizes as the difference between the mission of landscape design and the manner in which it is carried out in everyday practice, where it is heavily influenced by a market-based economy and the politics of multiple uses on public lands. This frequent omission of the design fields from ecologists' lists of necessary collaborators raises important questions that the design professions must face for their future in public practice (Chapter 5).

Teaching collaborative skills is as necessary as teaching new knowledge in designers' and ecologists' efforts to integrate fragmented understanding of how landscapes function as both ecological *and* cultural places.[2] To create a living bridge between design and ecology requires more than tweaking the margins of their respective educational curricula. It requires a deep reconceptualization of how designers, planners, and ecologists conceive of humans as members of Earth ecosystems. If we don't get real by tracking the ways in which humans affect those ecosystems, using consistent monitoring of built designs to detect failures and retrofitting them to improve their performance, we are teaching no more than good intentions. And good intentions alone will not address the problems we have created.

Following the Shire Conference, we reflected on the tremendous energy and sense of imperative brought by the participants. How could we encapsulate the pragmatic, ethical, and philosophical issues raised by our need to "get real" about the ecological performance of designed landscapes? What kind of transformation is needed to move from design *with* nature to a design that includes humans *in* nature? As editors, we decided that the term "landscape realism" captured much of what we desired. Realism, in our usage, is not a philosophical position that claims there is only one "true" reality; rather, in the sense of the artistic movement of social realism, it is the position that the diversity of life forms and lived experience matters and must be addressed in designing landscapes. This form of realism is necessary for the creative synthesis of human culture and ecological processes, and of science and art, that we envision. We offer the idea solely as a potential, knowing that it remains for future dialogues to determine what new words are truly needed.

Gardens and Scaffolding: Metaphors for Reenvisioning Design

Olmsted's founding vision of landscape architecture was that it would involve a melding of artistic, environmental, and social goals. Yet the field has experienced recurring schisms between proponents of art and proponents of science, and in the prioritization of beauty and function (Howett 1998). Even those who have advocated ecological priorities in design have asserted different roles for science. Consider the characterizations of design by McHarg or

Garrett Eckbo as preeminently guided by science, in contrast to Olmsted and others who saw design as an imaginative art informed by science, an art that harnesses "daring and original leaps of . . . insight . . . to express the mystery that lies at the heart of the human/nature dialectic" (Howett 1998, p. 90).

How can we envision new relationships between science and art, and ecology and design? James Corner (1997) argues that metaphors are an important source of visions for the future and that they are essential to imagining a future in which design and ecology enjoy a closer relationship. We agree, and note that the metaphors that have recurred frequently in the language of landscape architecture may offer the greatest potential for imagining its future. For example, we find that the metaphor of the garden can encourage shared understanding between designers and ecologists. In Chapter 2 of this book, Anne Spirn proposes that the garden, as a "well-tended region," is useful, both literally and metaporically, in working out a new relationship with nature. Similarly, Dan Janzen, a prominent ecologist, has also called attention to the potential of the garden. Janzen is a well-known and provocative tropical biologist who has devoted himself to protecting the extraordinary and endangered biodiversity of the tropics. In a recent article in *Science*, he asks how we can secure the enormity of natural diversity in an increasingly human-dominated world. His answer? In the "wildland garden" as an "unruly extension of the human genome" (Janzen 1998, p. 1312).

Garden design always has held a central place in the education of landscape architects. But if we wish to pursue the garden as a metaphor for landscapes at a larger scale, we must consider that gardens require more than design; they need people to nurture them. The concept of human landscapes as an ecological garden forces us to reconsider our own role, perhaps realizing that we might learn more from adopting the humility and observation skills of a good gardener than from implementing either untested theories or grand aesthetic schemes.

Ecological gardens must come in many shapes and sizes, from mosaics of small backyards that begin to cumulatively change the fabric of urban neighborhoods to gardens the size of large watersheds in which people harvest timber, grow crops, mine minerals, recreate and contemplate, and build houses and cities. We have little experience with this type of gardening. It must incorporate floods and fires. It requires science and design, planning and policy. It needs to allow and even celebrate human use, while comprehending that errors in judgment at any time could mean the extinction of species.

A second generative metaphor we consider useful for linking design and ecology is that of a scaffold as a structure that allows new forms to be constructed but does not determine those forms completely. Spirn (1988, 1998) extended earlier ideas of designer Lawrence Halprin for designing processes

(Howett 1998) when she proposed establishing an open-ended "framework congruent with the deep structure of a place" that could continue to evolve (Spirn 1988, p. 124). Corner (1997) echoed this earlier work when he wrote of creating "catalytic frameworks" rather than finished works. This metaphor operates in much the way that an ecologist may characterize an ecosystem as more a bundle of intertwined processes than a set of objects (Norton 1991). Design that tries to build ecological scaffolding could create enabling relationships instead of static patterns and accommodate "unpredictability, contingency and change" within a framework of "formal coherency and structural/material precision" (Corner 1997, p. 102).

In a similar way, Halprin's idea of scoring design as a stage for performance (Howett 1998) echoes the metaphor of ecologist G. E. Hutchinson for the landscape as an ecological theater, the living stage for the enactment of the evolutionary play (see Chapter 13). In the field of restoration ecology, William Jordan (1995) has stressed the need for restoration projects to encompass human interplay on multiple levels: product, process, experience, and performance. In the last category, Jordan proposes restorative ritual as a profound way to re-create deep relationships among people and with the natural world through the activity of restoration.

Howett (1987), Spirn (1988, 1998), and Mozingo (1997) advocate a blending of ecology and culture toward evocative and profoundly moving landscapes that reveal and celebrate ecological processes, and a special issue of *Landscape Journal* was devoted to a critique of recent "eco-revelatory" design (Brown, Harkness, and Johnston 1998). Meyer (1997) argues for a "systems aesthetic" for design that is "concerned with the relationships among things, not the things themselves" (p. 66). This emphasis on process seems healthy since design has often ignored how maintaining static designs may interfere with ecological change and associated processes. Yet ecologist James Karr (Chapter 6) reminds us that the "things"—owls and salmon, soils and water—are also important. The relationships among landscape form, function, change, and aesthetic perceptions are ripe for mutual investigations among designers and ecologists.

Common Ground for Education

What is it that makes something so fundamentally sensible as teaching ecology in design and planning so intrinsically challenging? During preparations for this book, we discovered that in spite of the availability of compelling metaphors, and a common perception among people we spoke to is that there is a need for greater ecological accountability in design education, there are still serious hurdles to overcome. As a precursor to the Shire Conference, we held panel sessions in 1997 at the annual meetings for the International

Association of Landscape Ecology (IALE, U.S. chapter) and the Council of Educators in Landscape Architecture (CELA) to begin identifying key issues and opportunities. We discovered that there are epistemological differences and skill differences that can make the divide between science and design seem difficult to bridge.

At the IALE session, seven panelists were asked to outline the principal issues of teaching ecology in design and planning programs. There were notable differences in the pedagogic challenges perceived by faculty whose primary academic training and research were in ecological science compared to those whose training was primarily in landscape design and planning. The first group seemed most concerned with the need for students to understand ecology as a rigorous search for cause-and-effect relationships, the creation of conceptual toolboxes and procedures that would allow students to go beyond "cookbook" approaches to ecology, and the empowerment gained by learning to read and interpret the ecological literature. The designers and planners, on the other hand, emphasized a different set of issues. They described the need for conceptual vehicles and tools, but more in the sense of "ordering frameworks" that could help transform ecological analysis into a design response. Faculty from both camps emphasized the need for designers to master strategic thinking about how to apply ecological knowledge to spatial design.

At the CELA session, a key issue that emerged was the degree to which institutional contexts determined the solutions most suitable for individual classes or departmental curricula. No single approach seemed to be applicable across the board. For example, some departments with strong bonds to applied science programs sent their students to ecologists and courses in other departments. Others had chosen, or felt compelled, to hire an ecologist as part of their faculty, while still others asked faculty who were principally trained in design to teach ecology. The character of the student body also emerged as a factor, in that students' expectations, basic skills, and priorities influenced how and when ecological understanding could be effectively communicated.

Following the CELA and IALE conferences, we convened an invited symposium of sixty landscape architects, planners, and ecologists, including both educators and practitioners at the forefront of efforts to integrate ecology with landscape design and planning. We also included in our discussions selected students who assisted in managing the logistics of the conference and shared essential insights from their perspectives as participants in the educational programs we discussed. The conference was held July 16–19, 1998, in the Columbia River Gorge between Washington and Oregon, in close proximity to the land John Yeon named The Shire. It became known as the 1998 Shire Conference, entitled "From Theory to Practice: Teaching Ecology in Landscape Design and Planning Programs."

To jump-start discussions at the Shire Conference, we invited ten distinguished scholars and practitioners to submit essays that all participants would read prior to the conference. The first day of the conference was spent in plenary sessions where these authors summarized their papers as a prelude to group discussions. The following three days were spent in working group sessions designed to develop synthetic papers on selected topics, interspersed with plenary discussions and a field trip. This format activated spirited exchanges and engaged the participants in the task of developing outlines for synthetic papers. Following the conference, we worked closely with all authors to incorporate the ideas and criticisms generated during the conference. As a result, this book contains selected essays authored by invited panelists from the Shire Conference, and six synthetic papers developed by conference working groups.

Common Challenges with Conservation Education

This book focuses on lessons for design and planning education. But are there relevant ideas for applied ecological education as well? We think so. We suggest the following reasons why applied ecologists can learn from the designers' and planners' educational approaches:

- Knowledge of spatially explicit design and planning is a necessary skill for many conservation practitioners.
- Design and planning disciplines offer important insights into how diverse human needs and aspirations can be addressed through design and physical planning.
- Design processes are operational models for linking program development to landscape analysis and spatial decision making. They have been developed and refined for well over a hundred years.
- The relative depth and breadth of skills needed for applied conservation work are similar to those needed for design and planning practice. Design programs have wrestled with these issues since their inception. Both their successes and shortcomings are instructive.

Decisions being faced in applied ecology programs aimed at training scientists skilled in spatial planning and human dimensions are a mirror image of decisions faced by design programs attempting to educate designers and physical planners well versed in ecology. Like design and planning, landscape ecology, conservation biology, restoration ecology, and ecosystem management all have been described as synthetic disciplines. Furthermore, because they focus on applied and often value-driven goals, their programmatic needs are leading them to chart educational directions substantially different from those of their "mother" programs in ecology and other natural sciences. As each has developed from bases in "pure" science toward action and advocacy that retain roots in sound science, they have recognized the need to incorporate landscape decision-making methodologies and knowledge of human cultures into their curricula. This has led to debates about the relative importance of different types of knowledge, especially those that are not traditional for scientists. We will address these issues in two main categories: interdisciplinary partnerships in education and practice, and real-world problem solving.[3]

Interdisciplinary Partnerships in Education and Practice

The impetus to create new partnerships in conservation applications stems from the need for "putting the science into practice and the practice into science" (Kohm et al. 2000, p. 593). In conservation education, this manifests as a need to "soften the boundaries" among academic departments to support synthetic disciplines that must draw from many fields (Meffe 1998, p. 259). In part, this can be achieved through institutional restructuring toward multidisciplinary and interdisciplinary degree programs (Jacobson 1995; Noss 1997; Meffe 1998), as well as other ways to encourage cross-disciplinary training and breadth (Jacobson 1990; Meffe and Carroll 1997; Richter and Redford 1999). In this book, Tamminga et al. (Chapter 14), Ahern et al. (Chapter 15), and Johnson et al. (Chapter 13) all consider related needs and strategies. In many cases, such structural changes require removing institutional barriers that result from the ways in which faculty are evaluated and rewarded (or not) for stepping outside of traditional roles (Meffe 1998) and for performing issue-driven research (Viederman, Meffe, and Carroll 1997).

A move toward broader educational foundations for students engenders a related challenge of enlarging educational scope without diluting core knowledge. Some of the answers may lie in developing different models for training a disciplinary specialist with sufficient breadth to communicate with specialists from other disciplines versus someone deeply trained in more than one discipline (Lidicker 1998; Chapter 9). Such issues can become central in considerations of whether to offer flexible, nontraditional degrees (design or science) for students with advanced training in other fields.

Meffe, Noss, and Jacobson (1998) suggest that the need for conservation biologists to apply scientific data toward societal changes means that educational programs may have to require more credit hours and additional time to complete degree requirements than has been traditional. This situation was faced long ago in landscape architecture, where a professional degree typically requires both undergraduate and masters students one year longer to complete their programs than their arts and sciences counterparts. A related and pressing need for both disciplines is to develop lifetime educational opportunities (Lidicker 1998; Chapter 18) so that professionals can renew and broaden their education in relation to on-the-job needs.

One of the most cited needs for breadth in conservation education is the inclusion of human dimensions (Cannon, Dietz, and Dietz 1996; Saberwal and Kothari 1996; Jacobson and McDuff 1998), defined as a "variety of people-oriented management considerations and a cross-disciplinary range of inquiry" (Jacobson and Mcduff 1998, p. 263). In none of these discussions, however, does there appear to be an appreciation for the relevance of physical design and planning. From a designer's point of view, this seems an odd and striking omission. As described throughout this book, design and planning skills are systematically focused on linking ecological and cultural knowledge toward prescriptions for the land. Moreover, they are clearly related to the ways conservation biologists proffer spatial proposals, such as those for reserve design. Jacobson and McDuff further identify the need for methods to identify problems, understand human cognition and value systems, and analyze alternatives. All these skills are central to design and planning education. We add a further need that these authors do not specifically articulate: applying knowledge of human responses to landscapes to creating and conserving places with desirable social and ecological qualities.

Real-World Problem Solving

The need to solve real-world problems is central to conservation practice and at the core of calls to broaden educational training, but it is not adequately addressed in most conservation programs (Jacobson 1990; Orr 1993; Noss 1997; Meffe 1998). Students must learn to craft solutions from among competing values and priorities (Richter and Redford 1999), to work effectively in teams (Touval and Dietz 1994; Mattingly 1997), and to link sound science to imaginative solutions (Meffe 1998; Kohm et al. 2000). All of these issues are familiar to landscape designers and planners and are echoed throughout this book.

Meffe (1998, p. 260) points out that "medical schools train their students in real hospitals with actual patients." Design and planning programs have long trained their students with actual problems in real landscapes. In a statement that seems to apply equally to design and conservation programs, Orr (1993) cautions " . . . we should introduce students to the mysteries of specific places and things before giving them access to the power inherent in abstract knowledge" (p. 11) and cites the experiences of Thoreau, Aldo Leopold, Annie Dillard, and others as precedents. He stresses the need for innovative approaches to help students encounter and learn from place—an idea that is at the core of models for place-based education proposed by Hough (Chapter 11) and Ahern et al. (Chapter 15).

If we expect students to learn how to imagine new solutions, then we must teach them imaginatively. The suggestion by Mattingly (1997) that conservation educators consider lessons from primary school teachers who engage multiple senses with active learning, and who use learning cycles of exploration, concept invention, and discovery (Orth 1995), will be familiar to design educators. More than a few of us have noticed that the opportunity to make colorful drawings, to build models with balsa wood and clay, and even to sculpt landscapes in sandboxes hearkens back to our students' first learning experiences in kindergarten. In this vein, design educators have long had an answer to the need for real-world problem solving that integrates multiple ways of knowing and different types of skills: They model their classroom on professional practice through the use of design studios.

As a learning mode, studios (Chapter 16) are at the heart of design education. They are small-enrollment classes in which students work as individuals or in teams with one or more faculty to fashion spatially explicit solutions to complex landscape problems. Each student has his or her own drafting table (and, increasingly, a personal computer) for an intensive learning experience that may require twelve hours per week of in-class time alone. Faculty target different skills and focal understandings by the choice of problem statements and landscapes in a sequence of studios that extend throughout the curriculum. Teaching demands can be high. Faculty who choose to engage real projects and stakeholders may essentially need to construct a new class every time they teach a studio. Yet studio is the key to design education. It is a dynamic and generative framework in which faculty guide students through the processes of discovery, analysis, idea generation, and proposal development. As described in later chapters, it is a prime vehicle for students to strategically engage ecological knowledge within the context of a cultural problem.

Overhauling higher education is not easy. But as David Orr (1999) queries: "Do we intend to perpetuate an academic system . . . or do we wish to preserve the Earth's biota?" (p. 1243). As he and other authors describe in this book, we believe the endeavor is worth the effort.

A Guide to the Book

We have organized the chapters in this volume in four main parts. The first, "Theories of Nature in Ecology and Design," is composed of three chapters that focus on the shifts that have occurred in theories of ecology and design that draw on nature as a source of authority. Part II, "Perspectives on Theory and Practice," is a set of four chapters that ask what designers and planners need to know about ecology from a range of professional, theoretical, practical, and philosophical perspectives. The third part, "Education for Practice," contains three chapters that explore the conceptual frameworks needed to realize an ecologically based approach to design and planning education and how that foundation can lead to more meaningful practice. Part IV, "Prescriptions for Change," includes six synthetic papers derived from conference working groups that offer recommendations for the integration of ecology within a landscape architecture curriculum and consider the links from education to professional practice. We conclude with a chapter that examines the key ideas presented in the book and identifies unresolved issues for future dialogue. In it we also consider shared ideas and points of contention that may evolve into different schools of thought in future design theory and practice.

Conclusion

With this book we have posed a question: What is an appropriate relationship between design and ecology, and in education and practice? Many voices need to be heard. They range from artists to scientists, from thoughtful academics to down-to-earth practitioners, and from visionary advocates to critical skeptics. We do not advocate a single solution, but rather a diversity of approaches sufficient to match the diversity of ecological, cultural, and institutional systems we inhabit.

Sometimes a single moment can capture the core of a complex issue. After two days of presentation and discussion at the Shire Conference, we all boarded a large bus and headed out into the landscape. Looking over the Columbia Gorge from a roadside perch hundreds of feet above the river, we were struck by the stunning beauty of the place: the brilliant blue of the sky meeting the shadowed rock of sheer cliffs, the dark green of the forested slopes and the bright green of agricultural fields cast against the brilliant blue of the wide Columbia River. And yet against this backdrop, Jim Karr was speaking of the Columbia Gorge as a collapsing ecosystem, of plummeting salmon populations, of agricultural chemicals dumping into the river, of an ecosystem on the verge of collapse from the combined weight of urban and rural development, pollution, dams, timber harvest, and poor agricultural practices. Collapsing ecosystems may be perceived as beautiful, and not everyone may see beauty in intact ones. But there seems to be no intrinsic conflicts between beauty and ecological health.

There are, however, important questions that must still be answered. How do we resolve the disparity between our aesthetic perceptions of beauty and the functional realities of ecosystems, which embody so many forces invisible to the uninformed eye? How can we align human aspirations and practices with new understandings of how ecosystems work? How do we make knowledge generated in the halls of academia relevant in the offices of professional practice, and what in turn do we need to learn from those who day after day face the challenges and insights of practice? The words stewardship, right living, care, and compassion come to mind as some of the most essential qualities for an ecological and humanistic approach to design and planning. Yet good attitudes and heartfelt goals are unlikely to succeed without skills, proficiency, critical thinking, and the creative interplay of ideas from far-ranging fields.

Education that aims to mold the ways people understand, imagine, and propose landscapes is an enthralling yet daunting enterprise. In the process of making this book, we have learned that our question is not "What from ecology should be incorporated in design?" or "What from design and planning is relevant to ecology?" Rather, from our historical position in the early years of the twenty-first century, we are asking, "What does it mean to be a designer? What does it mean to be an ecologist?" The answers are as much a process of learning as of teaching.

Citations

Brown, B., T. Harkness, and D. Johnston. 1998. Eco-revelatory design: nature constructed/nature revealed. Landscape Journal Special Issue.

Cannon, J. R., J. M. Dietz, and L. A. Dietz. 1996. Training conservation biologists in human interaction skills. Conservation Biology 10: 1277–1282.

Christensen, N. L., et al. 1996. The report of the Ecological Society of America committee on the scientific basis for ecosystem management. Ecological Applications 6: 665–691.

Collins, J., A. Kinzig, N. Grimm, W. Fagan, D. Hope, J. Wu, and E. Borer. 2000. A new urban ecology. American Scientist 88: 416–425.

Corner, J. 1997. Ecology and landscape as agents of creativity. Pages 81–108 in G. F. Thompson and F. R. Steiner, editors. Ecological design and planning. John Wiley and Sons, New York, New York, USA.

Grumbine R. E. 1994. What is ecosystem management? Conservation Biology 8: 27–38.

Hough, M. 1995. Cities and natural processes. Routledge, New York, New York, USA.

Howett, C. 1987. Systems, signs, sensibilities: sources for a new landscape aesthetic. Landscape Journal 6: 1–12.

———. 1998. Ecological values in twentieth-century landscape design: a history and hermeneutics. Landscape Journal, Special Issue: Eco-revelatory design: Nature constructed/nature revealed, pp. 80–98.

Jacobson, S. K. 1990. Graduate education in conservation biology. Conservation Biology 4: 431–440.

———. 1995. New directions in education for natural resources management. Pages

297–310 in R. Knight and S. Bates, editors. A new century for natural resource management. Island Press, Washington, D.C., USA.

Jacobson, S. K., and M. D. McDuff. 1998. Training idiot savants: the lack of human dimensions in conservation biology. Conservation Biology 12: 263–267.

Janzen, D. H. 1998. Gardenification of wildland nature and the human footprint. Science 279: 1312–1313.

Jordan, W. R. 1995. Good restoration. Restoration and Management Notes 13: 3–4.

Kohm, K., P. D. Boersma, G. K. Meffe, and R. Noss. 2000. Putting the science into practice and the practice into science. Conservation Biology 14: 593–594.

Lidicker, W. Z. 1998. Revisiting the human dimension in conservation biology. Letter in Conservation Biology 12: 1170–1171.

Lyle, J. T. 1985. Design for human ecosystems: landscape, land use, and natural resources. Van Nostrand Reinhold, New York, New York, USA.

MacMahon, J. A. 1997. Ecological restoration. Pages 479–511 in G. K. Meffe and C. R. Carroll, editors. Principles of conservation biology, second edition. Sinauer Associates, Sunderland, Massachusetts, USA.

Mattingly, H. T. 1997. Seeking balance in higher education. Conservation Biology 11: 1049–1052.

Meffe, G. K. 1998. Softening the boundaries. Conservation Biology 12: 259–260.

Meffe, G. K., and C. R. Carroll. 1997. What is conservation biology? Pages 3–27 in G. K. Meffe and C. R. Carroll, editors. Principles of conservation biology, second edition. Sinauer Associates, Sunderland, Massachusetts, USA.

Meffe, G. K., R. F. Noss, and S. K. Jacobson. 1998. Revisiting the human dimension in conservation biology, response to W. Z. Lidicker. Letter in Conservation Biology 12: 1171–1172.

Meyer, E. K. 1997. The expanded field of landscape architecture. Pages 45–79 in G. F. Thompson and F. R. Steiner, editors. Ecological design and planning. John Wiley, New York, New York, USA.

Mozingo, L. A. 1997. The aesthetics of ecological design: seeing science as culture. Landscape Journal 16: 46–59.

Nassauer, J. I., ed. 1997. Placing nature: Culture and landscape ecology. Island Press, Washington, D.C., USA.

Norton, B. G. 1991. Toward unity among environmentalists. Oxford University Press, New York, New York, USA.

Noss, R. F. 1997. The failure of universities to produce conservation biologists. Conservation Biology 11: 1267–1269.

Orr, D. W. 1993. The problem of disciplines/the discipline of problems. Conservation Biology 7: 10–12.

———. 1999. Education, careers, and callings: the practice of conservation biology. Conservation Biology 13: 1242–1245.

———. 2001. The nature of design. Oxford University Press, New York, New York, USA.

Orth, D. J. 1995. Pogo was right! Let's change the way we teach fisheries management. Fisheries 20: 10–13.

Parlange, M. 1998. The city as an ecosystem. Bioscience 48: 581–585.

Pickett, S. T., W. R. Burch Jr., S. E. Dalton, T. W. Foresman, J. M. Grove, and R. Rowntree. 1997. A conceptual framework for the study of human ecosystems in urban areas. Urban Ecosystem **1**: 185–199.

Richter, B. D., and K. H. Redford. 1999. The art (and science) of brokering deals between conservation and use. Conservation Biology **13**: 1235–1237.

Rowe, J. S. 1992. The ecosystem approach to forestland management. Forestry Chronicle **68**: 222–224.

———. 1996. Land classification and ecosystem classification. Environmental Monitoring and Assessment **39**: 11–20.

Saberwal, V. K., and A. Kothari. 1996. The human dimension in conservation biology curricula in developing countries. Conservation Biology **10**: 1328–1331.

Spirn, A. W. 1984. The granite garden: urban nature and human design. Basic Books, New York, New York, USA.

———. 1988. Poetics of city and nature: towards a new aesthetic for urban design. Landscape Journal **7**: 108–126.

———. 1998. The language of landscape. Yale University Press, New Haven, Connecticut, USA.

Thompson, G. F., and F. R. Steiner, eds. 1997. Ecological design and planning. John Wiley, New York, New York, USA.

Touval, J. L., and J. M. Dietz. 1994. The problem of teaching conservation problem solving. Conservation Biology **8**: 902–904.

Viederman, S., G. K. Meffe, and C. R. Carroll. 1997. The role of institutions and policymaking in conservation. Pages 545–574 in G. K. Meffe and C. R. Carroll, editors. Principles of conservation biology, second edition. Sinauer Associates, Sunderland, Massachusetts, USA.

Notes

1. The word "ecology" is derived from the Greek words *oikos* and *logos*, referring to household, and wisdom or knowledge, respectively. To ecologists it is a science founded on keen observation and measurement of the natural world. Designers and planners, along with other nonecologists, however, may use the term in a number of ways. Ecology may be invoked as a science, but it may also be considered as a framework for understanding, or as a metaphor or philosophical foundation for living. Each usage is important, but it is critical to know which meaning is being invoked in any given instance. In this book we have attempted to frame the term consistent with its use in ecological science, while recognizing other meanings. We note, however, that not all designers and planners, including some authors in this book, restrict themselves to a scientific definition of the word ecology.

2. We use the word "landscape" in the sense that Rowe (1992, 1996) describes ecosystems as spatially defined, three-dimensional volumetric units of the ecosphere—that is, geographic ecosystems that include all the components of life: atmosphere, land forms and earth layers (including water), and biota (organisms). At the same time, we include the understanding that because people

imbue landscapes with meaning, they also serve as symbolic referents and sources of nourishment for deep aspects of human culture and the human psyche.

3. Our categories reflect, in part, the way related issues have been addressed in design and planning education. Most of the ecological literature cited derives from an ongoing forum on conservation education in the journal *Conservation Biology*.

PART I

Theories of Nature in Ecology and Design

How have past theories of nature influenced the design and planning disciplines and ecology? What influences should and could current understandings have for the future of design and planning? This section is composed of three chapters that focus on the need for change in the education and practice of landscape architecture specifically, and more broadly in land planning and design professions. Each author explores the interplay between concepts of nature and disciplinary theory from different points of view. In Chapter 2, "The Authority of Nature: Conflict, Confusion, and Renewal in Design, Planning, and Ecology," designer Anne Whiston Spirn considers what we mean by nature and provides a history of how the ideas of several prominent design and planning professionals have influenced the definition of nature in American design and planning. Chapter 3, "Ecology's New Paradigm: What Does It Offer Designers and Planners?" by ecologist H. Ronald Pulliam and designer and ecologist Bart R. Johnson, explores the conceptual underpinnings of contemporary ecology and their implications for design and planning, including emerging ideas of landscape memory and ecosystem autonomy. Ecologist Richard T. T. Forman, in Chapter 4, "The Missing Catalyst: Design and Planning with Ecology Roots," envisions a marriage between design and ecology to produce bold practice with solid theoretical foundations.

CHAPTER 2

The Authority of Nature:
Conflict, Confusion, and Renewal in
Design, Planning, and Ecology

Anne Whiston Spirn

Landscapes are shaped by rain and sun, plants and animals, human hands and minds. Whether wild or clipped, composed of curved lines or straight, living plants or plastic, every garden and every region is a product of natural phenomena and human artifice. It is impossible to make a landscape without expressing, however unconsciously, ideas about nature. For thousands of years, nature has been both mirror and model for landscape design, has been looked to for inspiration and guidance.[1] Many practitioners who established landscape architecture as a profession in the nineteenth century accepted George Perkins Marsh's (1864) challenge: "In reclaiming and reoccupying lands laid waste by human improvidence or malice . . . the task is to become a co-worker with nature in the reconstruction of the damaged fabric" (p. 35). Landscape architects have explored and debated what it means to design with nature for well over a century. Clarifying conflicts and dispelling confusion over the nature of nature and of landscape design and planning are essential to renewing both the discipline and the landscapes themselves. Fields such as conservation biology, restoration ecology, and ecological engineering, which have emerged in recent years, may learn from this cautionary tale.

Designers and planners who refer to their work as "natural" or "ecological" make ideas of nature central and explicit, citing nature as authority to justify decisions, to select some materials or plants and exclude others, for example, to arrange them in particular patterns, and tend the result in certain ways. Appealing to nature as the authority for landscape architecture has pitfalls that are often overlooked by advocates of "natural" gardens and ecological planning and design. To describe one sort of landscape as natural implies that there are unnatural landscapes that are somehow different (and presumably wrong). Yet, over time and place, quite different sorts of landscapes have been claimed as natural, much the same way opposing nations claim to have God on their side. In fact, some designers invoke nature to call upon divine authority. To Frank Lloyd Wright, for example, nature was the manifestation of God: "Nature should be spelled with a capital 'N,' not because Nature is God but because all that we can learn of God we will learn from the body of God, which we call Nature."[2]

Now, too, the authority of science is cited to augment the authority of nature and God. Today, many landscape architects regard ecological science as an important source of principles for landscape design and planning. Indeed, the adoption of ideas from ecology contributed to a renewal of the discipline in the 1960s. Some, however, have embraced ecology as the primary authority for determining the "natural" (and therefore correct) way to design landscapes. To its most extreme practitioners, ecological design is deterministic, its "laws" couched in terms that recall religious dogma. This chapter should in no way be interpreted as a rejection of the approach launched at the University of Pennsylvania in the early 1960s and dubbed "ecological" design, but rather should be seen as an attempt to construct firmer ground for future discussions.[3] What I am attacking here is dogma, and what I am urging is a more reasoned, inclusive approach, well cognizant of the problems inherent in appeals to authority in general and to nature in particular.

Debates over what constitutes a "truly ecological landscape architecture" have escalated in recent years, with various groups accusing each other of "nonecological" behavior, and with the phrase "nonecological" used by authors with divergent views. Chapters by various authors in Thompson and Steiner 1997, for example, reveal some of the conflict and confusion in the field, as well as some pitfalls of appealing to "ecology" or "nature" for authority in landscape design. There have been bitter quarrels among landscape architects over the proper materials, styles, and methods of "ecological" landscape design. Some advocate the exclusive use of native, as opposed to naturalized, plants, while others advocate naturalistic plantings regardless of species composition or ecological function. Some urge the eradication of "exotic" "invaders" and condemn others for planting naturalized, nonnative plants. Some conceal the artifice of their works; others celebrate the human ability to transform the landscape. Some privilege the role of reason in design and promote science as the sole source of truth about nature, while others prefer personal revelation and reject science as a way of knowing.[4] Once confined to professional journals, such debates have spilled over into the popular press. "Attack of the Killer Weeds!" announced the cover of *The Philadelphia Inquirer Sunday Magazine* on June 30, 1996, alluding to the lead article,

"Aliens Among Us." Polemical rhetoric plays on fear of the alien, provokes opposition as well as support, and, ironically, obscures the real problems some species pose (Rodman 1993).

Such conflicts and the confusion they engender are about competing sources of authority and conflicting ideas of nature: whether humans are outside or inside nature, whether human impact is inevitably destructive or potentially beneficial, whether one can know an objective nature apart from human values. Anyone who invokes the authority of nature implies that he or she speaks for nature. But who confers that right and why? Some believe authority comes from traditional precedent: from the way things have "always" been done, or were done previously in some idealized period or exemplary models. Others derive authority from a rational system of rules or laws that can be proved or explained. Some are persuaded by the statements of a charismatic leader.[5] Differences in basic assumptions are so fundamental they may make it impossible to resolve the conflicts, but it is possible to clarify differences and dispel confusion. Much confusion comes from launching the debate without defining its terms. What is nature anyway?

The Nature of Nature

Nature is an abstraction, a set of ideas for which many cultures have no one name, "a singular name for the real multiplicity of things and living processes" (Williams 1980). The singular quality of the word masks this multiplicity and implies that there is a single definition, an impression that is grossly misleading. A. O. Lovejoy identified sixty-six different senses of the words "nature" and "natural" as used in literature and philosophy from the Ancient Greeks to the eighteenth century (Lovejoy 1935). The abstract quality of the word strips nonhuman features and phenomena of agency, of exerting an active force upon the world, on the one hand, yet invites personification ("Nature's revenge"), on the other.

Nature is both given and constructed. There is always a tension between the autonomy of nonhuman features and phenomena and the meanings ascribed to them. Nature is the word Raymond Williams (1983) called "perhaps the most complex word in the language" (p. 219).[6] It comes from the Latin *natura*, which comes in turn from *nasci*, to be born. Thus nature is linked to other words from the same root, such as nascent, innate, native, and nation. In English, as in French and Latin, the word "nature" originally described a quality—the essential or given character of something—then later became an independent noun. Williams identified two additional areas of meaning: "the inherent force which directs either the world or human beings or both" and "the material world itself, taken as including or not including human beings" (p. 219).

Nature is a mirror of and for culture. Ideas of nature reveal as much or more about human society as they do about nonhuman processes and features. Even as human cultures describe themselves as reflections of nature, their ideas of nature also mirror their culture. Lovejoy's review of the words "nature" and "natural" reveals how integral ideas of nature have been to religion, politics, and beliefs about what constitutes normal or abnormal, right or wrong behavior. Nor has science been immune to normative notions of nature. When ecologists once described the "harmony" of nature and the succession of plant "communities" from pioneers to stable climax forest, they were also describing a model for human society (see Worster 1979; Botkin 1990; Mitman 1992; Golley 1994). The shift in plant ecology from the dominance of Clements's association-unit theory prior to 1950 to that of Gleason's individualistic concept paralleled shifts in American culture (Barbour 1995). The idea of The Fall—of humanity expelled from paradise, a former state of grace within nature—has exerted a powerful influence on the imagination in Western cultures. Ecology, anthropology, and landscape architecture are laced with Edenic narratives, stories of an initial state of harmony, perfection, and innocence in which humans lived as one with other living creatures followed by the forced separation of humans from nature, often accompanied by nostalgia for the perfect past and a view of "native" peoples as living in a more worthy, morally superior relation to nature (Slater 1995).[7]

As products of culture, ideas of nature vary from people to people, place to place, period to period. Even in a particular time and place, what constitutes the "natural" way of doing things has been disputed. Frank Lloyd Wright and Jens Jensen, fellow residents of Chicago and Wisconsin, friends throughout most of their lives, agreed that nature was *the* authority for design and sought to express the moral messages or "sermons" they read in hills and valleys, rivers and trees (Jensen 1956). Despite this apparent common ground, the two men "argued incessantly about the nature of nature," about what form a "natural" garden should take (Tafel 1985, p. 152).[8]

Wright's understanding of nature was grounded in his family's Emersonian philosophy.[9] He had contempt for "some sentimental feeling about animals and grass and trees and out-of-doors generally," as opposed to reverence for nature as an internal ideal, the very "'nature' of God" (Wright 1937, p. 163). To Wright, landscape was often an imperfect manifestation of nature; the task of the architect was to bring its outer form in closer conformity with an inner ideal, its *nature*, or essential characteristics. Wright derived his principles for design from the underlying *structure* of flowers, trees, and terrain, and his landscape designs were often abstract versions of regional landscapes of prairie or desert.

If Wright's obsession was to extract and express an ideal inner nature, Jensen's was to protect and promote the "native" features of regional landscapes. Jensen believed there was a correspondence between a region's climate, physiography, and flora and its human inhabitants; landscape fostered, then symbolized, a relationship between people and place. Unlike Wright, Jensen (1956) gave no impression in his published works that he believed humans could improve upon the "native" landscape: "Nature talks more finely and more deeply when left alone" (p. 94). He revered what he called the "primitive" and found his "main source of inspiration . . . in the unadulterated, untouched work of the great Master" (p. 23). These ideas led Jensen to imitate the outward appearance of the local landscape, its meadows, woodlands, and riverbanks: "Through generations of evolution our native landscape becomes a part of us, and out of this we may form fitting compositions for our people" (p. 21).

Many of Jensen's ideas, such as the relation he saw between nature and nation and his advocacy of native plants, were common ideas in Europe and North America.[10] Contemporary ecological theories drew parallels between plant and animal "communities" and human communities and, in some cases, extended this analogy to justify certain human activities as "natural."[11] Ideas of the relationship between native plants and "folk," however, were carried to ideological extremes by German landscape architects under National Socialism.[12] The use of "native" plants and "natural" gardens to represent the Nazi political agenda should dispel forever the illusion of innocence surrounding the words nature, natural, and native and their application to garden design. Nature is one of the most powerfully loaded, ideological words in the English (and German) language.

"Nature" and "natural" are among the words landscape architects (and ecologists) use most frequently to justify their designs (or research) or to evoke a sense of "goodness," but they rarely examine or express precisely what the words mean to them, and they are generally ignorant of the ideological minefields they tread. Invoking nature, they imagine they are talking about a single phenomenon with universal meaning, when in fact their ideas may be entirely different from one another, even antithetical. At first, the abstraction of the word nature conceals differences. Then, when arguments inevitably ensue, it befuddles and confounds.

The Nature of Landscape Design and Planning

Landscape architects (and ecologists) hold strong ideas about nature; whatever it means to them, they tend to care about it, for the beliefs and values those ideas represent are usually at the heart of why they entered the profession. For the past thirteen years, I have asked my graduate students: What is nature? Their responses have included the following: Nature was given as a trust to humans by God; nature is trees and rocks, everything except humans and the things humans make; nature is a place where one cannot see the hand of humans, a place to be alone; nature consists of creative and life-sustaining processes that connect everything in the physical and biological worlds, including humans; nature is a cultural construct with no meaning or existence outside human society; nature is something that cannot be known; Nature is God. While this is a broad range of definitions, it does not represent the full spectrum of possible answers; the experiential and spiritual aspects of nature are cited frequently, for example, and nature as material resource is rarely mentioned. For several years I asked students for their personal definition of nature on the first day of class, and then at the end of the course asked them to write a short paper defining nature once again. Their answers were more articulate and reflective, but rarely changed in substance from the first brief statement. I have concluded that ideas of nature are deeply held beliefs, closely tied to religious values, even for those people who do not consider themselves religious.[13]

Tensions and contradictions in landscape architecture also stem from inherent, unresolved conflicts among the disciplines from which it draws. The roots of landscape architecture lie in several constellations of disciplines: agriculture (gardening, horticulture, forestry); engineering; architecture and fine arts; science (ecology, geology). These constellations are based on disparate ideas about the relationships of humans to nonhuman features and phenomena. Agriculture, engineering, and architecture are founded on the idea that nature can be improved upon, whereas ecologists tend to be observers of, rather than actors upon, nature. To gardeners (and, by extension, to horticulturalists and foresters), humans are stewards who manage plants, animals, and their habitats for human ends, for sustenance and pleasure; nature is both material and process, something to be reckoned with. To most engineers, nature consists of forces to be controlled or overcome. Engineers such as Ken Wright of Denver, who has devised drainage and floodways that deflect or adjust to flowing water, are exceptional; so are architects such as the Australians Glenn Murcutt and Richard Le Plastrier, who regard landscape processes as active agents and design their buildings to respond to wind, water, light, and heat; and so are artists such as Robert Smithson, James Turrell, Alan Sonfist, Newton and Helen Harrison, and Doug Hollis, who engage processes of erosion, water flow, light, wind, sound, and plant growth in their works. To most artists and architects, however, nature is generally not an active agent, though it is a source of inspiration, of symbolic forms to be drawn upon, a scene to be represented, a site to be occupied and transformed, something perceived. On the other hand, despite growing interest in applied fields such as restoration ecology and conservation biology, many ecologists consider humans to be interlopers in nature and have focused mainly on wild and rural landscapes.

These differences among disciplines are emphasized further by the fact that members of each discipline recognize the validity of a different type of authority to defend understanding of the world and justify their actions. While most derive authority to some degree from tradition, systems of rules, and charismatic leadership, they give more or less weight to each of these types. Modern science, for example, is based on the idea of rational, systematic studies whose results can be replicated. Historians of science have demonstrated that scientific practice is also tradition-bound (until the next paradigm shift), its course swayed by the ideas of powerful personalities; nevertheless, rational proofs are recognized as the only legitimate authority. Architecture, on the other hand, has long acknowledged the authority traditionally vested in certain styles (e.g., classicism, the vernacular) and exemplary buildings (the Pantheon, the Villa Savoye). Most architects seek legitimacy for their buildings through reference to a stylistic tradition or original model. Artists have more license to flaunt authority than do architects or scientists; society does not hold artists as accountable for their works. Particularly in the twentieth century, artists gained authority through originality, the production of works unlike anything seen before.

Landscape architects have drawn broadly from other disciplines without examining and reconciling the beliefs and traditions upon which they are based.[14] There is also a tendency to accord higher status to ideas generated in other disciplines, to cite authors from outside the field, but to ignore pertinent works in landscape architecture, and to draw freely from precedent without acknowledgment. Landscape architects fail repeatedly to build upon prior efforts and often reiterate ideas without advancing them significantly. The desire to be seen as original is typical of the field, and advocates of ecological design and planning are no exception. Ian McHarg ignored precedent when he asserted, as he has many times, "I invented ecological planning in the 1960s" (1997, p. 321). McHarg has made an enormous contribution to the theory and practice of landscape architecture, especially in the incorporation of ideas from ecology. The importance of his contributions is not diminished when seen in the context of work by others such as Phil Lewis, Angus Hills, and Artur Glickson (Glickson 1971) who pursued similar ideas from the 1950s and early 1960s, not to mention many prior figures, such as Patrick Geddes, Charles Eliot, and Warren Manning. This tradition was not acknowledged in the Department of Landscape Architecture and Regional Planning at the University of Pennsylvania when I was a student there in the early 1970s, nor did we draw from it in our work at Wallace McHarg Roberts and Todd during that period. Though both department and firm made numerous innovations, there were also many reinventions. Whether through honest oversight or deliberate disregard, failure to acknowledge precedents has prevented an appreciation for the evolution of ideas and practices in landscape architecture. Unless this habit is overcome, landscape architecture will not mature as a profession, and ecological design and planning will not mature as a field.[15]

The habit of borrowing theory and methods from other fields and applying them directly and uncritically to landscape architecture not only works against their integration, it often places these disparate ways of knowing and working in hostile juxtaposition. In graduate schools, it is not unusual to find students with backgrounds in horticulture, art, architecture, engineering, and ecology in the same class, and the faculty often includes members of several of these disciplines. At best, mating these fields in a single faculty is a rich marriage of ideas. At worst, it is a shotgun wedding where individuals cannot find common ground. Few have combined these roots successfully and inventively. The unresolved differences in academic departments over meanings of nature and ways of knowing have been played out in practice, producing a major muddle and too few built landscapes that fuse the contributions of art and science, gardening and engineering.

In 1957, Sylvia Crowe called landscape architecture a bridge between science and art, a profession whose greatest task was to "heal" the "breach between science and humanism, and between aesthetics and technology" (p. 4). Landscape architecture and its relation to allied disciplines was the subject of International Federation of Landscape Architects meetings during this period.[16] Many years later, landscape architecture is still caught in the breach, struggling to construct a core that integrates its diverse roots rather than privileging one over the others. Each root is distinctive: each has an important contribution to designing, planning, and managing landscape.

In 1969, Ian McHarg's *Design with Nature* led to fundamental changes in the teaching and practice of landscape architecture. McHarg advocated the systematic application of a set of "rules" derived from ecological science and demonstrated the value of this approach in professional projects. His charismatic personality and polemical language captured the attention of the profession and public, attracted a large following, and were instrumental in the acceptance of ideas that had also been explored by others. Nearly thirty years later, many innovations once seen as radical are now common practice. The claim that science is the only defensible authority for landscape design, however, proved particularly damaging to discourse and practice in landscape architecture. When McHarg, for example, continues to use the words nature and ecology interchangeably, as an "imperative" or "command" for design, he brooks no dissent: "I conceive of non-ecological design as either capricious, arbitrary, or idiosyncratic, and it is certainly irrelevant. Non-ecological design and planning disdains reason and emphasizes intuition. It is anti-scientific by assertion" (McHarg 1997, p. 321). Such aggressive overstatements have provoked equally dogmatic reactions from those who seek to promote landscape architecture as an art form. Provoked by such statements, many proponents of a new artistic thrust in landscape architecture chose to set this movement in opposition to "the ecological movement and its detrimental consequences for design." One article included gratuitous, unfounded attacks, some from critics who chose to remain anonymous, such as: "The so-called Penn School led by McHarg produced a generation of landscape graduates who did not build" (Boles 1989, p. 53). Statements such as these were retracted by the editors in a subsequent issue of the journal in response to letters to the editor.

Ecology as a science (a way of describing the world), ecology as a cause (a mandate for moral action), and ecology as an aesthetic (a norm for beauty) are often confused and conflated. McHarg (1997) does so when he calls ecology "not only an explanation, but also a command" (p. 321). As does his critic, 'James Corner,' when he offers an alternative "truly ecological practice of landscape architecture" and refers to "the processes of which ecology and creativity speak" as leading to "freedom" (1997, pp. 81, 102). It is important to distinguish the insights ecology yields as a description of the world, on the one hand, from how these insights have served as a source of prescriptive principles and aesthetic values, on the other. The perception of the world as a complex network of relations has been a major contribution of ecology, permitting us to see humans, ourselves, as but one part of that web. There has been a tendency, however, to move directly from these insights to prescription and proscription, citing ecology as an authority in much the same way that nature was employed in the past to derive laws for landscape design and to define a single aesthetic norm, in this case "the ecological aesthetic." Laurie Olin (1988) has criticized this approach as "a new deterministic and doctrinaire view of what is 'natural' and 'beautiful'" embodying a "chilling, close-minded stance of moral certitude" (p. 150).

Constructing Nature

Landscape architects construct nature both literally and figuratively, but the history of twentieth-century landscape architecture has been told largely as a history of forms rather than a history of ideas and rhetorical expression. This has been especially true of the history of "natural" or "ecological" design. Gardens of different periods built to imitate "nature" may appear similar, yet express different, even divergent, values and ideas. The Fens and Riverway in Boston and Columbus Park in Chicago, for example, were built to resemble "natural" scenery of their region, but the motivations that underlie them were quite different in several important respects. These projects are cited often as precedents and models for an ecological approach to landscape design without critically examining the values and motives that underlay them, thereby further confounding the current confusion around issues of nature and authority.[17]

Boston's Fens and Riverway, designed by Frederick Law Olmsted, were built over nearly two decades (1880s–1890s), the first attempt anywhere, so far as I know, to construct a wetland.[18] The function and the form of the Fens and Riverway were revolutionary; the "wild" appearance was in contrast to the prevailing formal or pastoral styles.[19] These projects, built on the site of tidal flats and floodplains fouled by sewage and industrial effluent, were designed to purify water and protect adjacent land from flooding. They also incorporated an interceptor sewer, a parkway, and Boston's first streetcar line. Together, they formed a landscape *system* designed to accommodate the flow of water, removal of wastes, and movement of people; Olmsted conceived them as a new type of urban open space that he took care to distinguish from a park. This skeleton of woods and wetland, road, sewer, and public transit structured the growing city and its suburbs. The Fens and Riverway were a fusion of art, agriculture, engineering, and science. Olmsted's contemporaries knew that these parks were constructed, for they had seen and smelled the stinking, muddy mess the Fens replaced; the recognition of the transformation was part of their social meaning and aesthetic power.

Jens Jensen designed Columbus Park (1916) in Chicago thirty years later to "symbolize" a prairie landscape (1956, p. 76). He made a large meadow, excavated a meandering lagoon, and planted groves of trees as a representation of the Illinois landscape: prairie, prairie river, and forest edge. All the plants used in the park were native to Illinois; they "belonged," as Jensen (1956, p. 77) put it. In outward appearance, the "prairie river" looked much like the Fens, as testified by photographs taken of each within about a decade of construction. Both Olmsted and Jensen intended their projects to expose townspeople to what they saw as the beneficial influence of rural scenery, particularly those people who were unable to travel to far-off places and were barred from "neighboring fields, woods, pond-sides, river-banks, valleys, or hills" (Olmsted 1895, pp. 253–254). Despite these similarities, the aims of the two men and the goals of their projects were very different in important ways.

Jensen's agenda at Columbus Park and elsewhere was to bring people, especially "the growing minds" of youth, into contact with their "home environment," for he believed that "[w]e are molded into a people by the thing we live with day after day" (1956, p. 83). Every region should display the beauty of its local landscape: "This encourages each race, each country, each state, and each county to bring out the best within its borders" (1956, p. 46). Jensen elaborated on these ideas of "environmental influences" in *Siftings* (1956), where he attributed certain characteristics among populations of European countries and American regions to the influence of their landscapes. While he stressed that each regional landscape has its own beauty, he repeatedly revealed his prejudice for the superiority of northern regions and peoples with such statements as: "Environmental influences of the hot south have almost destroyed the strong and hardy characteristics of . . . northern people" (1956, p. 35). Jensen drew parallels between people and plants and advocated the sole use of species native to a place: "To me no plant is more refined than that which belongs. There is no comparison between native plants and those imported from foreign shores which are, and shall always remain so, novelties" (p. 45).

Like many of his contemporaries, Olmsted thought that environment influenced human behavior, but his views and focus were different from Jensen's. He believed that contemplation of "natural scenery" had beneficial physical, mental, and moral effects, and that the lack of such opportunity could lead to depression and mental illness.[20] In constructing "natural" scenery, Olmsted advocated the practice of mixing native and hardy exotic plants, as described in William Robinson's *The Wild Garden* (1870), and argued with Charles Sprague Sargent, who opposed using nonnative plants in the Riverway. The upshot was that only native species were planted on the Brookline side of the Riverway (where Sargent had the authority of approval), while a mixture of native and nonnative species were planted on the Boston side![21] The primary purpose of the Riverway was "to abate existing nuisances, avoid threatened dangers and provide for the permanent, wholesome and seemly disposition of the drainage of Muddy River Valley" (Olmsted 1881). The Fens and Riverway are an application of the ideas proposed by George Perkins Marsh in *Man and Nature* (1864). In reclaiming polluted tidal flats and derelict floodplain, Olmsted planned to "hasten the process already begun" by nature, thereby achieving more than the "unassisted processes of nature" as he would also do a few years later at Niagara Falls (Olmsted 1887, pp. 21, 8). He must have been familiar with Marsh's well-known book, which was reprinted several times in the nineteenth century. The attempt to manage landscape processes to restore land and water polluted by human wastes and to promote human health, safety, and welfare was what made these projects so significant. Such goals were largely absent from Jensen's work.

The natural garden movement in the early part of the twentieth century, of which Jensen was a proponent, and the ecological design movement of the latter part seem to have much in common. Both have stressed native plants and plant communities as material and model for garden design. Beyond these and other similarities, however, there are deep differences in the ideas of nature underlying the two movements. In the United States, natural garden design in the early twentieth century was part of the larger context of regionalism expressed in art, literature, and politics. American regionalism was a populist movement that promoted the local roots of place and folk over the increasing power of the federal government, the growth of national corporations, and the influence of foreign styles (Dorman 1993). Jensen used regional landscapes and native plants to shape human society; he never discussed the value of plants, animals, or biological and physical processes apart from their significance for human purpose. This anthropocentric context is a contrast to late-twentieth-century environmentalism where animals, plants, and ecosystems may be accorded value, and even legal rights, not just for the present or future value they may have for humans, but also for themselves.[22]

Reconstructing Nature, Recovering Landscape and Language, Renewing Landscape Architecture

The features and phenomena we refer to as nature are both given and constructed; authors from Cicero to Karl Marx have distinguished between a "first" and a "second" nature, where the first represents a nature unaltered by human labor. Cicero defined second nature thus: "We sow corn, we plant trees, we fertilize the soil by irrigation, we confine the rivers and straighten or divert their courses. In short, by means of our hands we try to create as it were a second nature within the natural world" (Hunt 1993, p. 312). John Dixon Hunt has reminded us that gardens have been called a "third nature," a self-conscious re-presentation of first and second natures, an artful interpretation "of a specific place . . . for specific people" (p. 325).

Today, many people are struggling to redefine nature, and the landscape reflects this struggle. There is no consensus. Is nature a sacred entity where humans are one with all living creatures, or a wilderness refuge requiring protection from man? Or is nature just a bunch of resources for human use? Is nature a web of processes that link garden, city, and globe? These different natures and others all coexist in contemporary society. They underlie whether and how people value and shape landscapes and gardens. Despite this range of ideas about the nature of nature, there is widespread international concern about the future environment and a growing sense that we need to reconstruct our conceptions of nature, to find ways of perceiving and relating to nonhuman features and phenomena that assert the dynamic autonomy of the nonhuman while they also affirm the importance of human needs and dreams.

Landscape architects have a potential contribution to this exploration, and the landscapes we shape are part of our discourse, ideas in the archaic sense of the word as "a visible representation of a conception."[23] Gardens have been a medium for working out fresh ideas and forms of human habitation, and they are particularly fertile ground for exploring relationships between the human and nonhuman. In the garden, there is a recognition of constructedness and an attitude of beneficial management, as well as an acknowledgment that certain nonhuman phenomena are beyond human control. Gardens are never entirely predictable; one cultivates a garden with an acknowledgment of unforeseen circumstances. The garden, defined by *Webster's* as "a well-tended region," is a powerful metaphor for reconceiving cities and metropolitan regions; this metaphor was the inspiration for my first book, *The Granite Garden: Urban Nature and Human Design* (Spirn 1984).

A series of gardens on Bainbridge Island near Seattle makes the case for an artful, humanist environmentalism. These gardens, designed by Richard Haag for Bloedel Reserve, a place devoted to environmental education, transcend polemics and draw their power from ideas of nature, from past garden traditions, from art. The Bird Sanctuary, a dredged pond edged by trees, set between forest and meadow, is an "artful wilderness to perplex the scene," dug, planted, inhabited (Pope 1731). The pond's water is brownish-black; small islands make it appear larger. Red alders line the shore in clumps, like sprouts sprung up from the same old stump. The way that islands lie, poplars clump, and meadow slopes looks wild yet deliberate. A path in the sanctuary leads from pond through forest to the Anteroom, a wooded moss garden, where it winds around enormous tree stumps covered by thick moss. The Anteroom juxtaposes decay and rebirth; it celebrates both artful construction and organic regeneration, the power of life to renew a forest once cut for timber. From the Anteroom's moss garden, the path leads in to the Reflection Garden, a clearing in the forest bounded by a rectangular frame of clipped yew hedges, where a long rectangular pool reflects the dark forest and reveals the groundwater beneath the spongy turf. These gardens at Bloedel do not return the site to some imagined, ideal condition before humans cut the forest; rather, they create a garden where evidence of human use is incorporated into the whole. They are stories about the use, abuse, and renewal of landscape.

To call some landscapes "natural" and others "artificial" or "cultural" misses the truth that landscapes are never wholly one or the other. Cicero's "first nature" exists only as an ideal; there is no place unaltered by human activities. His "second nature" and Hunt's "third nature" are landscapes, the expression of actions and ideas in place. Landscape associates people and place. Danish *landskab*, German *landschaft*, Dutch *landschap*, and Old English *landscipe* combine two roots. *Land* means both a place and the people living there. *Skabe* and *schaffen* mean "to shape"; suffixes -*skab* and -*schaft*, as in the English -*ship*, also mean "association" or "partnership." Though no longer used in ordinary speech, the Dutch *schappen* conveys a magisterial sense of shaping, as in the biblical Creation.[24] Still strong in Scandinavian and German languages, these original meanings have all but disappeared from English. *Webster's Dictionary* defines landscape as static, "a picture representing a section of natural, inland scenery, as of prairie, woodland, mountains . . . an expense of natural scenery seen by the eye in one view"; the *Oxford English Dictionary* traces the word to a Dutch painting term (*landskip*).[25] But landscape is not a mere visible surface, static composition, or passive backdrop to human theater; dictionaries should be revised and the older meanings revived. Landscape connotes a sense of both the given and the purposefully shaped and recovers the dynamic connection between place and those who dwell there.

Language has consequences. It structures how one thinks and what kinds of things one is able to express. If the language one uses is purely formal, one is predisposed to engage the formal; the language itself does not prompt one to incorporate other dimensions of planning, design, and management. Similarly, if one's language is purely functional, then formal concerns are less likely to receive explicit attention. Without a language that fuses form, feeling, meaning, and human and ecological function, it is difficult to address all these dimensions. Landscape architects need a language of design that represents the scope and concerns of the discipline. Designers and planners need a language that integrates natural processes and human purpose, a language that will enhance how one experiences and reads the landscape, and how one designs it, a language that will permit us to assess how well the landscapes we make satisfy our fundamental physical, social, and spiritual needs; a language that will link the everyday with art, the past with the future, and the scale of the garden with the scale of the region. My recent book, *The Language of Landscape*, describes such a language and reflects upon the kinds of places it could inspire us to create (Spirn 1998).

To recover the language of landscape is to discover and imagine new metaphors, to tell new stories, and to create new landscapes. John Berger describes, and photographer Jean Mohr illustrates, a language of lived experience with which to interpret the common and the particular across the gulf of different cultures (Berger and Mohr 1983). Gregory Bateson says that humans must learn to speak the language "in terms of which living things are organized" in order to read the world not as discrete things but as dynamic relations and to practice the art of managing complex, living systems (Bateson 1991, pp. 310–311, 253–257). Aldo Leopold writes of the need for humans to "think like a mountain," to escape the shortsightedness that threatens the larger habitats of which humans are part (Leopold 1966, pp. 129–133). The language of landscape is such a language: in terms of it the world is organized and living things behave, and humans can think like a mountain, shape landscapes that sustain human lives and the lives of other creatures as well, and foster identity and celebrate diversity.

The language of landscape is derived from the core activity of landscape architecture: artful shaping, from garden to region, to fulfill function and express meaning. The roots of this theory are strong, deep, and varied, grounded in many fields, not just landscape architecture, but ecology, geology, geography, anthropology, history, art history, literature, and linguistics, among others. It is a radical theory: in the sense of being rooted in the basic elements of nature and human nature, in the sense of offering a perspective fundamentally different from any one individual root, and in demanding and enabling radical change in how we choose to think and act.

Human survival as a species depends upon adapting ourselves and our landscapes—settlements, buildings, rivers, fields, forests—in new, life-sustaining ways, shaping contexts that acknowledge connections to air, earth, water, life, and to each other, and that helps us feel and understand these connections, landscapes that are functional, sustainable, meaningful, and artful. Not everyone will be a farmer or a fisherman for whom landscape is livelihood, but all can learn to read landscape, to understand those readings, and to speak new wisdom into life in city, suburb, and countryside, to cultivate the power of landscape expression as if our life depends upon it.

Citations

Barbour, M. G. 1995. Ecological fragmentation in the fifties. Pages 233–55 in W. Cronon, editor. Uncommon ground: toward reinventing nature. W. W. Norton, New York, New York, USA.

Boles, D. 1989. The new American landscape. Progressive Architecture (July).

Botkin, D. 1990. Discordant harmonies: a new ecology for the twenty-first century. Oxford University Press, New York, New York, USA.

Boudon, R., and F. Bourricaud. 1989. A critical dictionary of sociology. University of Chicago Press, Chicago, Illinois, USA.

Collingwood, R. G. 1945. The idea of nature. Clarendon Press, Oxford, England, UK.

Corner, J. 1997. Ecology and landscape as agents of creativity. Pages 81–108 in G. F. Thompson and F. R. Steiner, editors. Ecological design and planning. John Wiley, New York, New York, USA.

Cronon, W. 1994. Inconstant unity. Pages 8–31 in T. Riley, editor. Frank Lloyd Wright: architect. Museum of Modern Art, New York, New York, USA.

———, editor. 1995. Uncommon ground: toward reinventing nature. W. W. Norton, New York, New York, USA.

Crowe, S. 1957. Presidential address. Journal of the Institute of Landscape Architects (November).

———, editor. 1961. Space for living. Djambatan, Amsterdam, NL.

Dorman, R. 1993. Revolt of the provinces: the regionalist movement in America. University of North Carolina Press, Chapel Hill, North Carolina, USA.

Eaton, L. 1964. Landscape artist in America: the life and work of Jens Jensen. University of Chicago Press, Chicago, Illinois, USA.

Emerson, R. W. 1836. Nature. James Munroe, Boston, Massachusetts, USA.

Evernden, N. 1992. The social creation of nature. Johns Hopkins University Press,

Baltimore, Maryland, USA.

Gill, B. 1987. Many masks. Ballantine, New York, New York, USA.

Glacken, C. 1967. Traces on the Rhodian shore: nature and culture in western thought to the end of the eighteenth century. University of California Press, Berkeley, California, USA.

Glickson, A. 1971. The ecological basis of planning. Lewis Mumford, editor. Martinus Nijhoff, The Hague, NL.

Golley, F. 1994. A history of the ecosystem concept in ecology. Yale University Press, New Haven, Connecticut, USA.

Grese, R. E. 1992. Jens Jensen: maker of natural parks and gardens. Johns Hopkins University Press, Baltimore, Maryland, USA.

Groening, G., and J. Wolschke-Bulmahn. 1992. Some notes on the mania for native plants in Germany. Landscape Journal **11**: 116–122.

Hunt, J. D. 1993. The idea of the garden, and the three natures. In *Zum Naturbegriff der Genegenwart* (Stuttgart).

Jensen, J. 1956. Siftings, a major portion of The Clearing, and collected writings. Chicago, Illinois, USA.

Leiss, W. 1974. The domination of nature. G. Braziller, Boston, Massachusetts, USA.

Lewis, C. S. 1948. Nature as aesthetic norm. Pages 69–77 in Essays in the history of ideas. Johns Hopkins University Press, Baltimore, Maryland, USA.

———. 1967.Nature. In Studies in words. Cambridge University Press, Cambridge, UK.

Lovejoy, A. O. 1935. Some meanings of "Nature." Pages 447–456 in A. O. Lovejoy et al. A documentary history of primitivism and related ideas. Johns Hopkins University Press, Baltimore, Maryland, USA.

Marsh, G. P. 1864. Man and nature. Harvard University Press, Cambridge, Massachusetts, USA.

McHarg, I. 1969. Design with nature. Natural History Press, Garden City, New York, USA.

———. 1997. Ecology and design. In G. F. Thompson and F. R. Steiner, editors. Ecological design and planning. John Wiley, New York, New York, USA.

Mitman, G. 1992. The state of nature: ecology, community, and American social thought 1900–1950. University of Chicago Press, Chicago, Illinois, USA.

Olin, L. 1988. Form, meaning, and expression in landscape architecture. Landscape Journal **7**: 2, 150. A special issue on "Nature, form, and meaning," edited by Anne Whiston Spirn.

Olmsted, F. L. 1865. The Yosemite Valley and the Mariposa Big Trees: a preliminary report. Landscape Architecture **43**: 12–25 (1952).

———. 1881. General plan for the sanitary improvement of Muddy River and for completing a continuous promenade between Boston Common and Jamaica Pond. Boston, Massachusetts, USA.

———. 1887. General plan for the improvement of the Niagara Reservation. New York, New York, USA.

———. 1895. Parks, parkways and pleasure grounds. Engineering Magazine **9**: 253–254.

Olwig, K. 1980. Historical geography and society/nature problematic: the perspective of J. F. Schouw, George Perkins Marsh, and E. Reclus. Journal of Historical Geography 6: 1.

———. 1984. Nature's ideological landscape. Allen and Unwin, London, England, UK.

Pedersen, P. 1995. Nature, religion, and cultural identity: the religious environmentalist paradigm. In A. Kalland and O. Bruun, editors. Asian perceptions of nature: a critical perspective. London, UK.

Pfeiffer, B. B., editor. 1987. Frank Lloyd Wright: his living voice. California State University Press, Fresno, California, USA.

Riley, R., and B. Brown. 1995. Analogy and authority: beyond chaos and kudzu. Landscape Journal 14: 1, 87–92.

Robinson, W. 1870. The wild garden. J. Murray, London, England, UK.

Rodman, J. 1993. Restoring nature: natives and exotics. Pages 139–153 in J. Bennett and W. Chaloupka, editors. The nature of things. University of Minnesota Press, Minneapolis, Minnesota, USA.

Ruff, A. 1979. Holland and the ecological landscape. Manchester, UK.

Scherer, D., and T. Attig, editors. 1983. Ethics and the environment. Prentice Hall, Englewood Cliffs, New Jersey, USA.

Slater, C. 1995. Amazonia as edenic narrative. Pages 114–131 in W. Cronon, editor. Uncommon ground: reinventing nature. W. W. Norton, New York, New York, USA.

Spirn, A. W. 1984. The granite garden: urban nature and human design. Basic Books, New York, New York, USA.

———. 1985. Urban nature and human design: renewing the great tradition. Reprinted in J. M. Stein, editor. 1995. Classic Readings in Urban Planning. McGraw-Hill, New York, New York, USA.

———. 1988. The poetics of city and nature: toward a new aesthetic for urban design. Landscape Journal 7: 2, 108–126.

———. 1991. The west Philadelphia landscape plan: a framework for action. Graduate School of Fine Arts, University of Pennsylvania, Philadelphia, Pennsylvania, USA.

———. 1995. Constructing nature: the legacy of Frederick Law Olmsted. Pages 91–113 in W. Cronon, editor. Uncommon ground: toward reinventing nature. W. W. Norton, New York, New York, USA.

———. 1996. Frank Lloyd Wright: architect of landscape. Pages 135–169 in D. DeLong, editor. Shaping an American landscape: 1922–1932. Abrams, New York, New York, USA.

———. 1998. The language of landscape. Yale University Press, New Haven, Connecticut, USA.

———. 2000a. Ian McHarg, landscape architecture, and environmentalism: ideas and methods in context. In M. Conan, editor. Environmentalism and landscape architecture. Dumbarton Oaks, Washington, D.C., USA.

———. 2000b. Reclaiming common ground: water, neighborhoods, and public spaces. Pages 297–313 in R. Fishman, editor. The American planning tradition: culture and policy. Woodrow Wilson Press and Johns Hopkins University Press, Washington, DC and Baltimore, Maryland, USA.

Tafel, E. 1985. Years with Frank Lloyd Wright: apprentice to genius. Dover, New York, New York, USA.

Thompson, G. F., and F. R. Steiner, editors. 1997. Ecological design and planning. John Wiley, New York, New York, USA.

Waugh, F. 1917. The natural style of landscaping. Boston, Massachusetts, USA.

Weber, M. 1978. Economy and society. G. Roth and C. Wittich, editors. University of California Press, Berkeley, California, USA.

Williams, R. 1980. Ideas of nature. Pages 67–85 in Problems in materialism and culture. Verso, London, UK.

————. 1983. Keywords: A vocabulary of culture and society. Oxford University Press, New York, New York, USA.

Wolschke-Bulmahn, J. 1993. The peculiar garden: the advent and the destruction of modernism in German garden design. In R. Karson, editor. Masters of American garden design III: the modern garden in Europe and the United States. Proceedings of the Garden Conservancy Symposium.

————. 1995. Jens Jensen: maker of natural parks and gardens (review). Journal of Garden History 15: 54–55.

————. 1997. Nature and ideology: Natural garden design in the twentieth century. Dumbarton Oaks, Washington, D.C., USA.

Worster, D. 1979. Nature's economy: the roots of ecology. Sierra Club Books, San Francisco, California, USA.

Wright, F. L. 1992. An autobiography. Page 163 in collected writings, Volume 2. Bruce Brooks Pfeiffer, editor. Rizzoli, New York, New York, USA.

Zaitzevsky, C. 1982. Frederick Law Olmsted and the Boston park system. Harvard University Press, Cambridge, Massachusetts, USA.

Notes

1. An earlier version of this essay was published in 1997 (Wolschke-Bulmahn 1997) and portions adapted in my book, The Language of Landscape (Spirn 1998). It is an extension of a chapter (Spirn 1995) for a book written with a group of scholars at the University of California Humanities Research Center in Irvine from January to June 1994 (Cronon 1995). For insights that inform this essay, I am indebted to my colleagues in the Irvine seminar: Michael Barbour, Ann Bermingham, Bill Cronon, Susan Davis, Giovanna Di Chiro, Jeff Ellis, Donna Haraway, Robert Harrison, Katherine Hayles, Carolyn Merchant, Ken Olwig, Jim Proctor, Jenny Price, Candace Slater, and Richard White, and to Mark Rose, Director of the Institute. I am also grateful to Paul Spirn, Carl Steinitz, Joachim Wolschke-Bulmahn, Kenneth Helphand, Bart Johnson, Kristina Hill, and anonymous readers for their comments on a preliminary version of this essay.

2. Quoted in Gill 1987, p. 22. Another version, almost word for word, is transcribed from a tape of August 4, 1957 (Pfeiffer 1987, p. 88). This is pure Emerson, who had written similar words more than 150 years earlier: " . . . the noblest ministry of nature is to stand as the apparition of God" (Emerson 1836, p. 77). Wright spoke with Mike Wallace in 1957 on the television program The Mike

Wallace Interview. "I've always considered myself deeply religious," said Wright. "Do you go to any specific church?" asked Wallace. Wright replied, "My church [pause], I put a capital 'N' on Nature and go there." Wright is a good example of a designer who appeals to divine authority through nature because he has written so extensively on the topic. Most designers who link the natural and the divine do so less explicitly. For a discussion of how ideas of nature are expressed in the work of Frank Lloyd Wright, see Spirn 1996.

3. Those familiar with landscape architecture will recognize that many of the figures quoted in this essay have been my colleagues at the University of Pennsylvania. I have great respect for each of them; all have made important contributions to the field through writing, teaching, and/or practice. Penn has been a center for the development and continuing evolution of this approach to landscape design.

4. This chapter does not discuss the full scope of the current controversy in landscape architecture over the conflicting authority of "nature" versus "culture," ecology versus art. I am just as critical of appeals to historical precedent or personal expression as authority for landscape design as I am of appeals to nature. Such a discussion is outside the scope of the essay, and I have treated the subject elsewhere (Spirn 1998).

5. See Max Weber's analysis of three forms of authority—traditional, legal-rational, and charismatic (Weber 1978). For an introduction to ideas of authority, see Boudon and Bourricaud 1989.

6. Many essays and entire books have been written on the origins, history, use, and significance of the word nature. See, for example, Williams 1980, Lewis 1967, Lovejoy 1935 and 1948, Collingwood 1945, Glacken 1967, Leiss 1974, Evernden 1992.

7. Slater traces Edenic narratives as they relate to the Biblical story of Genesis, but points out that such notions are not unique to the Judeo-Christian tradition. For a discussion of other religious traditions, see Pedersen 1995.

8. Also personal communication, Cornelia Brierly of the Taliesin Fellowship. Brierly was assigned to assist Jensen when he visited Taliesin.

9. See Cronon 1994 for an excellent discussion of the roots of Wright's philosophy.

10. Jensen was born of a Danish-speaking family in the Slesvig region—a border zone of northern Germany and southern Denmark. This region was politically and culturally contested ground for more than a century. Despite two good books on Jensen's life and work (Eaton 1964 and Grese 1992), his complex relationship to Danish and German ideas of nature and nation has not been fully explored (Wolschke-Bulmahn 1995), and it is beyond the scope of this chapter to do so. For perspectives on contemporary Danish and German ideas of nature and natural gardens, see Olwig 1980 and 1984, Wolschke-Bulmahn 1993. See also Waugh 1917 for a North American perspective, and Ruff 1979 for an introduction to the work of Jacques Thijsse, J. Landwehr, and the Dutch "Heem" (home) parks.

11. See Mitman 1992 for a history of the Chicago school of ecology and the interplay between science and a social philosophy that stressed the value of cooperation over conflict.

12. See Groening and Wolschke-Bulmahn 1992 for a discussion of ecological theory in Germany during the nineteenth and early twentieth centuries and parallels between the eradication of nonnative plants in Nazi Germany and the extermination of non-Aryan human populations. There is some evidence that Jensen was sympathetic to at least some of these ideas; see Wolschke-Bulmahn 1993.

13. By the age of twenty-five, most students' ideas of nature seem set or at least not modified greatly by a single course on the subject (they ranged in age from twenty-two to fifty; most were in their mid to late twenties). While largely North American, approximately one-third have been from other parts of the world, including Europe, the Middle East, Africa, Asia, South America, and Australia. Of the North Americans, most grew up in the suburbs or in rural areas; a higher proportion of foreign students are from cities.

14. Robert Riley and Brenda Brown have addressed this topic in an editorial (Riley and Brown 1995).

15. McHarg did always acknowledge his debt to Lewis Mumford and, late in life, spoke of the contributions of Charles Eliot and his peers Lawrence Halprin and Roberto Burle Marx.

16. Geoffrey Jellicoe portrayed the situation as "A Table for Eight," where the landscape architect shares concerns for shaping the environment with seven others: the philosopher, the town and country planner, the horticulturalist, the engineer, the architect, the sculptor, and the painter. Another essay by Francisco C. Cabral outlined a curriculum for landscape architecture that stresses the importance of science (ecology, geology, climate) and agriculture (horticulture, forestry), as well as architecture and fine arts (Crowe 1961).

17. Grese 1992 presents a useful comparison of the work of Olmsted and Jensen in this and other respects, but emphasizes similarities and does not probe their ideological differences.

18. My essay "Constructing Nature" treats material presented here in more detail (Spirn 1995).

19. The Ramble at Central Park was planted to appear "wild," but it was only a small part of the park. William Robinson, an English acquaintance of Olmsted, published his book *The Wild Garden* in 1870. Olmsted was undoubtedly also aware of Martin Johnson Heade's contemporary paintings depicting marshes along Boston's North Shore.

20. Such views were common at the time, and Olmsted discussed them frequently in relation to his work. See, for example, Olmsted 1865 and 1887.

21. See Zaitzevsky 1982, page 196, for quotations of Olmsted's and Sargent's disagreement on this subject.

22. See Scherer and Attig 1983. At its most extreme, contemporary environmentalism can even sound antihuman. Ian McHarg still refers to humans as a "planetary disease," the phrase he used in *Design with Nature* (1969). Despite misanthropic rhetoric, McHarg has close links to the earlier regionalist movement through his mentor Lewis Mumford.

23. From *idein*, "to see." *Webster's New International Dictionary*, second edition, unabridged (Merriam, 1955).

24. V. Dahlerup, *Ordbog over det Danske Sprog* (Nordisk, Copenhagen, 1931); J.

Grimm and W. Grimm, *Deutsches Worterbuch* (Verlag von S. Hirzel, 1885); A. R. Borden Jr., *A Comprehensive Old English Dictionary* (University Press of America, 1982). For a review of the histories of the words *landscape, nature, land,* and *country* in English, German, and Scandinavian languages, see Olwig 1996. See also Jackson 1984, 3–8. I am grateful to Andre Wink for the translation and interpretation of J. Heinsios, *Woordenboek der Nederlandsche Taale* (Martinus Nijhoff, A. W. Sijthoff, 1916).

25. *Webster's New Universal Unabridged Dictionary* (New York: Simon and Schuster, 1983) and *Oxford English Dictionary* (Oxford: Oxford University Press, 1989).

CHAPTER 3

Ecology's New Paradigm: What Does It Offer Designers and Planners?

H. Ronald Pulliam and Bart R. Johnson

Ecology is in the midst of a major paradigm shift (Pickett, Parker, and Fiedler 1992; Fiedler, White, and Leidy 1997). The transition has come about gradually, but nonetheless has resulted in a radically different view of how the natural world works and has profoundly changed not only the discipline of ecology but also what ecology has to offer applied disciplines such as landscape design and planning.

In a nutshell, the paradigm shift involves two primary changes: (1) a shift from an equilibrium point of view where local populations and ecosystems are viewed as in balance with local resources and conditions, to a disequilibrium point of view where history matters and populations and ecosystems are continually being influenced by disturbances; and (2) a shift from considering populations and ecosystems as relatively closed or autonomous systems independent of their surroundings, to considering both populations and ecosystems as "open" and strongly influenced by the input and output or "flux" of material and individuals across system borders. Taken together, these two changes result in a new view of natural systems that emphasizes how local ecological conditions are greatly influenced by events that occur at other times and in other places. Since local conditions depend on what happens elsewhere and at other times, how humans design and use landscapes is of paramount importance both locally and more broadly. These changes also point toward the importance of emerging concepts such as landscape memory and ecosystem autonomy, which we propose as conceptual tools to help landscape designers, planners, and managers apply these new understandings.

Roots of the Paradigm Shift

The 1960s were intellectually heady times for ecology. Led by the late Robert MacArthur, ecologists developed a view of the world that encapsulated the notion of the "balance of nature" into simple but elegant mathematical theory (e.g., MacArthur 1958, 1960, 1970; MacArthur and Pianka 1966). MacArthur and his followers reasoned that the abundance and distribution of organisms was controlled by the abundance and distribution of the limiting resources used by those organisms. MacArthur argued that if we know enough about exactly how each species utilizes resources and how those resources are distributed in time and space, we should be able to deduce patterns of the abundance and distribution of species. At the heart of this theory was the implicit assumption that given enough time, an equilibrium between consumers and the resources they consumed would occur and that this equilibrium condition was deterministic and predictable. Many practitioners of this brand of theoretical ecology enthusiastically set out to show how MacArthur's theories explained many of nature's patterns. As we discuss later, although he was the most eloquent defender of this equilibrium view of nature, it was MacArthur himself who sowed many of the seeds that would eventually grow into a new, disequilibrium paradigm.

This "old" or equilibrium paradigm was built on three central propositions:

1. Species, and the biological communities and ecosystems in which they are found, are usually at, or close to, equilibrium with their resources.
2. Disturbances such as fire, storms, floods, and the like temporarily dislodge biological communities from their natural equilibrium condition, but such perturbations are relatively short-lived, and the strong biological interactions among species quickly restore equilibrium.
3. Although nature is spatially heterogeneous, each local patch can be viewed as a homogeneous, relatively autonomous ecological system in equilibrium with its own local resources. This proposition did not deny the movement of individuals, energy, and materials among patches, but rather argued that this interchange is relatively unimportant compared to the strong nature of interactions among species and resources within the separate patches.

These simple propositions led directly to a number of nontrivial conclusions. For example, since communities come into equilibrium with their resources, species that utilize resources more efficiently will displace other species requiring the same resources. This reasoning leads to the conclusion that complete competitors cannot coexist (Gause's law) and the corollary that there is a limit to the similarity of coexisting species. These propositions also led to what we might call the "hand-in-glove" view of nature. If we view species as fingers and the pouches of a glove as niches, then the equilibrium view is that there will be a finger in every pouch and a pouch for every finger. In other words, the theory predicted a one-to-one relationship between species and the niches provided by the resources. Using this analogy, one could characterize the new paradigm, which would eventually come to replace the equilibrium viewpoint, as one in which there is still a general relationship between the number of fingers and the number of pouches, but some pouches of the glove are empty and others are frequently stuffed with more than one finger (Pulliam 1997). These propositions also led ecologists, designers, and the public to generally assume that the best care for natural systems, including those degraded by humans, was to leave them alone. People also tended to greatly overestimate the autonomy of local ecosystems and to underestimate the impacts of actions and conditions at a distance.

Philosophers of science often discuss paradigm shifts in terms of scientific "revolutions" (Kuhn 1962) that occur when the old models and old ways of looking at things no longer match the new experimental and observational results. Because natural systems are extremely "noisy," like a radio signal with static, high degrees of intrinsic variability often obscure even strong causal patterns or processes. Therefore, it is not surprising that no one experiment at odds with theory can be pointed to as the cause of the paradigm shift in ecology. Rather, many researchers working on a host of different organisms in different ecosystems gradually came to realize that the equilibrium view of a balance of nature was not consistent with what they were observing. There were, in our opinion, also a number of new and competing theoretical concepts that contributed to the paradigm shift. We summarize a number of these and why they matter to designers and planners.

Ecosystem Theory

Unfortunately, ecologists working on issues of evolutionary ecology, population dynamics, and species diversity have often had surprisingly little contact with ecosystem ecologists working on biogeochemical cycles and the role of species in regulating the flux of energy and matter in nature. In contrast to theory that focuses on species and populations, ecosystem ecology emphasizes the transfer and processing of matter and energy (Figure 3-1).

Unlike population ecologists, who often treated local populations as autonomous units, ecosystem ecologists recognized very early on that ecosystems are open to the flux of living and nonliving matter and organisms, and that ecosystem dynamics could not be understood unless ecosystems were treated as open systems subject to sometimes massive movement of materials across their boundaries. By tracking the exchange and storage of such "common currencies" as nitrogen and organic carbon among biotic and abiotic system components and their throughflow across system boundaries, ecosystem

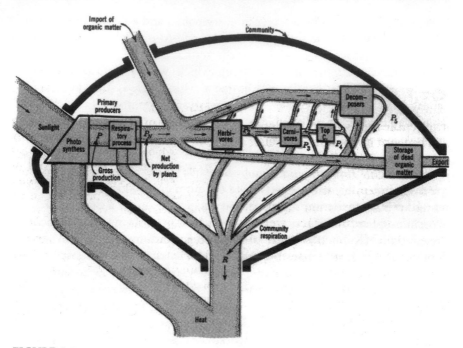

FIGURE 3-1.
The hydraulic analogy of H. T. Odum. In this analogy the energy cascade
of an ecosystem is imagined as being channeled through pipes whose thick-
ness is proportional to the rates of energy flow. (Adapted from Odum
1956 by Collinvaux 1993.)

ecologists demonstrated how ecosystems functioned as highly interconnected
networks. Moreover, key studies, such as the Hubbard Brook watershed
experiment (Likens et al. 1970; Bormann and Likens 1979), demonstrated
how management activities could dramatically affect such landscape flows.
As population ecologists became more aware of ecosystem theory and began
to consider the open nature of the ecosystems where they worked, they, in
turn, began to ask how important the flux of individuals across habitat
boundaries is in influencing the local abundance and diversity of organisms.

Because any given piece of the landscape is an open system, receiving and
contributing matter, energy, organisms, and information from and to nearby
and even distant locations, designers and planners must consider individual
sites in the context of broader landscape dynamics. Whether this is the quality and flow regime of urban stormwater or the population dynamics of a
species of concern, no site is sufficiently independent that it can be treated in
isolation from its surroundings. Moreover, because cultural boundaries,

including political jurisdictions and property lines, often bear little relationship to biophysical boundaries such as those of drainage basins, ecosystems, or species territories, maintenance of flows across site boundaries may be particularly important. Two of the key questions that designers and planners need to ask, then, are "What are the key landscape flows that affect our site, and in what ways might they be altered by our design?" and "What might be the consequences, direct and indirect, as well as intentional and unintentional, of those alterations, both on the site and elsewhere?"

Predation Theory and Keystone Species

In a paper that was to become an "ecological classic," Robert Paine (1966) reported that the predatory starfish *Pisaster ochraceous* strongly influenced the outcome of competition between various species of invertebrates in a marine intertidal community. Following experimental removal of the starfish, species diversity plummeted as a single species of mussel began to monopolize available space. Both the complexity of the food web and the structure of the ecosystem became dramatically simplified. On the basis of his experimental work, Paine introduced the concept of a "keystone species" that could drastically alter the interactions among other species and, in doing so, cause substantially different patterns of species abundance and diversity. Unlike species that have major direct effects on community structure and ecosystem processes by virtue of sheer numbers and weight of living matter, keystone species influence systems through indirect effects that belie their numbers or biomass. Certain tropical bats are keystone species in that they perform critical fruit pollination or dispersal without which forest fruit production would crash, affecting myriad other species whose survival depends on the fruits during times of the year when other foods are scarce. Sea otters are also keystone species in that they prey on herbivorous sea urchins, preventing them from decimating the kelp beds that serve as important nurseries for many fish species. A number of different types of keystone species can be usefully recognized; among them are keystone predators, keystone food resources, and keystone habitat modifiers (Meffe, Carroll, and Pimm 1997). While Paine's results and the concept of keystone species were not at all incompatible with the notion of the balance of nature, they did provide strong evidence against the widely held view that competition for limited resources was the dominant mechanism that structured biological communities, and this opened the door for consideration of other mechanisms that limited the importance of resource competition.

For designers and planners, it is important to understand that certain species may have critical functional roles affecting community composition and that their loss or addition can have cascading effects across the system.

One of the questions that designers and planners need to ask, then, is "Are there keystone species that may be affected by our plans, and if so, what would be the consequences for the ecosystem?"

Island Biogeography

Robert MacArthur and E. O. Wilson (1967) introduced the idea that the numbers of species on islands is not controlled solely by the habitats and resources present on the islands but also by the balance of immigration and local extinction. They further argued that patterns of immigration and extinction are primarily determined by island size and isolation. Immigration is a function of distance from the mainland (or in theoretical extensions, from other sources of potential colonists). Extinctions are a function of island size, since larger islands will support larger populations, which are less vulnerable to extinction.

The emphasis by MacArthur and Wilson on the role of immigration and emigration as determinants of local diversity was a radical departure from previous theory that had emphasized the importance of resources and habitat heterogeneity as the principal determinants of island diversity patterns. It was also one of the first times islands were explicitly treated as open ecological systems where fluxes of individuals across the island boundaries, as compared to resource conditions on particular islands, were viewed as the most important factor determining biological diversity.

In an important test of the theory, the resident arthropod populations on a series of small mangrove islands were exterminated, and species recolonizations were monitored over time (Wilson and Simberloff 1969; Simberloff and Wilson 1969). Following rapid recolonization, species numbers stabilized at levels that were consistent with expectations based on island size and distance from source islands, providing experimental validation of the model. Moreover, although the number of species on each island then remained constant over time, individual species appeared to come and go with rapid turnover. Island biogeography has become one of the fundamental paradigms for conservation reserve design and for understanding the biological consequences of habitat fragmentation, that is, the splitting of contiguous habitat into smaller, isolated fragments. The principal assumption that has allowed its broad application is that reserves or other habitat fragments may be considered as "islands" in a "sea" of other habitat or land-use types. Even the difficulties associated with this simplistic assumption have led to richer theory, since the ability of species to reach the island depends on how inhospitable the intervening habitat "ocean" is, leading to ideas of *contrast* between a reserve and the surrounding landscape matrix and to the *permeability* of the matrix (Forman 1995).

Further developments have included the concept of the *rescue effect* (Brown and Kodrik-Brown 1977), in which small populations vulnerable to extinction may be "rescued" by periodic immigrants from nearby populations without ever going extinct, as well as the SLOSS debate about whether a "Single Large Or Several Small" reserves of the same total size constituted the optimal design. Conservation applications have focused on reducing the insular quality of habitat patches by maintaining or restoring connectivity through the use of habitat stepping-stones or corridors (Figure 3-2). Corridors continue to be a topic of considerable debate (Simberloff et al. 1992; Beier and Noss 1998), including whether they actually increase movement among habitat patches as intended. Robust empirical evidence regarding the effectiveness of corridors is still a fairly new phenomenon (Haddad 1999; Haddad et al. 2000). Moreover, corridors are a two-edged sword, and potential benefits must be weighed against the dangers of increased transmission of disease or exotic species. Notwithstanding these concerns, corridors have become a staple of conservation planning, albeit one to be used thoughtfully and with caution.

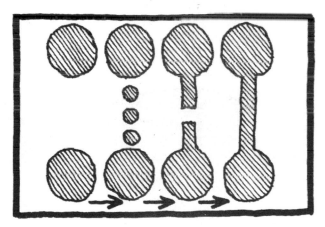

FIGURE 3-2.

Degrees of habitat connectivity. Conservation applications of island biogeography have focused on reducing the insular quality of habitat patches by maintaining or restoring connectivity through the use of habitat stepping-stones or corridors. All else being equal, *functional* connectivity is expected to increase with progressively greater degrees of *physical* connectivity. From left to right, the diagram shows two habitat patches with no apparent connectivity, followed by the same patches connected by stepping-stones, a broken corridor, and a complete corridor, respectively. (From Dramstad, Olson, and Forman 1996.)

Key questions that designers and planners need to ask, then, are "Will our plans further isolate important habitats or degrade them by fragmentation?" "Are there ways to maintain or restore meaningful levels of habitat connectivity by the use of stepping-stones and corridors or by enhancing the quality of the surrounding habitat?" and "How can we design corridors so that the benefits outweigh the risks?"

Intermediate Disturbance Hypothesis

The Intermediate Disturbance Hypothesis (Connell 1978) posits that the greatest biological diversity occurs at intermediate levels of disturbance (Figure 3-3). A "disturbance" in this sense is defined as any relatively discrete event in time that disrupts ecosystem or community structure and changes resource availability or the physical environment (White and Pickett 1985). Frequently, disturbances cause some level of mortality or reduction of biomass, such as when a tree falls in a windstorm or fire burns across a prairie. The hypothesis is based on two assumptions: (1) In the absence of periodic

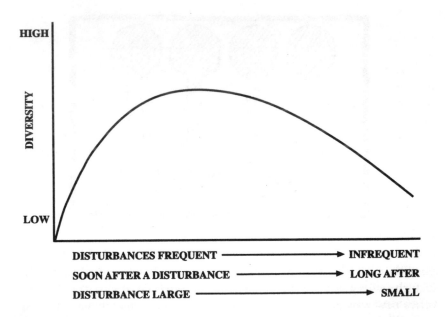

FIGURE 3-3.
The intermediate disturbance hypothesis. The greatest biological diversity is expected to occur at intermediate levels of disturbance. (Modified from Connell 1978.)

disturbances, competition among species will result in superior competitors displacing less capable species and thereby reducing diversity, and (2) very few species can tolerate extremely frequent or intense disturbances. As a consequence, diversity is greatest at intermediate levels of disturbance between these two extremes. There is now considerable evidence in favor of the Intermediate Disturbance Hypothesis, and the new paradigm in ecology explicitly recognizes disturbance as an integral part of natural systems. The extension of the Intermediate Disturbance Hypothesis to the landscape level results in the view that the greatest diversity occurs in landscapes large enough to contain various seral, or successional, stages as the result of a number of localized disturbance events at different times in the past. The resulting habitat heterogeneity at the landscape level will help maintain conditions suitable for a greater number of species, including both opportunistic species that invade newly created patches and highly competitive species that predominate in older patches.

Further developments concerning the role of disturbance followed this early theoretical work. One of the most important is the understanding that conserving both species diversity and community composition often requires maintaining historic *disturbance regimes*, which are characterized by the type, frequency, intensity, and spatial extent of disturbances. Conversely, the introduction of novel disturbances or a change from the historic regime typically will redirect the community along a new successional trajectory. This is true whether one considers a riparian forest, a prairie, or an urban lawn. In addition, most ecosystems, far from being in a state of equilibrium interrupted infrequently by disturbances followed by rapid reequilibration, are typically in the process of recovering from the last disturbance (Reice 1994).

A basic understanding of these concepts is implicit for anyone who has seen what happens to a lawn that isn't mown for a long period of time or a field that hasn't been plowed—pioneer species (a.k.a. "weeds") move in, followed over time by shrubs and trees. The utility of disturbance theory for designers and planners is that it can encompass a wide array of processes that may or may not be initiated by humans—mowing, plowing, weeding, harvesting, grazing, fire, floods, windstorms, insect outbreaks, and disease—and place these disparate mechanisms within a common conceptual framework of disturbance regimes characterized by type, frequency, intensity, and extent. It offers new ways for designers to respond to natural disturbances, to employ unfamiliar tools such as fire, or to consider the interaction of management disturbances with natural ones.

In certain parts of the country, prairie restoration has become an important component of urban and rural park management in recent years. This not only requires managers to understand how to use fire to achieve specific ends, but opens opportunities for designers to consider how to design settings so that fire can be safely and effectively employed (Figure 3-4). Disturbance theory also offers conceptual possibilities for mimicking historic disturbance regimes as part of traditional land management. For instance, researchers and managers collaborating on the Augusta Landscape Project in the Willamette National Forest in Oregon developed maps of historic fires and landslides and used them as the basis for a long-term forest management plan in which harvest rotation lengths, tree retention rates, size distribution of harvest units, and geographic patterns have been set to mimic the frequency and severity of historic disturbances (Swanson, Jones, and Grant 1997; Cissel, Swanson, and Weisburg 1999).

One of the principal goals was to maintain a wide range of system characteristics within the "historic range of variability" (Landres, Morgan, and Swanson 1999), a concept that can guide an understanding of how much variability is desirable for maintaining an ecosystem over time. For instance, the system characteristic may include the amount of closed-canopy or old-growth forest present at any given time, or the frequency and severity of flooding that renews riparian habitats. Forest planners for the Augusta project believe that their plan for the area is more likely to retain native species

FIGURE 3-4.
Prairie restoration in urban and rural parks creates opportunities to design settings so that fire can be safely and effectively employed. (Photo courtesy of Bob Grese.)

diversity than the riparian buffer/critical habitat strategy of the current Northwest Forest Plan because it is based on temporal variability rather than static patterns.

In these ways, disturbance theory forms a powerful conceptual model to understand all management activities as disturbance regimes and to explore ways to link natural disturbance regimes with human management activities. Two questions that designers and planners need to ask, then, are "Are there key historic disturbance regimes that can be maintained, restored, or mimicked so as to conserve native species diversity, community structure, or ecosystem function?" and "What was the historic range of variability in this system, and how much of that might be maintained in the present landscape?" Moreover, for a design profession too often obsessed with preserving planting designs essentially unaltered over time, disturbance theory offers a fresh framework and set of principles to foster and facilitate change while retaining a plant community with desired characteristics.

The Importance of Unpredictable Events

With mounting evidence for the importance of disturbance in structuring ecological systems came a further recognition that the inherent uncertainty around disturbances meant that many alternative community types could potentially exist at any given location, in stark contrast to the nearly inevitable march toward a predetermined climax community as envisioned by Clements (1916, 1936). Although many disturbances were not wholly unpredictable (e.g., winter floods, spring tornadoes, summer droughts), the specifics of what, when, where, and level of intensity were contingent on a host of factors that made them impossible to predict in advance. Moreover, infrequent major disturbances—a 100-year flood, a stand-replacing wildfire, or a devastating hurricane—were not only the least predictable but could also have extremely long-term consequences for communities and ecosystems.

Under the equilibrium paradigm, the overwhelming importance of deterministic processes meant that even if unusual events might temporarily displace a system, it would soon return to its former trajectory. The idea that stochastic events (those that are either random or indistinguishable from chance by organisms that experience them) played a major role in structuring ecosystems began to overturn long cherished beliefs in the orderly, gradual, and predictable progression of events. While some such events may be catastrophic and singular, others may be small and cumulative. The role of random or unpredictable events in structuring communities and its implications for the conservation and planning disciplines can be illustrated by considering the risk of extinction faced by small populations.

Small population size poses a number of potential problems for most species and has become a key concern of conservation biology. The basic risks are related to increased vulnerability to extinction due to environmental, demographic, and genetic stochasticity. Environmental stochasticity, that is, random fluctuations in environment, can drive a small population to extinction more easily than a large one. In part, this is because local fluctuations or disturbances may wipe out a small, concentrated population; and in part, it is because small populations may have less environmental heterogeneity or genetic variability to buffer them against environmental fluctuations. On the other hand, broad-scale catastrophes may decimate even very large populations. Demographic stochasticity is the random variation in growth and survivorship of individuals and is most likely to influence only very small populations that, as a rule of thumb, are smaller than a few hundred individuals. Genetic stochasticity is expressed in terms of "genetic drift"—random changes in gene frequency that are particularly pronounced in small populations and may limit the ability of a population to track environmental changes via selection and adaptive evolution. While there are no magic numbers, some researchers believe genetic effects are likely to become more important in much larger populations than are demographic effects.

Gilpin and Soulé (1986) proposed four types of "extinction vortices" that model ways in which deterministic and stochastic factors can interact in sequences or feedback loops to cause population extinction. For example, rapid habitat loss or degradation due to development, which can result in reduced population size as well as reduced habitat heterogeneity, may leave a population of wildflowers more vulnerable to a local drought. After drought-related population decline, the floral display may not be large enough for effective pollination, and genetic inbreeding may increase. If inbreeding leads to inbreeding depression, reproduction may be suppressed and cause further population decline. At some point, genetic drift may begin to take hold and increase the loss of genetic diversity, finally reducing the population to such a small size that demographic stochasticity finishes it off. Although this may sound like the plot of a Hollywood movie, it is the type of scenario that ecologists believe may drive many population and species extinctions.

Unpredictable events thus play important—and inevitable—roles in structuring ecological communities. To a great extent, they are critical to maintaining structural and compositional diversity over time. Human activities that reduce the sizes of previously large landscape mosaics, habitat patches, or species populations, however, increase the chances that these same unpredictable events will lead to local, regional, or global extinctions. An important question for designers and planners, then, is "How likely are our plans to reduce population sizes of sensitive species so that they are highly vulnerable to stochastic events?"

Hierarchy Theory

The application of hierarchy theory, an outgrowth of general systems theory, to ecology (Allen and Starr 1982; O'Neill et al. 1986) opened new doors to understanding the complexities of ecological systems and was particularly important to the growing emphasis on large landscapes and on how to integrate information from studies at different scales of interest. In particular, most traditional ecological research had been conducted at fine scales, whereas the advent of tools for broad-scale analysis, particularly remote sensing and Geographic Information Systems (GIS), made it critical that information could be transferred and integrated across scales.

Concepts of scale are intrinsically linked to those of hierarchy. Fundamentally, they represent ways of measuring or perceiving the world (scale) and of system organization (hierarchy), respectively. Scale represents the spatial or temporal dimensions of an object or process, characterized by both grain (the finest level of spatial or temporal resolution) and extent (the size of area or temporal boundaries of the system under consideration). Hierarchy, on the other hand, refers to ranked levels of organization. Levels may be defined by physical or spatial structure, interactions, rates, or other selected characteristics, and hierarchies are typically distinguished as structural or control hierarchies. Structural hierarchies (Figure 3-5) are orderings of subsystems within systems, which in turn are components of systems at the next higher level, and so forth. For instance, a tree is composed of leaves organized on twigs, twigs organized on branches, branches organized on limbs, and limbs organized on a trunk. Structural hierarchies typically are nested, mean-

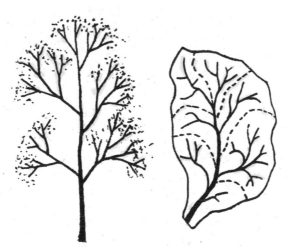

FIGURE 3-5.
Many different phenomena, from trees to watersheds with their subbasins and stream networks, constitute structural hierarchies.

ing that each subsystem is contained within the system at the next higher level. Other examples of structural hierarchies are river systems with their different orders of tributary streams, watersheds and their subbasins, and the hierarchy of gene-organism-population-community-ecosystem-landscape-biome-region-biosphere. Control hierarchies, on the other hand, represent such phenomena as military command structures or biological trophic levels (e.g., plants, herbivores, carnivores) in which components at one level exert control on components at lower levels and are in turn controlled by components at higher levels. Control hierarchies are typically nonnested. Moreover, control may be exerted on components that are not subsystems of the controlling unit.

Much recent theory and experimentation on control hierarchies in ecological systems has focused on the question of "bottom-up versus top-down" control. The basic idea is simple. If the only species in an ecosystem are plants (primary producers), the plant populations are necessarily controlled by the resources (nutrients) available to them. If there are also herbivores in the ecosystem, the herbivores may reduce the biomass of plants to a level far lower than could be supported by the available nutrients in the absence of herbivory. When the plants are controlled by herbivores, we say there is top-down control because the herbivores are "above" the plants in a food web. By contrast, when the plants are controlled by the nutrient supply, we call that bottom-up control because the nutrients can be thought of as "below" the plants in a food web.

Now consider what happens when a predator that eats herbivores is added to the ecosystem. The predators may consume so much of herbivore population that the herbivores no longer are abundant enough to control the plants. The result is that control of the plant population may revert from top-down control by herbivores to bottom-up control by the nutrient supply. A practical implication is that addition or removal of predators from a system may have unpredicted consequences that "cascade" through the system. For example, removal of a top predator, such as raptors or large cats, may result in increases in numbers of herbivores, such as rabbits and rodents, resulting, in turn, in decreases in plant populations and even extinctions of the plant species most preferred by the herbivores. Obviously, in large, complex ecosystems with several trophic levels and many species, these cascading impacts may be very difficult to predict and have many undesirable consequences.

Concepts of hierarchy and scale also provide a practical means to structure and organize information about landscapes and to analyze it. Moreover, hierarchy and scale can be used to understand and interpret system dynamics—why systems behave the way they do— based on characteristics of internal system structure, rates of key processes, and so on. For instance, the ability to link phenomena at the landscape scale to those within individual watersheds, subbasins, and parcels is critical for managing the complexity of such systems as well as for organizing complex spatial data. Moreover, hierarchical structuring is a practical necessity for broad-scale planning projects, since the plan will be meaningless unless it can be implemented at finer scales. Conversely, plans for small parcels need to be coordinated to work in concert for protection of large landscape resources (Figure 3-6). A key concept for

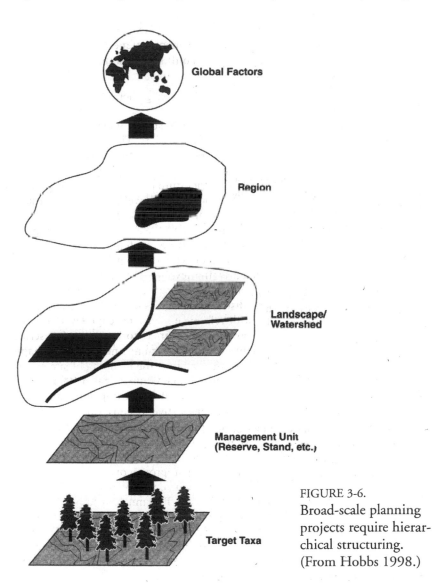

Global Factors

Region

Landscape/
Watershed

Management Unit
(Reserve, Stand, etc.)

Target Taxa

FIGURE 3-6.
Broad-scale planning
projects require hierarchical structuring.
(From Hobbs 1998.)

designers and planners in this regard is that of the triadic structure of hierarchical systems (O'Neill 1989), which posits that to understand the system at a particular level of interest, you must examine levels immediately above and below (Figure 3-7). Dynamics at the focal level are the result of the behaviors and interactions of components at the next lower level. On the other hand, the significance of behavior at the focal level is explained through reference to the next higher level. Moreover, higher levels impose constraints or boundary conditions on the focal level. For instance, the dynamics of a forest stand are created through the interactions of individual trees as they grow, compete, and die, creating a mosaic of forest gaps with new cycles of growth and competition. At the same time, the forest biome in which the stand is located, whether a mixed-hardwood forest of the southeastern United States or a Douglas fir forest of the Pacific Northwest, constrains what types of trees will grow there and which species are favored through its particular species pool, climate, and disturbance regime.

Some key questions for designers and planners, then, are "What are the most useful hierarchies to employ in classifying and analyzing this landscape?" and "Will my design for a large landscape result in the loss of large carnivores, and, if so, what unintended consequences might result from their loss?" A key lesson for designers and planners is to always examine the levels above and below the project scale to understand site dynamics, constraints as well as opportunities, and significance of the project. In their own work, designers have long understood the importance of looking up and down at least one scale, the former to consider the influences of site context and the

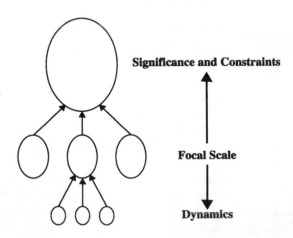

FIGURE 3-7.
A key concept for designers and planners is that of the triadic structure of hierarchical systems.

latter to develop detailed design elements that influence the specific ways in which a site is used. Hierarchy theory formalizes such relationships in ways that can be generalized to human and nonhuman systems and can offer deeper insights into both.

Metapopulation Theory

Metapopulation theory (Levins 1969, 1970; Hanski and Gilpin 1991, 1997; Hanski 1999), an extension of island biogeography from habitat islands to population islands, recognizes that clusters of populations may interact over time through the exchange of individuals or genetic material (e.g., pollen) and that individual populations frequently may go extinct only to be recolonized at a later time by immigrants from extant populations. The causes of local extinction may vary, ranging from local vagaries of weather, to local outbreaks of disease, to random drift in population size due to stochastic events. Because of the dynamic nature of local extinction and recolonization, any particular patch of suitable habitat may or may not be occupied at a given point in time, but the metapopulation as a whole persists because some patches are always occupied. The situation has been likened to that of a lighted Christmas tree (Kareiva and Wennergren 1995). In the old paradigm of stability and equilibrium, all the lights were continuously burning, although some might occasionally burn out, not to be replaced. In the new paradigm, the lights are constantly blinking on and off, representing populations going extinct and being recolonized over time. Among the more important implications of this view of natural populations is that local extinction may be frequent and that unoccupied suitable habitat may be a common occurrence.

In a metapopulation, large populations are less likely to go extinct than small populations so that, at any point in time, large habitat patches, which are more likely to support large populations, are more likely to be occupied than are small patches. Likewise, habitat patches that are relatively near to other patches are more likely to be occupied than are more isolated patches, because they are likely to be recolonized sooner after a local extinction event. There are now many documented examples of metapopulation dynamics in natural populations of organisms, ranging from shrews (Hanski 1993) to butterflies (Hanski, Kuussaari, and Nieminen 1994). In these examples, both patch size and patch isolation had clearly discernible effects on the probability of occurrence of the species. Although metapopulations may persist for very long periods of time, the entire metapopulation may go extinct if the number of suitable patches is reduced.

Perhaps the best-studied metapopulation is that of the checkerspot butterfly (*Euphydryas editha bayensis*) at Stanford University's Jasper Ridge pre-

serve in northern California and studied by Paul Ehrlich and his collaborators. These butterflies are restricted to native grasslands occurring on serpen-
tine soils, and patches of such habitat are rare and widely dispersed. In a long
series of papers (including Ehrlich et al. 1980; Ehrlich and Murphy 1987;
Harrison, Murphy, and Ehrlich 1988; Murphy, Freas, and Weiss 1990),
Ehrlich and his colleagues have shown that the checkerspot butterflies have
occupied a number of distinct habitat patches at Jasper Ridge and that local
patch extinctions are often followed by recolonization from nearby patches.
They also discovered what appears to be a relatively stable large population
surrounded by smaller, satellite populations whose likelihood of occupancy
increases with proximity to the larger population (Figure 3-8). Moreover,
which sites showed the highest reproductive success varied over time, as pop-

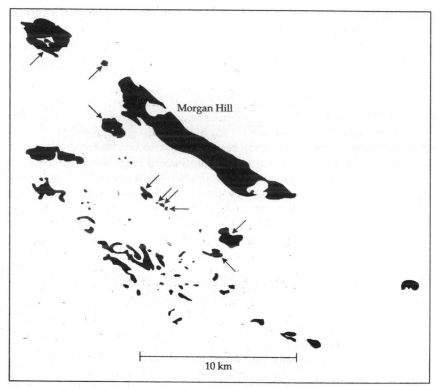

FIGURE 3-8.
Metapopulation dynamics. In this metapopulation of the checkerspot butterfly, a relatively stable large population at Morgan Hill is surrounded by
smaller, satellite populations whose likelihood of occupancy increases with
proximity to the larger population. Arrows indicate occupied patches.
(From Harrison, Murphy, and Ehrlich 1988.)

ulations on warm south-facing slopes experienced the most successful egg hatching and larval emergence in cool years while those on cool north-facing slopes did best on hot years. Unfortunately, many of the patches once suitable for this species outside of the Jasper Ridge preserve have been destroyed by development, and this habitat destruction has resulted in fewer source areas for the recolonization of empty patches on the preserve. In 1997, Ehrlich reported that, for the first time since he began his studies in 1960, no checkerspot butterflies were found at Jasper Ridge.

The key questions that designers and planners need to ask, then, are "Are there other nearby habitats for species of concern that may affect or be affected by the status of populations on the project site or planning area?" and "How can this overall system, including the ability of species to move among these areas, be maintained over time?" Moreover, like island biogeography, metapopulation theory points to the need for designers, planners, and managers to think across a broad range of spatial scales about issues of dispersal and connectivity, and about flux over time due to both deterministic and chance events, in ways that one would never have imagined under the equilibrium paradigm.

Source-Sink Theory

Just as metapopulation theory leads to the notion that suitable habitat might frequently be unoccupied, source-sink theory (Pulliam 1988; Pulliam and Danielson 1991) leads to the conclusion that unsuitable habitat might often be occupied. Source habitat is defined as that habitat where local reproductive success is greater than local mortality, resulting in the production of surplus individuals that emigrate from the source area. On the other hand, sink habitat consists of areas where local mortality is greater than local reproductive success, resulting in a local reproductive deficit. Clearly, in the absence of immigration from more productive areas, or sources, whatever population exists in sinks will gradually disappear. However, when sink areas are relatively close to source areas, continued immigration into the sinks may maintain large and stable sink populations. In fact, under a number of reasonable conditions, sink populations may be larger than source populations, even though sinks represent unsuitable habitat inasmuch as sink populations will disappear in the absence of continued immigration from source areas. In an example from the southeastern United States, Bachman's sparrows (*Aimophila aestivalis*) occupy both old growth and clearcuts. In fact, because much more forestland is in clearcuts than old growth, the majority of the sparrows are found within these very young stands. However, sparrow reproduction in clearcuts is less than replacement rates, and it appears that clearcut populations are maintained only through recruitment from old-growth stands, which, although much smaller in area, provide a consistent reproductive surplus. If managers were to base forest management decisions solely on bird counts and habitat associations, they could easily assume that harvesting remaining old growth would have no effect on Bachman's sparrow populations, when in fact it could lead to local extinction.

One of the key questions that designers and planners need to ask, then, is "Are there habitat types that may serve as critical source areas for species of concern, without which a population may be destined to extinction?" In particular, source-sink theory provides a strong precautionary principle that the habitat type where a majority of organisms of any given species resides is not necessarily the key or sole habitat type required for population persistence. One of the particular challenges of source-sink and metapopulation theories is that they require knowledge of demography, habitat-based reproductive success, and dispersal—characteristics that can be obtained only through intensive study of individual species. As such, empirical evidence may be available in only isolated instances, and yet the simple knowledge that such phenomena exist establishes further reason for designers and planners to be cautious in their evaluations of design outcomes and to seek out expert knowledge when concerns arise.

Emerging Applications of the New Paradigm

Arising from these various shifts in theory, the new paradigm that has emerged in ecology over the past twenty years emphasizes a disequilibrium point of view, where flux and change are the norm and where local areas are strongly influenced by conditions and events that occur elsewhere in the landscape as well as earlier in time. Although we have attempted to highlight the implications of these developments for designers and planners on a case by case basis, it has been the cumulative effects of these shifts that have truly changed ecology and that offer new ways for those in land conservation and development disciplines to think about and act toward landscapes that range from small properties to entire bioregions. This new paradigm can be said to be the conceptual basis of two of ecology's new subdisciplines: landscape ecology and conservation biology. In reality, landscape ecology is not so much a new subdiscipline as it is a shift in perspective that is gradually changing how ecologists approach and study the natural world around them. For example, landscape ecology by no means has resulted in population ecologists abandoning the point of view that patterns of diversity are influenced by the local availability of resources, but rather it has embedded the idea that resources are important in a broader perspective that takes into account local resources, land-use and disturbance history, and the importance of flux of individuals into and out of the local patch from elsewhere in the landscape. Thus, the landscape perspective has resulted in a more complete view of the factors influencing the current abundance, distribution, and diversity of organisms.

The field of conservation biology constitutes the application of the new paradigm in ecology to the preservation and management of biological diversity. Conservation biology embraces the new perspective of the importance of events and conditions at other places and times in determining patterns of diversity. Thus, conservation biology attaches considerable importance to such topics as habitat fragmentation, the size and isolation of habitat patches, and the role of corridors. To date, conservation biology has not put as much emphasis on the role of history in determining patterns of diversity. Assuming that most landscape designers and planners are familiar with the basic concepts from landscape ecology (Forman and Godron 1986; Forman 1995; Dramstad, Olson, and Forman 1996) and conservation biology (Soulé 1986; Meffe and Carroll 1997) concerning habitat fragmentation and the interconnectedness of habitat patches, in the following sections we place particular emphasis on the somewhat newer ideas of landscape memory and ecosystem autonomy. In particular, we address the following questions:

- *How does landscape history constrain the options available to landscape designers, planners, and managers? Moreover, how can a consideration of landscape memory lead to better design and planning decisions?* We introduce the concept of "landscape memory" and give examples of how forests and other habitats that look superficially the same but that have had different landscape histories provide different potential for conserving biological diversity.
- *What factors control whether dynamics inside the landscape boundaries are more important than conditions outside its boundaries in controlling the fate of the landscape?* We discuss the relationship between system autonomy and landscape flows, size, and management, and how different design and management goals may dictate different landscape sizes to achieve those goals. We offer some simple guidelines about how to use concepts of landscape autonomy to judge what can and cannot be controlled on landscapes of a particular size or context.

Landscape Memory

One commonly accepted meaning of memory is the property of retaining or reflecting the past. Clearly, in this sense of the word, landscapes have memory. The tens of thousands of lakes and wetlands in Minnesota and Saskatchewan are memories of past glaciers, and the worn-out, red clay soils of Georgia are a memory of past farming practices and land uses and abuses. Even the species composition of an area can be thought of as the memory of past extinction and colonization events.

Because we have not thought a lot about landscape memory, we tend to overlook its importance. Consider, for example, two woodlots that appear, at first glance, nearly identical based on the size and types of trees and the degree of canopy coverage. On closer examination, however, one site is found to have a diverse assemblage of spring wildflowers and soil arthropods while the other is found to be depauperate in both respects. Most ecologists noticing such a situation would think first of how present-day differences between the two sites, such as slope, aspect, elevation, soil nutrients, and the like, might explain the observed differences in flora and fauna. But the differences between the sites may have little to do with slope, aspect, and elevation except inasmuch as a farmer, fifty or a hundred years before, may have chosen not to plow a steeper slope. The ultimate difference between the two sites may be that one was formerly used to grow row crops, while the other was once harvested for timber but never plowed. The present-day correlates of low biological diversity, such as loss of soil structure and nutrients, fragmented landscapes, and the invasion of weedy exotics, can be thought of as how landscapes remember their pasts.

Although they may not have referred to it as landscape memory, numerous authors have documented ways in which the present-day condition of ecosystems and landscapes reflect the past. One of the most striking is the physical mosaic of patch types reflecting not natural landscape boundaries but past land-use practices. Flying over any part of North America, for example, one sees an obvious pattern of often rectangular and even circular habitat patches reflecting the current and historic location of property boundaries and land-use treatments. Although the same aerial view may show natural boundaries and gradients, human-induced boundaries are often much more pronounced and tend to obscure or entirely obliterate natural ones.

A long history of studies documents how human land use and management influence both ecosystem processes and the distribution of organisms. A number of recent investigations (e.g., Foster 1988; Abrams and Nowacki 1992; Foster 1992; Nowacki, Abrams and Lorimer 1990) have emphasized how land-use practices such as farming and logging interact with natural disturbances such as fires and storms to determine patterns of nutrient abundance, plant diversity, and distribution. As a case in point, Abrams, Orwig, and Demeo (1995) studied one of the few remaining primary forests in the Blue Ridge province of West Virginia. The forest is currently dominated by Eastern white pine (*Pinus strobus*), red maple (*Acer rubrum*), and three species of oaks (*Quercus alba, Q. rubra, Q. velutina*). By coring and aging trees, Abrams et al. determined that the oldest trees were pines and oaks, but that there was little recruitment of pines and oaks after 1900. By contrast, most of the smaller, younger trees were maples (*Acer rubrum* and *A. saccharum*), beeches (*Fagus grandifolia*), and hemlocks (*Tsuga canadensis*). They concluded that periodic disturbances, particularly fire, in the eighteenth and nineteenth centuries maintained the pines and oaks and that fire suppression in the twentieth century has favored an increase in maples, beeches, and hemlocks. The fact that the forest still has substantial numbers of pines and oaks attests to its memory of fires that occurred more than a hundred years ago.

More and more studies are showing that past land-use changes and management practices have left a long-lasting legacy or landscape memory. Pearson, Smith, and Turner (1998), for example, have demonstrated that the forest fragmentation and land-use practices of the late nineteenth century and early twentieth century have influenced current conditions in southern Blue Ridge forests in North Carolina. They found that wind-dispersed plant species were unaffected by forest fragmentation, but that ant-dispersed plants were less likely to be found in small, isolated forest patches than in large patches or in small patches near to large contiguous areas of forest. They also found that past land-use practices affected current soil conditions, such as nutrient concentrations and percent humic matter, and that these factors in turn influenced the diversity of forest plants.

In a study of stream invertebrates and fish in the southern Appalachian mountains, Harding et al. (1998) found that present-day patterns of river basin sediments and fish and invertebrate species distribution and diversity strongly reflect past land-use histories. Although stream biota may quickly recover from short-term disturbances, such as floods, logging, and construction projects, Harding et al. demonstrated that streams may remember sustained anthropogenic disturbance, such as long-term agricultural use, many decades after the disturbance has ceased. They refer to these long-term effects as the "ghost of land use past," and they argue that recovery of stream invertebrate and fish faunas may require restoration of entire watersheds rather than just streamside protection.

Several key questions for designers and planners, then, are "What has been the landscape's history?" "What types of long-lasting effects may past or proposed human activities have on hydrology, soils, species composition, and the like?" "How do they affect conservation and management strategies?" and "Should they be mitigated?"

Ecosystem Autonomy

Autonomy is defined as the condition or quality of being independent or self-determining. As open, flow-through systems, all ecosystems are dependent on their surroundings. But autonomy, like purity and honesty, comes in many shades of gray; some ecosystems are more independent of their inputs and surroundings than others.

The concept of degrees of ecosystem autonomy can be illustrated by consideration of a lake or pond. The water level and many other physical attributes depend on the balance of inputs and outputs of energy, water, and other materials. Water enters a lake or pond by precipitation, stream flow, and surface and subsurface inputs, and leaves by evaporation, transpiration, stream output, and seepage into groundwater. The water in a pond or small lake fed by a large creek may turn over and be replaced by new water in a matter of days or weeks. The water in a deep, spring-fed pond or lake, on the other hand, may require months or years to be replaced by new inputs. When the turnover of water and other materials is rapid, physical conditions such as temperature, pH, or nutrient content are likely to be very similar to that of the input water, but when turnover is very slow, physical conditions are more likely to result from internal ecosystem processes. Similarly, conditions in a lake or pond receiving large amounts of overland surface flow from nearby agricultural or residential areas may be greatly influenced by the nutrient and contaminant content of the water flowing into it. However, conditions in a similar-sized lake or pond in a similar agricultural or residential landscape that is configured to receive less of its input from surface flows may be more or less immune from the activities in the landscape around it.

Although the concept of ecosystem autonomy is relatively new and undeveloped, a few authors have attempted to quantify or model the degree of autonomy of an ecosystem or components of an ecosystem. For example, Pulliam et al. (1995) developed a simulation model called MAP to explore the consequences of long-term land-use change and management decisions on the population trends of birds, such as the red-cockaded woodpecker and Bachman's sparrow, in forested landscapes of the southeastern United States. In particular, they considered how the size and management of relatively small biological reserves determined how vulnerable endangered and threatened species on the reserve were to land-use changes and forest management practices in the larger landscapes surrounding the reserves. They found that core population densities were most sensitive to the size of the core reserve and the management practices used there but that management activities in the surrounding landscape could also influence the population dynamics in the core. They concluded that large, well-managed core areas were more autonomous than smaller, less well-managed core areas. This is because large area and good management practices, such as periodic burning of mature pine stands, result in larger bird populations, and, consequently, migration into and out of the areas has less impact on overall population dynamics.

In both the pond example and the bird population example, the degree of autonomy is largely determined by the relative magnitude of inputs and outputs in relationship to the size of the systems under consideration. For example, the flow of water into and out of the pond in relationship to the amount of water in the pond helped determine the relative impact of the outside environment.

The question of exactly how large an area has to be for it to be relatively independent of outside disturbance and to protect its internal biological diversity can never be fully answered. One estimate is that if an area is large enough for large carnivores, then it is likely large enough for most other species. Although this is a helpful generalization, like all generalizations, it is far from exact. An area may be very large but not have the diversity of habitats necessary to support many species. Designers and planners concerned about preserving biological diversity should consult with local experts, such as Fish and Wildlife Service biologists and university-based ecologists, about the particular habitat requirements of species of local management concern. Even then, such efforts will require thoughtful collaboration among designers and ecologists to apply incomplete knowledge of species to the inherent uncertainties of predicting the effects of specific actions in specific landscapes (Chapter 13).

Designers, planners, and managers have two ways of influencing ecosystem autonomy. By influencing the flows into and out of the system or by influencing the internal state of the system. One way to explore how flows from and to the outside environment might influence, for example, a bird population is to use what is known as a BIDE, or birth-immigration-death-emigration model (see Pulliam 1988 for details). The basic model builds from the observation that any change in population size is due to one of the four BIDE factors. The basic equation is $\Delta N = B + I - D - E$, where ΔN means change in population size. In other words, the change in population size is equal to the number of births (B), plus the number of immigrants (I), minus the number of deaths (D) and number of emigrants (E). A resource manager could influence population growth either by changing what happens within the system (B or D) or by changing flows into and out of the system (I or E). Birth and death might be influenced by changing how much food or cover is available to the birds on the reserve; immigration and emigration might be influenced by changing how the reserve is connected to other reserves or by changing conditions in other sites in the surrounding landscape.

Although the term "autonomy" has not been used very much in the past, the same idea has been considered under other rubrics. For example, a recent report of the National Research Council (1999) entitled "Ecological Indicators for Monitoring Aquatic and Terrestrial Ecosystems" has proposed indicators quantifying the degree of "independence" of ecosystems. The report maintains that when a biological reserve is isolated from its surrounding landscape, immigration of biological populations into the area is reduced and some species will vanish from the area as a result. The report also argues that small, isolated reserves are more likely to lose native species and have them replaced by exotic species. The report proposes a quantitative "indicator of independence," which is a measure of autonomy at the species level, and suggests that low independence can result either from isolation or from degradation due to disturbance or poor management practices.

In the above examples, more attention has been given to ecosystem autonomy at the species level than at the level of ecosystem function. However, the concept is readily applied to the ecosystem function level as can be illustrated by returning to our lake example. Some simple measures of the autonomy of a lake are given by ratios of physical conditions, such as temperature, acidity, and nutrient concentration, in the lake to the value of the same condition in the input waters. For example, the pH of high-elevation Adirondack streams and lakes is strongly influenced by the pH of the acid rain falling on the lakes, while the pH of streams and lakes of the southern Appalachian Mountains are more buffered by the soil and bedrock conditions of their surrounding landscapes and are therefore less affected by the pH of rainwater. In this case, the ratio of the pH of lake water to the pH of rainwater would presumably be closer to 1.0 for Adirondack lakes, showing low autonomy of these lakes, and the ratio for southern Appalachian lakes would be far from 1.0, showing greater autonomy. Since the biotic community of a lake can be substantially influenced by changes in pH, a measure of autonomy or an indicator of independence at the species level would show a similar pattern, but measurement of the biotic response would be more difficult and expensive to obtain than measurement of the pH response. Understanding the sensitivity of a system to inputs is another facet of ecosystem dynamics that can help designers and managers predict and manage the impacts of human activities.

Although ecosystem autonomy is a new concept, many of the factors that contribute to autonomy are well known and familiar to students of landscape ecology. For example, it is now generally agreed that larger reserves are likely to contain more species and have larger, less-extinction-prone populations of those species. This is, in part, due to larger reserves being more autonomous; that is, events outside of large reserves are less likely to influence what happens within them than is the case with smaller reserves. The way by which reserve size contributes to autonomy can be quite different than the way by which reserve size is generally thought to influence extinction probability, however. Consider two reserves, both old deciduous woodlands surrounded by pine plantations but one ten times as large as the other. The usual reason given for why the larger reserve has less-extinction-prone populations is that population size is proportional to reserve area, so the larger reserve will have larger populations that are less likely to go extinct due to demographic or environmental stochasticity. Although this is likely to be true, the probability of extinction may also be influenced by other mechanisms, more related to reserve autonomy. In our example, if similar-sized tracts of pine plantation adjacent to both the large and small reserves are clearcut and displaced bird and mammal populations disperse into the reserves, the impact of that immigration is likely to be greater on the smaller reserve because the immigrants will constitute a larger fraction of all birds and mammals on the smaller reserve than on the larger reserve. If the immigrants are different species than the residents, the influx of competitors may contribute to higher probabilities of extinction on the smaller reserve, but via a mechanism other than stochastic population fluctuations.

Reserve isolation and the presence or absence of corridors may also influence ecosystem autonomy by different mechanisms than those by which these factors are usually thought to influence population persistence. In fact, proximity to other reserves and corridors between reserves may actually result in less autonomy inasmuch as individual reserves may be more influenced by what occurs outside their boundaries. In this and other examples, ecosystem autonomy cannot be seen, in and of itself, as a desirable attribute, because greater autonomy could result in greater extinction probabilities and lower species diversity. Rather than be seen as a necessarily positive or negative attribute, autonomy should be viewed as an ecosystem characteristic that should always be considered when designing and managing reserves. The more autonomous a reserve is, the less attention needs to be paid to its landscape context, but, just as no individual is an island, to some extent, no ecosystem is fully autonomous, and activities in the surrounding landscape may always have some impact on the integrity of an individual reserve.

Landscape memory and ecosystem autonomy interact in a number of ways. Recall the effect of "ghosts of land use past" on fish and invertebrate diversity in Appalachian streams. In this example, current streams remember past agricultural practices. This is because the streams are not autonomous; rather, conditions in the stream are strongly dependent on land use in the surrounding landscape. Sediments washed into the stream accumulate and may influence the suitability of the streams as habitat for decades to come. However, streams may have different degrees of autonomy and they may vary in their degree of memory of past land-use practices. For example, the "high-gradient" streams on steep slopes may be more prone to sedimentation because of higher erosion from plowing of nearby steep slopes. On the other hand, the sediments in these high-gradient streams may wash out more quickly and have less long-term impact on stream faunal diversity. In this case, low autonomy may be counteracted by a shorter memory. Also, sediments may accumulate in low-gradient streams on flatter terrain and, despite greater autonomy, result in greater memory and longer-lasting impacts on biodiversity.

A landscape that has been severely degraded in the past is likely to reflect that past degradation in its current physical and biological condition, and therefore any remnant "undisturbed" patch in such a landscape is more likely to be influenced by conditions of neighboring degraded patches. The extent of flow of energy and materials from the surrounding landscape into the remnant patches may have profound impacts on the condition of remnants. Small patches of original forest that have never themselves been plowed or clearcut are often depleted in native species and full of invasive exotics if they are surrounded by highly disturbed landscapes. For example, small woodlots surrounded by farmland typically have lower native bird diversity and larger incidence of nest parasitism by cowbirds, while small remnant forest patches surrounded by residential lands often are depauperate in native species and full of weedy exotics. Further, such influence tends to accumulate with time, with native species gradually being replaced by exotics.

Landscape memory can also work to the advantage of land managers. In the Cascade Mountains of Oregon, the loss of upland meadows in forested areas due to fire suppression may reduce local biodiversity and contribute to habitat homogenization. Because grasslands develop very different soil structure than forests, an examination of soil horizons can distinguish historic meadows that have recently been taken over by coniferous forest. Such areas may be the best targets for meadow restoration because the remnant soil structure, a function of landscape memory, may confer greater autonomy following restoration.

Ecosystem autonomy has important ramifications for landscape dynamics and is not easily measured. Designers and planners should use it as a conceptual basis to ask and answer "How important are flows into and out of the landscape in relation to within-system dynamics?" "How can design and management affect that balance in ways that support desired outcomes?" and "How do landscape memory and ecosystem autonomy interact, and how can that inform design and planning decisions?"

Using Concepts of Landscape Memory and Ecosystem Autonomy in Design and Planning

What implications do landscape memory and ecosystem autonomy have for landscape design and planning? A few messages are clear. First, history counts. Aerial photographs are available for many parts of the country going back to the late 1930s or early 1940s. A simple check of historic aerial photography can quickly reveal whether current forests, wetlands, or natural areas were in such a natural condition fifty years ago. An area that had mature forest in 1940 probably had not been plowed or clearcut earlier in this century or else it would not so quickly have returned to a mature forest. An area that was mature forest in 1940 and still is in 2000 should be considered a prime target for conservation, because it is likely to be more intact and diverse than more recently developed forest patches. Moreover, larger patches of intact natural areas are more likely than smaller ones to have retained more of their original species, to be more self-maintaining, and to require less human intervention.

Even a large, relatively intact patch of natural habitat is subject to degradation from overdevelopment along its edges. Commercial, industrial, and high-density residential areas are a primary source of disturbance, weeds, sediments, and pollution that degrade natural areas. Buffer zones, such as recreational areas and low-density housing around remnant patches of older-growth forest and other natural areas, are one approach to minimizing the cumulative impacts of development on the natural amenities of natural areas. Often such protection can be sold to local citizens, planning boards, and even developers by pointing out the positive consequences of such protection, such as a source of clean water, wildlife, and recreation, and an increase in surrounding property values. Planning boards may not understand the concepts of landscape memory and ecosystem autonomy, but they do understand land values.

Landscape designers and managers can directly influence the flow of materials across a landscape and therefore influence its degree of autonomy. Strips of natural vegetation along streamsides protect streams from receiving sediments and contaminants from nearby agricultural fields and development. Confining construction and other soil-disturbing activities to relatively flat terrain outside of the reach of floodwaters can be thought of as a means of managing ecosystem autonomy and memory at a landscape level. The goal, from an ecological perspective, ought to be to reduce the flow from highly disturbed areas into natural areas and to increase the flow from one natural patch to another. Obvious examples come from consideration of water, but many substances besides water flow across the landscape, and terrestrial habitats as well as aquatic ones can be greatly degraded by improper flows. A mature forest patch next to an agricultural field is more likely to be impacted by inputs of pesticides and nutrients than is a mature forest patch buffered from the agricultural fields by a commercial pine plantation that uses far less pesticide and fertilizer than a cornfield. Pest species also flow from highly disturbed areas into natural areas, and a natural area next to an agricultural, industrial, or commercial area is likely to receive more weed seeds and visits by parasitic cowbirds, crows, rats, dogs, and cats than a natural area next to a buffer zone.

In addition to protecting natural areas from unwanted visitors, desirable visitors can be encouraged by interconnecting high-quality patches and, thereby, encouraging the flow of, for example, foxes and owls between remnant natural areas. Such connections between patches not only help maintain populations of large mobile animals but also result in the positive benefits of their visitations, such as pest population control. On the other hand, managing autonomy by connecting a high-quality patch with a lower-quality one containing invasive species might enhance the poor-quality site but degrade the higher-quality one. Finally, in an interaction of memory and autonomy, conservation corridors that follow historic species migration routes, including riparian corridors and ridgelines, are more likely to serve their intended purpose than newly created corridors that have no such lineage.

Concluding Remarks

Ecology, like other disciplines, is an ever-changing body of knowledge and ideas. It is our hope that this chapter will encourage discussion between ecologists and landscape designers about how new and changing ideas from ecology can be usefully incorporated into landscape designs and plans. Many of the concepts being explored by ecologists are extremely complex, and as a young science, ecology is not able to predict nearly all of the consequences of human activities in the landscape. Nonetheless, progress is being made, and there are useful lessons from ecology for design and planning.

Ecologists also have much to gain by working with landscape designers. Thirty years ago, most ecologists sought out pristine, untouched landscapes far from the influence of humans. Now most ecologists understand that if their work is to be relevant to real-world problems and issues, they need to focus some, if not most, of their research on human-dominated landscapes. As ecologists seek to apply their work to the conservation of species and ecosystems, they must work closely with those who are designing and managing landscapes heavily populated by humans. Just as the fate of biological diversity is becoming increasingly intertwined with the fate of human communities, the fate and effectiveness of ecologists is becoming increasingly intertwined with that of designers, planners, and managers of the human-dominated landscape.

Citations

Abrams, M. D., and G. J. Nowacki. 1992. Historical variation in fire, oak recruitment, and post-logging accelerated succession in central Pennsylvania. Bulletin of the Torrey Botanical Club 119: 19–28.

Abrams, M. C., D. A. Orwig, and T. E. Demeo. 1995. Dendroecological analysis of successional dynamics for a presettlement-origin white-pine-mixed-oak forest in the southern Appalachians, USA. Journal of Ecology 83: 123–133.

Allen, T. F. H., and T. B. Starr. 1982. Hierarchy: perspectives for ecological complexity. University of Chicago Press, Chicago, Illinois, USA.

Beier, P., and R. F. Noss. 1998. Do habitat corridors provide connectivity? Conservation Biology 12: 1241–1252.

Bormann, F. H., and G. E. Likens. 1979. Pattern and process in a forested ecosystem. Springer-Verlag, New York, New York, USA.

Brown, J. H., and A. Kodrik-Brown. 1977. Turnover rates in insular biogeography: effect of immigration on extinction. Ecology 58: 445–449.

Cissel, J. H., F. J. Swanson, and P. J. Weisburg. 1999. Landscape management using historic fire regimes: Blue River, Oregon. Ecological Applications 9: 1217–1231.

Clements, F. E. 1916. Plant succession: an analysis of the development of vegetation. Carnegie Institution of Washington, Washington, D.C., USA.

————.1936. Nature and structure of the climax. Journal of Ecology 24: 252–284.

Colinvaux, P. 1993. Ecology 2. John Wiley and Sons, New York, New York, USA.

Connell, J. H. 1978. Diversity in tropical rain forests and coral reefs. Science 199: 1302–1310.

Dramstad, W. E., J. D. Olson, and R. T. T. Forman. 1996. Landscape ecology principles in landscape architecture and land-use planning. Island Press, Washington, D.C., USA.

Ehrlich, P. R., and D. D. Murphy. 1987. Conservaton lessons from long-term studies of checkerspot butterflies. Conservation Biology 1: 122–131

Ehrlich, P. R., D. D. Murphy, M. C. Singer, C. B. Sherwood, R. R. White, and I. L. Brown. 1980. The response of checkerspot butterfly (Euphydryas) populations to the California drought. Oecologia: 46: 101–105.

Fiedler, P. L., P. S. White, and R. A. Leidy. 1997. The paradigm shift in ecology and its implications for conservation. Pages 83–92 in by S. T. A. Pickett, R. S. Ostfeld, and G. E. Likens, editors. The ecological basis of conservation: heterogeneity, ecosystems and biodiversity. Chapman and Hall, New York, New York, USA.

Forman, R. T. T., 1995. Land mosaics: the ecology of landscapes and regions. Cambridge University Press, Cambridge, UK.

Forman, R. T. T., and M. Godron. 1986. Landscape ecology. John Wiley and Sons, New York, New York, USA.

Foster, D. R. 1988. Disturbance history, community organization, and vegetation dynamics of the old growth Pisgah forest, southwestern New Hampshire, USA. Journal of Ecology 76: 105–134.

————. 1992. Land-use history (1730–1990) and vegetation dynamics in central New England, USA. Journal of Ecology 80: 753–772.

Gilpin, M. E., and M. E. Soulé. 1986. Minimum viable populations: processes of species extinction. Pages 19–34 in M. E. Soulé, editor. Conservation biology: the science of scarcity and diversity. Sinauer Associates, Sunderland, Massachusetts, USA.

Haddad, N. M. 1999. Corridor and distance effects on interpatch movements: a landscape experiment with butterflies. Ecological Applications 9: 612–622.

Haddad, N. M., D. K. Rosenberg, and B. R. Noon. 2000. On experimentation and the study of corridors. Conservation Biology 14: 1543–1545.

Hanski, I. 1993. Dynamics of small mammals on islands. Ecography 16: 372–375.

————. 1999. Metapopulation ecology. Oxford University Press, New York, New York, USA.

Hanski, I., and M. E. Gilpin. 1991. Metapopulation dynamics: brief history and conceptual domain. Biological Journal of the Linnean Society 42: 3–16.

————. 1997. Metapopulation biology: ecology, genetics, and evolution. Academic Press, San Diego, California, USA.

Hanski, I., M. Kuussaari, and M. Nieminen. 1994. Metapopulation structure and migration in the butterfly *Melitaea cinxia*. Ecology 75: 747–762.

Harding, J. S., E. F. Benfield, P. V. Bolstad, G. S. Helfman, and E. B. D. Jones III. 1998. Stream biodiversity: the ghost of land-use past. Proceedings of the National Academy of Sciences U.S. 95: 14843–14847.

Harrison, S. D., D. Murphy and P. R. Ehrlich. 1988. Distribution of the bay checkerspot butterfly, *Euphydryas editha bayensis:* evidence for a metapopulation model. American Naturalist 132: 360–382.

Hobbs, R. J. 1998. Managing ecological systems and processes. Pages 459–484 in D. L. Peterson and V. T. Parker, editors. Ecological scale: theory and applications. Columbia University Press, New York, New York, USA.

Kareiva, P., and U. Wennergren. 1995. Connecting landscape patterns to ecosystem and population processes. Nature 373: 299–302.

Kuhn, T. S. 1962. The structure of scientific revolutions. University of Chicago Press, Chicago, Illinois, USA.

Landres, P., P. Morgan, and F. Swanson. 1999. Overview of the use of natural variability concepts in managing ecological systems. Ecological Applications 9: 1179–1188.

Levins, R. 1969. Some genetic and demographic consequences of environmental heterogeneity for biological control. Bulletin of the Entomological Society of America 15: 237–240.

————. 1970. Extinction. Pages 77–107 in M. Gesternhaber, editor. Some mathematical problems in biology. American Mathematical Society, Providence, Rhode Island, USA.

Likens, G. E., F. H. Bormann, N. M. Johnson, D. W. Fisher, and R. S. Pierce. 1970. Effects of forest cutting and herbicide treatment on nutrient budgets in the Hubbard Brook watershed-ecosystem. Ecological Monographs 40: 23–47.

MacArthur, R. H. 1958. Population ecology of some warblers of northeastern coniferous forests. Ecology 39: 599–619.

————. 1960. On the relative abundance of species. American Naturalist 94: 25–36.

————. 1970. Species packing and competitive equilibrium for many species. Theoretical Population Biology 1: 1–11.

MacArthur, R. H. and E. R. Pianka. 1966. On optimal use of a patchy environment. American Naturalist 100: 603–609.

MacArthur, R. H., and E. O. Wilson. 1967. The theory of island biogeography. Princeton University Press, Princeton, New Jersey, USA.

Meffe, G. K., and C. R. Carroll. 1997. Principles of conservation biology, second edition. Sinauer Associates, Sunderland, Massachusetts, USA.

Meffe, G. K., C. R. Carroll, and S. L. Pimm. 1997. Community-level conservation: species interactions, disturbance regimes and invading species. Pages 235–267 in

G. K. Meffe and C. R. Carroll, editors. Principles of conservation biology, second edition. Sinauer Associates, Sunderland, Massachusetts, USA.

Murphy, D. D., K. E. Freas, and S. B. Weiss. 1990. An environment-metapopulation approach to population viability analysis for a threatened invertebrate. Conservation Biology 4: 41–51.

National Research Council. 1999. Ecological indicators for monitoring aquatic and terrestrial ecosystems. National Academy Press, Washington, D.C., USA.

Nowacki, G. J., M. D. Abrams, and M. D. Lorimer. 1990. Composition, structure, and historical development of northern red oak stands along an edaphic gradient in north central Wisconsin. Forest Science 36: 276–292.

Odum, H. T. 1956. Efficiencies, size of organisms, and community structure. Ecology 37: 592–597.

O'Neill, R. V. 1989. Perspectives in hierarchy and scale. Pages 140–156 in J. Roughgarden, R. M. May, and S. A. Levin, editors. Theoretical ecology. Princeton University Press, Princeton, New Jersey, USA.

O'Neill, R. V., D. L. Deangelis, J. B. Wade, and T. F. H. Allen. 1986. A hierarchical concept of the ecosystem. Princeton University Press, Princeton, New Jersey, USA.

Paine, R. T. 1966. Food web complexity and species diversity. American Naturalist 100: 65–75.

Pearson, S. M., A. B. Smith, and M. G. Turner. 1998. Forest patch size, land use, and mesic forest herbs in the French Broad River Basin, North Carolina. Castanea 63: 382–395.

Pickett, S. T. A., V. T. Parker, and P. L. Fiedler. 1992. The new paradigm in ecology: implications for conservation biology above the species level. Pages 65–88 in P. L. Fiedler and S. K. Jain, editors. Conservation biology: the theory and practice of nature conservation preservation and management. Chapman and Hall, New York, New York, USA.

Pulliam, H. R. 1988. Sources, sinks, and population regulation. American Naturalist 135: 652–661.

———. 1997. Providing the scientific information conservation practitioners need. Pages 16–22 in S. T. A. Pickett, R. S. Ostfeld, M. Schak, and G. E. Lickens, editors. The ecological basis of conservation. Chapman and Hall, New York, New York, USA.

Pulliam, H. R., and B. J. Danielson. 1991. Sources, sinks and habitat selection: a landscape perspective on population dynamics. American Naturalist 137: S50–S66.

Pulliam, H. R., J. B. Dunning, D. J. Stewart, and T. D. Bishop. 1995. Modelling animal populations in changing landscapes. Ibis 137: S120–S126.

Reice, S. R. 1994. Nonequilibrium determinants of biological community structure. American Scientist 82: 424–435.

Simberloff, D., J. A. Farr, J. Cox, and D. W. Mehlman. 1992. Movement corridors: Conservation bargains or poor investments? Conservation Biology 6: 493–504.

Simberloff, D. S., and E. O. Wilson. 1969. Experimental zoogeography of islands: the colonization of empty islands. Ecology 50: 278–296.

Soulé, M. E. 1986. Conservation biology: the science of scarcity and diversity. Sinauer Associates, Sunderland, Massachusetts, USA.

Swanson, F. J., J. A. Jones, and G. E. Grant. 1997. The physical environment as a basis for managing ecosystems. Pages 229–238 in K. A. Kohm and J. F. Franklin, editors. Creating a forestry for the 21st century: the science of ecosystem management. Island Press, Washington, D.C., USA.

White, P. S. and S. T. A. Pickett. 1985. Natural disturbance and patch dynamics: an introduction. Pages 3–13 in S. T. A. Pickett and P. S. White, editors. The ecology of natural disturbance and patch dynamics. Academic Press, Orlando, Florida, USA.

Wilson, E. O., and D. S. Simberloff. 1969. Experimental zoogeography of islands: Defaunation and monitoring techniques. Ecology 50: 267–278.

CHAPTER 4

The Missing Catalyst: Design and Planning with Ecology Roots

Richard T. T. Forman

The breadth of knowledge demonstrated by Frederick Law Olmsted, Charles Eliot, and other pioneers in designing and planning the land continues to amaze me. They seemed to have studied and rigorously understood biology, the physical environment, aesthetics, and socioeconomics. They successfully tied together nature protection, recreation, sewage treatment, transportation, land restoration, visual quality, solid waste disposal, and water quality (McHarg 1969; Eliot 1971; Spirn 1984; Hough 1995; Nassauer 1997; McHarg and Steiner 1998). Amazing!

Today society and the land desperately need designers and planners, or some other knowledgeable group, to step forward onto the broad solid shoulders of those giants. Ecologists, economists, and engineers obviously contribute in major ways. But today each has too few of the tools needed to create a sustainable synthesis of nature and culture. Meanwhile most landscape designers are inspired by and primarily focused on important aesthetic dimensions, leaving society's other major objectives to secondary status (Muschamp et al. 1993; Smith and Ferguson 1994; Seddon 1995). And most planners today highlight important economics or public policy dimensions, leaving lesser status to other key societal objectives (Duany and Plater-Zyberk 1991; Tjallingii 1995; Diamond and Noonan 1996; Lynch and Hack 1996; Beatley and Manning 1997). Not surprisingly, I salute the impressive exceptions to these general patterns. Also, clearly, each field evolves over time, and today each has its vision of sustainability.

Nevertheless, this leaves society with tough questions: Is landscape design now largely peripheral to the major concerns of society? Is planning now largely enmeshed in the economic and governmental status quo? Is society now degrading landscapes and land at an accelerating pace? Do we hear the environmentalists' crescendo calling for a sustainable future for land and people?

"Yes" resounds for all four questions. Yet I believe that design and planning have the potential to make a difference for land and people. Indeed, the footsteps of a vanguard of emerging leaders, outlining a new design and planning field, can be heard. Some have their names in this book.

Three key steps are needed to reach this new level. First, the science of ecology must become a central foundation of design and planning. This will noticeably strengthen the field. But it also makes this the only discipline with a palette of expertise effectively embracing both natural systems and human culture.

Second, theory must become clearly stated and put to use in design and planning. That is, the central body of principles needs to be delineated and refined, both to solidify the field and to underpin dependable practice.

Third, boldness must become the norm. Boldness is an alternative to tinkering or the status quo, not a license that "anything goes." Rather, to alter the dominant direction of human land-use change that is so detrimental to long-term nature and culture, boldness requires either a multitude of minor changes or a major new vision. In either case, the proposed solution must be sufficiently understandable and beneficial to society that it has real potential to spread widely in the near future. Neither a few minor changes nor an idiosyncratic new vision holds promise for accomplishing this objective. Assuming a major role in society requires stepping forward with bold new solutions (Leopold 1949; McHarg 1969; Wilson 1984; Forman 1995; Dramstad, Olson, and Forman 1996).

To develop the core thesis of this chapter, I briefly introduce and link ecology, design and planning, culture, and landscape ecology. Key strengths, opportunities, and even shortcomings are highlighted. I hasten to add that my design and planning background is peppered with conspicuous lacunae. However, as an ecological scientist I have been fortunate to be deeply involved in both landscape architecture and ecological planning (plus learning about regional and urban planning from the sidelines) for the past fifteen years.

Three key themes emerge in this chapter: (1) The design and planning field could, with some specific feasible strengthening, become a key to solving the accelerating degradation of land; (2) a serious incorporation of theory, the science of ecology, several dimensions of human culture, and bold solutions is required to create a land where both nature and people thrive; and (3) at least at present, landscape ecology offers the most promising foundation for sustainably meshing nature and culture on land. Together, these themes form a vision for design and planning of the future.

Trend and Opportunity

Worldwide statistics show an increase, often an acceleration, in soil erosion, freshwater scarcity, human population, habitat loss, habitat fragmentation, transportation impacts, suburban sprawl, air and water pollution, and more (McKibben 1989; Cairns, Niederlehner, and Orvos 1992; Gore 1993; Barrow 1994; Noss and Cooperrider 1994; Beatley and Manning 1997; National Research Council 1997). Virtually all these trends also pertain to the United States. Each trend has major implications for socioeconomic costs as well as for the web of life on Earth. Specialists in the various fields seem to be in surprising agreement that, without effective action, a cascading and coalescence of crises will occur within approximately three decades. What will it be like after such an unraveling of land and people? Will you be here?

A shrinking opportunity exists to prevent this coalescence. Consider those major society objectives addressed by the pioneer designers and planners. And consider the spaces that so often receive little or no design and planning, such as farmland, greenways, logging areas, strip development, nature reserves, lakeshores, road networks, parks, ranchland, and stream corridors. These are extensive, rapidly changing areas. Who today can best incorporate the diverse societal objectives into designs and plans for these areas? An ecologist, an economist, or a designer?

To appreciate how different the solutions of these experts would be, let us consider a particular tract of land, such as in a rapidly suburbanizing area of Maryland (USA). The visible portion (Figure 4-1) is representative of an extensive landscape where farmland with fields, farmsteads, streams, and woodland is progressively replaced by housing, commerce/industry, roads, vehicles, and people. Patches of woods, agriculture, and housing are habitats, as well as sources of species, water, and chemical substances. Corridors (strips) of woods, houses, and roads/roadsides are partial barriers that subdivide the landscape, but they also provide routes for movement. Species, water, chemicals, and people move through this land mosaic in routes determined by the pattern of patches and corridors. Furthermore, the pattern of the land mosaic changes over time, as patches and corridors appear and blink out. This dynamic mosaic pattern therefore alters the amounts and locations of ecological resources, as well as the directions of flows and movements across the landscape.

Now consider the contrasting solutions for the land by the ecologist, the economist, and the designer. (Ignore the obvious point that what's around the tract is often more important than what's in it [context over content].) The ecologist might propose creating a large forest on the left side of the area to provide the many biodiversity, soil, and water values of a large natural ecosystem, plus protecting the large grassland on the right for butterflies and grassland birds and establishing a moratorium on development until nitrogen and

FIGURE 4-1.
A rapidly suburbanizing landscape in Carroll County, Maryland, where ecologists, economists, and designers/planners would provide contrasting solutions. Former agricultural landscape with woodlots now has scattered single-family-home development, plus cluster development in center and limited industrial/commercial activity in upper right. Town center is off the photo from right foreground. (Photo courtesy of U.S. Department of Agriculture.)

organic chemicals from the housing areas no longer pollute the stream ecosystems. The economist might see the grassland as a good area for a shopping center, the upper right as a prime location for a light industry park, the need for "upgrading" of some roads, and the consequent skyrocketing values of the surrounding land ripe for immediate residential development. The designer might emphasize the view from the grassland to create an attractive park, trail system, and welcoming entrance to town, as well as adding more street trees as an amenity in the built areas. The future of this landscape will be fundamentally different depending on whose ideas get implemented, or in the typical case, if it continues to be altered with no overall planning and design.

A focus on short-term economics would miss most of society's major objectives, as would an aesthetics or design focus, though opportunities for the latter will be further explored in this chapter. But what would a plan by ecologists look like? They have never really studied roads, culverts, detention basins, grading, septic systems, aesthetics, economics, housing, and human culture (Odum 1989; Forman 1995; Smith 1996). Like many scientists, they tend to wrap themselves in jargon, and some shun projects labeled "applied ecology." The major strengths of ecologists' plans cannot hide their weaknesses.

Huge land areas, including the ex-urban fringe, numerous rural towns, and many counties, are "up for grabs." Spreading development here largely proceeds without overall plans. Designers and planners are uniquely poised to play an effective role, with more of the needed foundations than economists and ecologists have. But can design and planning expand, and deepen, their central foundation of expertise faster than natural scientists, including some landscape ecologists, who are clearly expanding theirs (Odum 1989; Smith and Hellmund 1993; Noss and Cooperrider 1994; Forman 1995, 1999b)? This is an opportunity, the "last great land grab." Furthermore, these areas are where the Tivoli Gardens and Emerald Necklaces of the future (Spirn 1984; Palazzo 1997) should appear.

Ecology, Design, and Culture

To address this huge societal challenge of altering the direction of land transformation to provide a hopeful future, three key components are presented: (1) the science of ecology, (2) design and planning, and (3) culture and ecology in design.

Science of Ecology

Ecology, with a body of theory developed over a century, focuses on "interactions among organisms and the environment" (Odum 1989; Forman 1995; Smith 1996). For instance, ecological theories provide understanding for protecting or enhancing: natural processes, such as succession and water flow; biodiversity, including rare species, fish and wildlife populations; and landscape elements, such as wetlands and stream/riparian corridors. Thus the science of ecology is especially promising as a core foundation of design and planning, including the stated interest in stewardship of natural resources (McHarg 1969; Spirn 1984; Steiner 1990; Hough 1995; Kehm and Yokohari 1995; Steinitz 1995).

The physical sciences—geology, soil science, hydrology, and microclimatology—have long been at least lightly studied and used by landscape architects (McHarg 1969; Spirn 1984; Steiner 1990; Marsh 1991; Hough 1995). The science of ecology includes key portions of these subjects. Specifically, it focuses on how physical factors affect plants, animals, and ecosystems and how animals and plants in turn affect the physical environment.

Much of the traditional core of the science of ecology applies well to small spaces. Indeed, site ecology, or the ecological science of small spaces, needs to have a much higher profile, especially in design. Landscape ecology has developed a rather distinct body of theory for large heterogeneous spaces (Zonneveld and Forman 1990; Saunders and Hobbs 1991; Turner and Gardner 1991; Forman 1995; Klopatek and Gardner 1999), such as seen from an airplane window or in an aerial photograph. In contrast, the subject of small-space ecology remains diffuse.

Site ecology seems particularly important for areas from patio size to football-field size (or window box to airport). A single small patch in a landscape could be usefully analyzed with site ecology (Turner and Gardner 1991; Forman 1995). A site often has distinct boundaries, and its shape, from rounded or square to multilobed (with coves), strongly affects its ecology. Edge conditions, which usually vary widely on different sides, cover much of the small space. The internal structure of the small space is commonly formed by the arrangement of individual trees, shrubs, water, soil, and rock, as well as many kinds of human structures. Since the number of objects is limited in a small space, the presence and the arrangement of each object may play a important ecological role. A distinctive subset of species can thrive in a small space, and while some animals live within it, many others move through it. Being small, the site normally has major interactions with adjacent sites, as well as with more distant sites. These incoming and outgoing flows and movements mean that the site is heavily dependent on, and indeed impacted by, surrounding conditions. Fluctuations within a site with little inherent stability tend to be high, so human maintenance costs may be high. Indeed, many small spaces subject to design are heavily used by people. In short, the distinctive combination of characteristics of a small space or site makes its ecology distinctive, differing from the core of both general ecology and landscape ecology.

At present, the science of site ecology must draw from wildlife biology, aquatic ecology, vegetation structure and dynamics, microclimatology, soil science, hydrology, urban ecology, human ecology, landscape ecology, and more (Marsh 1991; Gilbert 1991; Shepard 1991; Forman 1995). In effect, the major components of a rigorous small-space or site ecology exist, but are scattered over several fields. A great opportunity beckons for someone to make the synthesis. It will become a leg of the future design and planning field.

Design and Planning

Design and planning may be said to shape space by integrating human structures with protection of natural resources (Spirn 1984; Hough 1995;

Nassauer 1997). Within this is a role for earth art and beautiful small spaces that can inspire people. The creative greening of city squares and shopping centers, as well as the design of delightful gardens, commonly makes urban or suburban sites appealing or pleasant (Muschamp et al. 1993; Smith and Ferguson 1994; Van der Ryn and Cowan 1996; Nassauer 1997). In addition, creating symbols or representations of nature can be great education. However, caution is warranted and care must be manifest so the designs themselves do not destroy nature or significantly damage natural ecosystems. Imagine creating images of rare species instead of restoring the habitats where they live, or removing natural habitat to locate symbols of nature and people coexisting. The land degrades and the wrong message is highlighted.

But society looks beyond such green-spot projects to those broader major issues, where the alarming trends outlined above come into focus. Slowly but inexorably, society is turning to ecologists and others of various stripes with expertise in natural systems for planning solutions (Marsh 1991; Smith and Hellmund 1993; Noss and Cooperrider 1994; Forman 1995, 1999a; Diamond and Noonan 1996; Canters 1997). To avert the crises at the end of our dreams of development, we must seriously address stewardship of natural resources, and over large spaces. Designers and planners should and can become part of the solution.

Let us look more closely at the field and profession of landscape design and planning. It reveres its predecessors and elders (McHarg 1969; Spirn 1984; Steiner 1990; Hough 1995; Palazzo 1997). But paradoxically, overall it seems highly constrained by their paradigm, often avoiding rather than welcoming the hybrid vigor of major new ideas and adaptations to new conditions. Of course, delightful exceptions exist, offering bases for hope. In general, the field seems vibrant in areas such as landscape design, aesthetics, and landscape architectural history. Yet there appears to be unusually wide variation in the importance and understanding required for site ecology, landscape ecology, landscape planning, technology, perception/visual quality, physical environment, and electronic representation. This conspicuous mixture of strength and weakness from place to place and person to person in landscape architecture, like a chain with strong and weak links, dilutes the usefulness of the field in the eyes of society. The influence of the profession rolls along nearly in steady state, while development rampages over the land.

Theory lies at the core of a major field. The body of theory—the important principles, models, and concepts developed over time—defines the central expertise of those in the discipline. Architectural theory, geological theory, music theory, economic theory, and ecological theory are the solid roots of the respective fields. In addition, theory underpins a profession, providing foundations for effective practice, projects, and solutions. Society can depend on the practitioners to provide solutions well grounded in theory.

For landscape architecture, the development of theory is both a special problem and a special opportunity. Although individual theories have been periodically identified or developed (McHarg 1969; Lyle 1985; Thompson and Steiner 1997; Nassauer 1997; Palazzo 1997), the central body of generally accepted theory apparently has not been clearly articulated in print for many years. I'm told that some think it is waiting to be extracted from history, others consider it to be essentially modified architectural theory, and some consider the subject of tangential interest. I sense that this central gap is a great opportunity awaiting lucid thinkers and writers.

Working with and without theory produces contrasting results. Consider an artist guided mainly by inspiration and imagination creating a unique or distinctive design for a space such as a canvas or site. The distinctive design is associated with the person, and over time the artist produces an array of unique spaces. Basic theory or principles are used, but frequently are minimized as constraints to be overcome by imagination.

Now consider designing sites such as house lots or small parks, or even wheels as a metaphor, with and without theory. An array of beautiful wheels may emerge, but the wheel is reinvented on each site. Without using the basic theories of wheel design, or of ecology and technology in design, the beautiful wheels are apt to be oblong, wobbly, single-spoked, or square. Such wheels do not work. House lots and small parks are multiplying faster than are designers and planners. Without theory, generic approaches or solutions for lots and parks covering the land are unlikely to emerge. Design and planning would continue to create scattered attractive green spots that have little coherent positive impact on nature and on society. In short, make the wheel right, make it replicable, and then make it unique, distinctive, imaginative, inspirational, or beautiful.

A landscape architect, rather than transforming a homogeneous canvas, analyzes and alters a site's existing rich texture, plus its myriad interactions with the surrounding mosaic. Indeed, consider these spatial patterns and natural processes so central to the designer's and planner's work. Ecological flows (or natural processes) crossing the landscape are numerous, concurrent, and interacting (Figure 4-2a) (Forman 1995, 1999a; Harris, Hoctor, and Gergel 1996; Forman and Hersperger 1997). Although short stretches may be nearly straight, the flows are overwhelmingly curvilinear and irregular. Most important, they create visible spatial patterns.

These spatial patterns of nature—the forms, shapes, and structures—are very distinctive, and normally contrast dramatically with those produced by people (Forman 1995, 1999a; Forman and Hersperger 1997; Klopatek and Gardner 1999). Nature's forms are primarily aggregated, curvy, convoluted with lobes and coves, elongated, variable in size, irregular, fractal or dendritic,

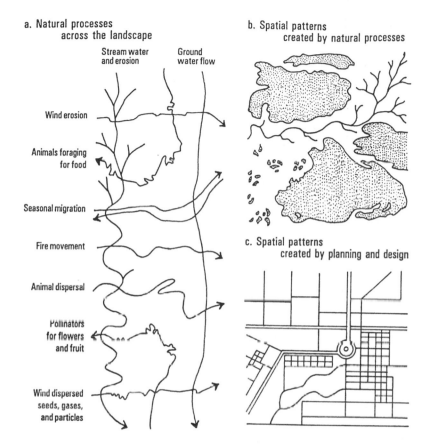

a. Natural processes
 across the landscape

Stream water
and erosion

Ground
water flow

Wind erosion

Animals foraging
for food

Seasonal migration

Fire movement

Animal dispersal

Pollinators
for flowers
and fruit

Wind dispersed
seeds, gases,
and particles

b. Spatial patterns
 created by natural processes

c. Spatial patterns
 created by planning and design

d. Common spatial patterns produced by:

NATURE	DESIGN AND PLANNING
Aggregated	Square
Curvy	Rectangle
Convoluted with lobes & coves	Grid
Elongated	Two parallel lines
Variable in size	Smooth curves
Richly textured	Circle with radiating lines
Fractal or dendritic	

FIGURE 4-2.

Common ecological flows and spatial patterns in landscapes. (a) Typical trajectories for ecological flows or natural processes across a landscape. For illustration, routes are superimposed on but independent of one another. (b) Spatial patterns resulting from natural processes illustrated. (c) Spatial patterns resulting from design and planning illustrated. (d) Summary list of common spatial patterns produced by the two processes. See Forman and Hersperger 1997; Forman 1999a.

and richly textured (Figure 4-2b and d). These are not amorphous forms; they are well-documented spatial patterns with known causative mechanisms from the scientific literature.

In contrast, the spatial patterns of design and planning by people seem limited by Euclidean geometry (Duany and Plater-Zyberk 1991; Smith and Ferguson 1994; Lynch and Hack 1996; Van der Ryn and Cowan 1996). Squares, grids, rectangles, two parallel lines, smooth curves, and circles with radiating lines seem to predominate (Figure 4-2c and d). Forms determine functions, and vice versa (Forman 1995). Thus areas with mainly natural patterns function differently from planned and designed areas. This stark contrast highlights the major need for ecological knowledge in design and planning.

Two processes in addition to nature and design/planning produce spatial patterns important to landscape architects and planners (Forman 1995, 1999a; Pietrzak 1989). Based on qualitative observations on three recent visits to central Tuscany (Toscana, Italy), it appears that long-term unplanned development, essentially the product of trial and error, results in: rather small, evenly sized squarish patches; a low variance in shape; an abundance of short, thin corridors; and a fine-grained landscape. In long-inhabited landscapes where access is along a coastline, river, canal, or road, the patches are long and narrow rather than squarish.

In contrast, recent unplanned development, such as in new American suburbs (Diamond and Noonan 1996), seems to produce a high variance in patch size, high variance in patch shape, predominance of "mixed forms" (with boundaries partly rectilinear and partly irregular), and relatively fine-grained landscape, with all land types fragmented and seemingly somewhat dysfunctional. In short, these patterns produced by long-term versus short-term unplanned development differ markedly.

Because of the relationship between form and function (Forman 1995), the four types of patterns must function or work very differently. Excluding nature's case, I hypothesize that ecologically the long-term unplanned pattern is best, the short-term unplanned pattern intermediate, and the design/planning pattern poorest. Clearly, that needs evaluation, pinpointing the ecological gains and losses. Irrespective, the comparison again emphasizes the importance of ecology in design and planning. Furthermore, the recent unplanned areas represent that great opportunity highlighted near the outset for design and planning to have an impact on the globe.

Thus to plan and design the land for a more sustainable future, catalyzing and publishing high-quality research in the design field and clearly articulating the field's body of theory and expertise, is a *sine qua non*. For example, research publications, especially by designers and planners, could have a major impact on any of the following:

- Unloved places and "loving a place to death"
- Meshing the ecology and design of roads in a landscape
- Neighborhood configurations making good sense both ecologically and for people
- Regional ecology, and incorporation of landscape ecology into metropolitan land-use planning
- Integrating ecology and visual quality or human culture
- Long-term monitoring and evaluation of completed projects

In effect, high-quality research and publication could quickly raise the stature of the field, and thus the profession.

Culture and Ecology in Design

Imagine designing a city plaza as a meeting place for local artisans and shoppers, which also attracts the sequential waves of migrating songbirds in season. The design of a beautiful garden can also provide habitat interspersion and convergency points (junctures of three or more habitats) for key wildlife (Forman 1995). A park can be designed for both intensive recreation and no increased soil deposition into an adjacent stream with threatened snails or fish. Each example combines a key cultural and ecological objective. Successfully combining two or more such objectives should be easy and the norm.

Also, because so many design projects are currently at this fine scale, a clear presentation of site ecology theory and principles for design is needed (Kress 1985; Bormann, Balmori, and Geballe 1993). It could have an immediate impact.

Clearly, aesthetics, as well as economics, are important to society. Yet if a site is ecologically unsound, designing it just for beauty or scenic views is largely a waste of time and society's resources. The site simply floods out, erodes, becomes species impoverished, is overrun by invasive exotics, degrades a stream, lowers a wetland water table, blocks a major wildlife corridor, or fragments a critical large natural-vegetation area (Marsh 1991; Gilbert 1991; Smith and Hellmund 1993; Bormann, Balmori, and Geballe 1993). Even a site dominated by the urban fabric has some natural patterns and processes.

Instead, after understanding the surrounding ecology and culture, plus the existing values within a project area, it is important to first design and create a solid ecological foundation for the area. Then enhance its aesthetics or economic viability as appropriate (Forman 1995). A possible alternative approach would be to first develop solid ecological knowledge of an area, and then design iteratively for ecological function along with aesthetics (Lyle

1985; Nassauer, pers. comm.). The results of these two design processes should be compared to determine the optimum model. Nevertheless, in some cases ecological integrity itself will provide the appropriate aesthetics or economics.

Indeed, ecology and aesthetics are linked in diverse ways. For instance, combining the perceptions of different groups of people with the array of ecological components (water, wildlife, rare plants, and so on) and at varied spatial scales produces a cornucopia of potential design and planning solutions. In a few cases the most aesthetic and the best ecological solution may coalesce. But, probably, the optimum solution usually would not be ranked number one for beauty. Often that is not an easy message to absorb and implement.

Aesthetics can also be a powerful cultural influence on how humankind perceives nature, and on how individual people experience ecological processes in daily life (Nassauer 1992; Muschamp et al. 1993; Im 1995; Nassauer, pers. comm.). This suggests that aesthetics could be quite important for establishing environmental policy, a useful way to help achieve the ecological objectives of society.

Short-term economics dictates the key design parameters of far too many projects today (Morgan and King 1987; Duany and Plater-Zyberk 1991; Van der Ryn and Cowan 1996; Diamond and Noonan 1996; Beatley and Manning 1997). For the human half of the design and planning equation, I would rather see longer-term culture be the lead factor. Here I use culture in the narrow traditional sense of expressing language, art, customs, education, morals, and literature passed by groups from generation to generation (Buell 1995; Park 1995; Forman 1995; Nassauer 1997; Seddon 1997). Aesthetics of course is a part of this cultural core.

More important, though, is the explicit pinpointing and highlighting of the other varied cultural dimensions, such as specific traditions, literature, language idioms, music, and so forth. These could be manifest in every design and planning project (Seddon 1997; Nassauer 1997). Society would love it. Much has been written on this subject, but it remains a puzzle why seriously addressing this whole core of culture (like ecology) is not a central foundation of the design profession.

Note that this refers to the culture of the people, for example, who will use the place, not that of the designer. Again there is a limited place for designs reflecting the designer's culture. But for the profession to have an impact over the land, it must go further, clearly attempting to understand and highlight the culture of the people. Furthermore, analogous to the range of ecological conditions present, the diversity of cultures encourages a richness of designs.

Indeed, culture and ecology are not unrelated. If I were a designer or planner, I would find it exhilarating to try to seriously mesh and highlight these two major objectives of society in projects. Compared with transportation, economics, housing, public policy, and the like, ecology and culture reach deeply and persistently into the very fabric of humans and the planet (Leopold 1949; Wilson 1984; Shepard 1991; Park 1995; Buell 1995; Seddon 1997). Culture and ecology provide stability and underpin sustainability.

Landscape Ecology

In contrast to the preceding subjects, landscape ecology, which elucidates the ecological patterns, processes, and change of land mosaics (Zonneveld and Forman 1990; Saunders and Hobbs 1991; Turner and Gardner 1991; Forman 1995; Klopatek and Gardner 1999), has rapidly developed in only the past fifteen years. By explicitly integrating science and spatial pattern, it dovetails especially well with planning and design (Forman 1999b). Landscape ecology has mushroomed in our midst because it opened conspicuous new scholarly frontiers of research and theory, forged synergistic linkages among people and disciplines, and provided novel solutions for persistent environmental and societal problems (Dramstad, Olson, and Forman 1996). Its patch-corridor-matrix model has become an effective handle for analyzing the structure, functioning, and change of a landscape as a specific object, a

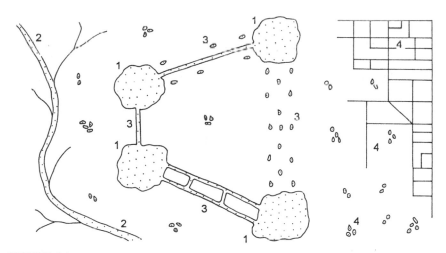

FIGURE 4-3.
Top-priority "indispensable patterns" in planning a landscape based on landscape ecology. 1 = a few large patches of natural vegetation; 2 = major vegetated stream or river corridor; 3 = connectivity with corridors and stepping-stones between large patches; 4 = heterogeneous "bits of nature" across the matrix. See Forman 1995; Forman and Collinge 1995, 1997; Forman and Hersperger 1997.

living system (Figure 4-3) (Saunders and Hobbs 1991; Forman 1995; Klopatek and Gardner 1999). The model highlights the commonalties among landscapes from urban to pristine, as well as the linkages between natural and human places within a landscape.

Yet landscape ecology is no panacea. Its theory needs tuning, empirical bases need strengthening, applicability to small sites needs development, and problem-solving applications should proliferate. The field needs stronger mutual linkages with several adjoining fields.

Nevertheless, students love the subject because it is spatially focused, a frontier field, at the broad scale, obviously important, rife with applications, and a drawing card to interact with students of related disciplines. More than eighty theories and principles of landscape ecology have been articulated (Forman 1995). Examples and case studies of applications have been published for many of them (Zonneveld and Forman 1990; Saunders and Hobbs 1991; Smith and Hellmund 1993; Noss and Cooperrider 1994; Harris, Hoctor, and Gergel 1996). Almost all are directly usable by designers and planners (Dramstad, Olson, and Forman 1996).

By placing a site in the context of, and interacting with, its surrounding mosaic of land uses, landscape ecology provides ecological understanding for the individual house lot, pasture, natural area, garden, and so on (Figure 4-4) (Forman 1987, 1995). More important, the field focuses directly on the whole landscape, ecologically explaining its patterns, how it works in terms of flows and movements, and how it changes spatially over time.

Landscape ecology is rapidly becoming important both in natural resource management and in land-use planning (Nassauer 1997; Forman 1999a, 1999b). Indeed, its large-area and long-term focus provide an obvious foundation for how we can design and plan the land for a more sustainable future (Forman 1990, 1995, 1999b).

Several specific trends largely developed in landscape ecology are quite likely to become important in design and planning practice of the future (Forman 1995, 1999a; Forman and Collinge 1995, 1997; Forman and Hersperger 1997). Seven of these, described in more detail in the literature, are briefly introduced here.

First, several recognized "indispensable spatial patterns" of nature will be among the top priorities in almost all projects (Figure 4-3). At present four such patterns have been identified, a few large natural-vegetation patches, connectivity among the patches, vegetation along major streams, and "bits of nature" scattered over a less hospitable matrix. The patterns are considered indispensable because no known or technologically feasible alternative exists to provide the ecological benefits each provides.

Second, an "aggregate-with-outlier" model portrays an effective way of fitting diverse land uses together (Figure 4-4). This indicates that one should

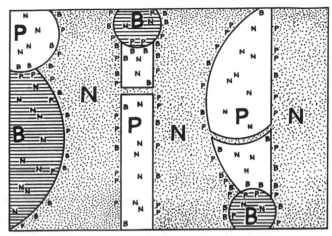

FIGURE 4-4.
Aggregate-with-outlier model for ecologically meshing land uses. N = natural vegetation; P = pasture; B = built area; large letters indicate a large patch; small letters indicate locations of small patches. See Forman 1995; Forman and Collinge 1995; Forman and Hersperger 1997.

aggregate land uses, yet maintain small patches and corridors of nature throughout developed areas, as well as outliers of human activity spatially arranged along major boundaries. The distance between human outliers increases with distance from a source, such as a large built area. The model effectively incorporates several important ecological characteristics: risk spreading; genetic variation; boundary zone; corridors; large patches of natural vegetation; and small patches of natural vegetation.

Third, providing for horizontal ecological flows or natural processes, that is, the flows and movements across the land, will be a key objective (Figure 4-2a). The interruption of these ecological flows by human activities is often a major reason for high maintenance costs, as well as serious damage and repair. Conversely, building in consort with these horizontal flows and movements is more sustainable.

Fourth, comparing the spatial patterns produced by natural processes, by design/planning, and by lack of planning will highlight the ecological impoverishment in most design and planning, and reveal rich opportunities for the designer and planner (Figure 4-2b, c, and d). Creating anthropogenic spatial patterns that mimic those produced by nature, rather than the rather rigid geometric patterns characteristic of planning and design, should, for example, increase protection of streams, aquifers, and habitats for biological diversity. Indeed, the ecological flows across the land, such as groundwater flow and animals foraging and dispersing, should be enhanced (Harris, Hoctor, and Gerge 1996). Of course, the patterns or objects are best oriented relative to the predominant directions of flows (Forman 1995).

Fifth, incorporating at least one "bio-rich place" in every project will begin to produce impressive ecological benefits (Forman 1995). Such a concentration of native species, even in a relatively small spot, serves as a source of seeds and animals for the surroundings. It is often alive with animals moving across the landscape. Moreover, a bio-rich place can serve as a magnet for people to observe, appreciate, and understand nature.

Sixth, an ecologically optimum sequence for changing a landscape will provide the directional and time dimensions for planning and design. Thus the "jaws-and-chunks" model has currently emerged as the best way for a large landscape to change or be changed (Figure 4-5). Its specific sequence of mosaics results from comparing, and improving on, the basic ways that a large, more ecologically suitable land type is progressively transformed to a less suitable land type in the real world. This proposed optimum pattern of landscape change not only portrays the whole landscape over time, but permits one to identify the best and worst location for, for example, the next shopping mall or nature reserve. Two analyses indicate that spatial planning has the greatest ecological impact if done during the first 40 percent of the land transformation (Forman 1995; Forman and Collinge 1997).

Seventh, the omnipresent conspicuous road network will be thoroughly integrated ecologically in the landscape (Figure 4-6) (Cairns, Niederlehner,

JAWS - AND - CHUNKS MODEL

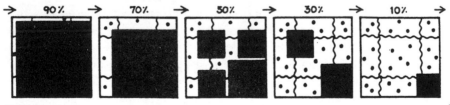

FIGURE 4-5.
Jaws-and-chunks model, the ecologically optimum sequence of mosaics in landscape change. The landscape begins all black and, if allowed to proceed, ends all white; black land type is better than white land type for ecological characteristics, such as biodiversity, animal movement, water flows, and erosion/sediment flows. See Forman 1995; Forman and Collinge 1995; Collinge and Forman 1998.

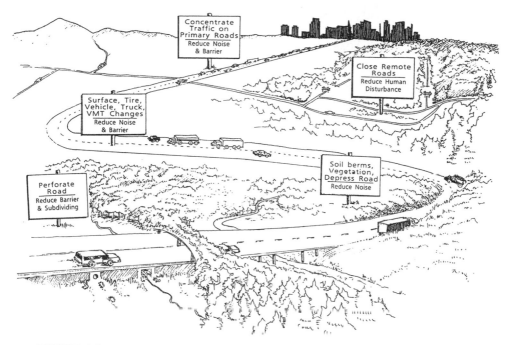

FIGURE 4-6.

Changes in the road-effect zone to reduce ecological impacts on the land. The five categories of change summarized in the signs are designed to reduce ecological effects that extend outward furthest from roads, namely, barriers to animal crossing, subdividing large persistent populations into small threatened subpopulations, noise impacts inhibiting the avian community, and human disturbance in remote areas. See Forman 1995, 1999; Forman and Alexander 1998; Forman and Deblinger 1998. Drawing courtesy of L. Giersbach.

and Orvos 1992; Forman and Alexander 1998; Forman 1999a; Forman and Deblinger 2000). The road system used by vehicles permeates widely through the land and causes an array of major ecological effects. Key ecological impacts include traffic noise, a barrier subdividing species into relatively isolated small populations more subject to local extinction, hydrologic effect on stream systems, network structure effects on large mammals, and access from remote roads resulting in human disturbance. Five categories of policy change could significantly reduce this cumulative impact of the road system (Figure 4-6).

These trends of today are apt to develop into the principles of tomorrow. Adding an anthropogenic focus points to three categories or dimensions as keys to design and planning of the future:

THE SPATIAL SOLUTION

Incorporate indispensable patterns.
Use aggregate-with-outlier model.

NATURAL PATTERNS AND ECOLOGICAL FLOWS

Provide for horizontal natural processes.
Use the spatial patterns of nature.
Provide "bio-rich places."

CULTURE AND SOCIOECONOMICS

Make core cultural values manifest.
Make economically and socially viable.
Provide high visual quality.

The spatial solution provides spatial patterns that make good ecological sense anywhere for land and people (Forman 1995; Forman and Collinge 1995, 1997). Natural processes and patterns will strengthen the ecological fabric for both the near term and long term (Forman 1995, 1999a; Harris, Hoctor, and Gergel 1996; Forman and Hersperger 1997). Similarly, the anthropogenic focus will provide short- and long-term solutions for society (McHarg 1969; Buell 1995; Forman 1995; Diamond and Noonan 1996; Seddon 1997).

Planning for Nature and Culture: Two Examples

I know of no place that effectively integrates core cultural values with modern landscape ecology principles. Thus today's scholarship and scientific evidence are outlining a vision for the future, which could be accomplished by planners and designers. Although such a proposed sustainable meshing of nature and culture may not currently exist, there are places where culture and nature coexist in positive and distinctive ways. We can learn from such places. Two U.S. towns, Concord, Massachusetts, and The Woodlands, Texas, are briefly introduced (see references for their temporal and spatial contexts). Here I focus on listing key attributes that were designed or planned to link nature and culture, and identifying some important integrative themes.

Concord is a product of lack of overall planning, plus numerous small design and planning decisions over time, whereas The Woodlands began and is controlled by an overall master plan. In reflecting on these two examples, we should consider whether they are bizarre outliers, unreplicable in essence, or alternatively, whether they may reflect promising steps toward reaching a sustainable mesh of nature and culture.

Consider Concord, Massachusetts (Wheeler 1967; Eaton 1974; Garrelick 1985). Its long-range open space plan, spearheaded by a landscape designer (J. Ferguson) and a landscape ecologist (R. Forman), focuses on large patches of natural vegetation (plus built and agricultural areas) (Ferguson et al. 1993). These areas are connected by major water and wildlife corridors, and peppered with small, special sites of geologic, scenic, ecological, historical, infrastructural, and recreational value. Several designed or planned features in town explicitly link local nature and culture:

- "Open space" areas of importance to the naturalist Henry David Thoreau and to the pioneer writer Louisa May Alcott
- A public path in the English tradition, and a creative boardwalk connecting midtown areas through a marsh
- Savanna-like areas maintained to resemble seventeenth-century wood pastures, and roadside clearing to highlight views of farmland
- Miles of nearby stone walls recently constructed to help portray the Minutemen's landscape of the American Revolution
- A tower for bird watching
- A vegetable garden with old varieties right where Thoreau dug a garden as a wedding gift to Sophia and Nathaniel Hawthorne
- Historic districts to maintain architectural integrity, and a pervasive backdrop of woods and swamps, reminding residents of the rugged individualistic character of earlier residents

These visible features benefit, indeed highlight, both culture and ecology. Two of the major landscape ecology recommendations (see preceding section) are evident in Concord: (1) the four indispensable patterns (a few large natural-vegetation patches, connectivity among them, vegetation corridors along major streams, and bits of nature over the matrix); and (2) providing for ecological flows across the landscape (large mammals, species assemblages, groundwater, surface water, and walkers/skiers/horseback riders). Several core cultural values, including literature, art, architecture, customs, and high visual quality, are manifest for all to see. Of course, all is not well; misplaced houses, polluted streams, unprotected aquifers, traffic jams, exotic plant invasions, and demolished historic structures are also manifest. Nevertheless, Concord has begun to implement landscape ecology principles together with culture in planning and on the ground.

The planned community of The Woodlands, Texas, where certain ecological characteristics were important in planning, design, and construction, offers a very different example (Spirn 1984; Morgan and King 1987; Hough

1995; McHarg and Steiner 1998). Superficially, the place looks similar to an upper-middle-class, outer American suburb. But the ecological planning makes it fundamentally different. For example, flooding was nearly eliminated by linking housing density to soil type and by maintaining natural vegetation in front and back yards of house lots. Wildlife movement was provided for in continuous wetlands. Further distinctive natural and cultural attributes include the following:

- Relatively small house lots helping to form neighborhoods of people and requiring less loss of natural habitat
- Fingers of golf courses interdigitating with housing clusters, reinforcing a strong cultural message throughout neighborhoods
- Walking and cycling paths connecting everything
- Concentrated commercial, religious, and cultural centers, which are separated from clusters of homes
- Streams and ponds providing placid scenes and rich habitat

Design and planning of The Woodlands, an ecologically remarkable community, did not benefit from landscape ecology that coalesced a decade later. For instance, natural vegetation patches are relatively small, and forest edges that favored generalist edge species at the expense of interior species were promoted (Spirn 1984; Morgan and King 1987; Claus 1994; Forman, pers. observ.). Connectivity with a large adjacent state park was largely severed, and pervasive road and trail networks interrupt animal movement routes. The important red-cockaded woodpecker was selected against by emphasizing hardwoods protection. Natural groundwater and surface water flows were squeezed and interrupted. And a fine-grained, geometry-dominated pattern was imposed on the natural landscape patterns. In short, this suburban town would look very different if planned and designed today using the landscape ecology principles presented in this chapter.

Ecology and culture are linked in The Woodlands, yet both appear very different than they do in Concord. Indeed, the richness of possible designs and plans for a town or a county, all solidly based on principles and portraying the synthesis of nature and culture, is stupendous.

Conclusion

As design and planning, ecology, and landscape ecology develop in the years ahead, all offering solutions to conservation and development conflicts, we could take the time-tested approach of working to improve the current trajectory. An alternative, more speculative and more powerful approach is to provide a vision. Sketch out the form or appearance of a desirable future. Add a framework or trajectory for getting there. If alternative visions arise, fine. Accumulating evidence and testing alternative hypotheses should be our bread and butter. Society gets engaged with people offering vision.

My vision of future planning and design has both ecology roots and a major role in society. Current positive traditions plus inertia will keep many elements of today's design and planning intact, so my focus is on the exciting area of coalescence where visionaries in the field will lead. In this vision, designers and planners of the future will continue as leaders in visual quality (or aesthetics) expertise, will maintain a solid foundation in the science of ecology, will build from a body of theory or principles, and will incorporate a range of core cultural values in projects. This is a distinctive set of expertises. No other discipline or profession provides that combination. Society will increasingly turn to planners and designers, not just to counteract the degradation of landscapes, but to provide inspired solutions to big problems of large areas.

The standard usage of this array of principles will also demonstrate to society that landscape architects take seriously, and have the expertise to provide solutions for, protection of natural resources (McHarg 1969, Kehm & Yokohari 1995; Kiemstedt 1995; Steinitz 1995; Nassauer 1997; Seddon 1997; Andropogon Associates, pers. comm.). Furthermore, more projects will be more sustainable in meshing nature and culture. In mundane terms they will require a lower maintenance budget. More significant, these projects will approach the sustainability target of lasting over human generations (Forman 1990, 1995).

A serious knowledge and use of the science of ecology is within reach and can be accomplished. Taking this critical extra step is completely consistent with the design and planning goal of excellence in stewardship of natural resources or minimizing environmental problems. Transportation planning in The Netherlands and Australia provides a useful model, where engineers and natural scientists both sign off on plans and projects (Saunders and Hobbs 1991; Canters 1997; Forman and Alexander 1998). This approach will protect society from projects that are essentially just aesthetics, just economics, or just ecology.

Ecology and culture at the core. A solid underpinning of theory. Boldness, not tinkering. Imagine how useful designers and planners could become to society, and to the solution of major environmental and societal issues! I am convinced that this vision is within reach. The history of the fields should record that a cohort of today's forward-thinking planners and designers stepped forward and made such an expansive leap ahead. Many authors in this book represent the vanguard. Imagine. This offers a rare but concrete basis for optimism.

The future is not just what lies ahead; it is what we create.

Acknowledgments

I am delighted to thank: my students and faculty colleagues over fifteen years at the Harvard University Graduate School of Design for helping frame and sharpen the ideas herein; Carl Steinitz, who has continually made me think of the applications of landscape ecology in landscape planning; Vittorio Ingegnoli for pinpointing long-term unplanned trial-and-error development as a process producing distinctive spatial patterns; Lisa Giersbach for the transportation change artwork; and Frederick Steiner, Joan I. Nassauer, Rossana Vaccarino, Gary Hilderbrand, Bart Johnson, and Kristina Hill for educating me with their valuable insights on the manuscript. I hope my friends in these fields pardon my transgressions while pushing the envelope; I want to see the land actually improve, and that can't wait forever.

Citations

Barrow, C. J. 1994. Land degradation. Cambridge University Press, Cambridge, UK.

Beatley, T., and K. Manning. 1997. The ecology of place: planning for environment, economy, and community. Island Press, Washington, D.C., USA.

Bormann, F. H., D. Balmori, and G. T. Geballe. 1993. Redesigning the American lawn: a search for environmental harmony. Yale University Press, New Haven, Connecticut, USA.

Buell, L. 1995. The environmental imagination: Thoreau, nature writing, and the formation of American culture. Belknap Press of Harvard University Press, Cambridge, Massachusetts, USA.

Cairns, J., Jr., B. R. Niederlehner, and D. R. Orvos, editors. 1992. Predicting ecosystem risk. Princeton Scientific Publishing, Princeton, New Jersey, USA.

Canters, K., editor. 1997. Habitat fragmentation and infrastructure. Ministry of Transport, Public Works and Water Management, Delft, NL.

Claus, R. C. 1994. The Woodlands, Texas: A retrospective critique of the principles and implementation of an ecologically planned development. Masters thesis, Massachusetts Institute of Technology, Cambridge, Massachusetts, USA.

Collinge, S. K., and R. T. T. Forman. 1998. A conceptual model of land conversion processes: predictions and evidence. Oikos **82**: 66–84.

Diamond, H. L., and P. F. Noonan. 1996. Land use in America. Island Press, Washington, D.C., USA.

Dramstad, W., J. D. Olson, and R. T. T. Forman. 1996. Landscape ecology principles in landscape architecture and land-use planning. Harvard University Graduate School of Design, American Society of Landscape Architects, and Island Press, Washington, D.C., USA.

Duany, A., and E. Plater-Zyberk. 1991. Towns and town-making principles. Harvard University Graduate School of Design, and Rizzoli, New York, New York, USA.

Eaton, R. J. 1974. A flora of Concord: an account of the flowering plants, ferns and fern-allies known to have occurred without cultivation in Concord, Massachusetts, from Thoreau's time to the present day. Special Publication 4. Harvard University Museum of Comparative Zoology, Cambridge, Massachusetts, USA.

Eliot, C. W. 1971. Charles Eliot, landscape architect: a lover of nature and of his kind, who trained himself for a new profession, practised it happily and through it wrought much good. Books for Libraries Press, Freeport, New York, USA.

Ferguson, J. D., M. Connelly, R. T. T. Forman, M. Kellett, C. Mackenzie, D. H. Monahan, S. Schnitzer, J. Sprott, and B. Stokey. 1993. Town of Concord 1992 open space plan. Concord Natural Resources Commission, Concord, Massachusetts, USA.

Forman, R. T. T. 1987. The ethics of isolation, the spread of disturbance, and landscape ecology. Pages 213–229 in M. G. Turner, editor. Landscape heterogeneity and disturbance. Springer-Verlag, New York, New York, USA.

———. 1990. Ecologically sustainable landscapes: the role of spatial configuration. Pages 261–278 in I. S. Zonneveld and R. T. T. Forman, editors. Changing landscapes: an ecological perspective. Springer-Verlag, New York, New York, USA.

———. 1995. Land mosaics: the ecology of landscapes and regions. Cambridge University Press, Cambridge, UK.

———. 1999a. Horizontal processes, roads, suburbs, societal objectives, and landscape ecology. Pages 35–53 in J. M. Klopatek and R. H. Gardner, editors. Landscape ecological analysis: issues and applications. Springer-Verlag, New York, New York, USA.

———. 1999b. Landscape ecology, the growing foundation in land-use planning and natural-resource management. Pages 13–21 in P. Kovar, editor. Nature and culture in landscape ecology. The Karolinum Press, Prague, Czech Republic.

Forman, R. T. T., and L. E. Alexander. 1998. Roads and their major ecological effects. Annual Review of Ecology and Systematics 29: 207–231.

Forman, R. T. T., and S. K. Collinge. 1995. The "spatial solution" to conserving biodiversity in landscapes and regions. Pages 537–568 in R. M. DeGraaf and R. I. Miller, editors. Conservation of faunal diversity in forested landscapes. Chapman & Hall, London, UK.

———. 1997. Nature conserved in changing landscapes with and without spatial planning. Landscape and Urban Planning 37: 129–135.

Forman, R. T. T., and R. D. Deblinger. 2000. The ecological road-effect zone of a Massachusetts (USA) suburban highway. Conservation Biology 14: 36–46.

Forman, R. T. T., and A. M. Hersperger. 1997. Ecologia del paesaggio e pianificazione: una potente combinazione. Urbanistica 108: 61–66.

Garrelick, R. 1985. Concord in the days of strawberries and streetcars. Town of Concord Historical Commission, Concord, Massachusetts, USA.

Gilbert, O. L. 1991. The ecology of urban habitats. Chapman & Hall, London, UK.

Gore, A. 1993. Earth in the balance: Ecology and the human spirit. Penguin Books, New York, New York, USA.

Harris, L. D., T. S. Hoctor, and S. E. Gergel. 1996. Landscape processes and their significance to biodiversity conservation. Pages 319–347 in O. Rhodes Jr., R. Chesser, and M. Smith, editors. Population dynamics in ecological space and time. University of Chicago Press, Chicago, Illinois, USA.

Hough, M. 1995. Cities and natural process. Routledge, London, UK.

Im, S.-B. 1995. Skyline conservation and management in rapidly growing cities and regions. Process Architecture 127: 88–93.

Kehm, W., and M. Yokohari. 1995. Toward the recovery of nature. Process Architecture 127: 130–140.

Kiemstedt, H. 1995. Landscape planning in Germany. Process Architecture 127: 54–63.

Klopatek, J. M., and R. H. Gardner, editors. 1999. Landscape ecological analysis: issues and applications. Springer-Verlag, New York, New York, USA.

Kress, S. W. 1985. The Audubon Society guide to attracting birds. Charles Scribner, New York, New York, USA.

Leopold, A. 1949. A Sand County almanac and sketches here and there. Oxford University Press, London, UK.

Lyle, J. T. 1985. The alternating current of design process. Landscape Journal 4: 7–13.

Lynch, K., and G. Hack. 1996. Site planning. MIT Press, Cambridge, Massachusetts, USA.

Marsh, W. M. 1991. Landscape planning: environmental applications. John Wiley and Sons, New York, New York, USA.

McHarg, I. 1969. Design with nature. Doubleday, Garden City, New York, USA.

McHarg, I. L., and F. R. Steiner. 1998. To heal the earth: selected writings of Ian L. McHarg. Island Press, Washington, D.C., USA.

McKibben, B. 1989. The end of nature. Doubleday, New York, New York, USA.

Morgan, G. T., Jr., and J. O. King. 1987. The Woodlands: new community development, 1964–1983. Texas A&M University Press, College Station, Texas, USA.

Muschamp, H., S. B. Warner, Jr., P. Phillips, E. Ball, and D. Balmori. 1993. The once and future park. Princeton Architectural Press, New York, New York, USA.

Nassauer, J. I. 1992. The appearance of ecological systems as a matter of policy. Landscape Ecology 6: 239–250.

———, editor. 1997. Placing nature: culture and landscape ecology. Island Press, Washington, D.C., USA.

National Research Council. 1997. Toward a sustainable future: addressing the long-term effects of motor vehicle transportation on climate and ecology. National Academy Press, Washington, D.C., USA.

Noss, R. F., and A. Y. Cooperrider. 1994. Saving nature's legacy: protecting and restoring biodiversity. Island Press, Washington, D.C., USA.

Odum, E. P. 1989. Ecology and our endangered life-support systems. Sinauer Associates, Sunderland, Massachusetts, USA.

Palazzo, D. 1997. Sulle spalle de giganti: le matrici della pianificazione ambientale negli stati uniti. FrancoAngeli/DST, Milano, Italy.

Park, G. 1995. Nga Uruora: The groves of life: ecology and history in a New Zealand landscape. Victoria University Press, Wellington, New Zealand.

Pietrzak, M. 1989. Problemy i metody badania struktury geokompleksu. Seria Geografia Nr 45, Uniwersytet Im. Adama Mickiewicza W Poznaniu, Poznan, Poland.

Saunders, D. A., and R. J. Hobbs, editors. 1991. Nature conservation 2: the role of corridors. Surrey Beatty, Chipping Norton, Australia.

Seddon, G. 1995. Commentary on the Perth foreshore competition by a citizen of Perth. Process Architecture **127**: 94–104.

———. 1997. Landprints: reflections on place and landscape. Cambridge University Press, Cambridge, UK.

Shepard, P. 1991. Man in the landscape: A historic view of the esthetics of nature. Texas A&M University Press, College Station, Texas, USA.

Smith, D. S., and P. C. Hellmund, editors. 1993. Ecology of greenways: design and function of linear conservation areas. University of Minnesota Press, Minneapolis, Minnesota, USA.

Smith, E. A. T., and R. Ferguson. 1994. Urban revisions: current projects for the public realm. Museum of Contemporary Art, Los Angeles, California, USA.

Smith, R. L. 1996. Ecology and field biology. HarperCollins, New York, New York, USA.

Spirn, A. W. 1984. The granite garden: urban nature and human design. Basic Books, New York, New York, USA.

Steiner, F. 1990. The living landscape. McGraw-Hill, New York, New York, USA.

Steinitz, C. 1995. A framework for planning practice and education. Process Architecture **127**: 42–53.

Thompson, G. F., and F. R. Steiner, editors. 1997. Ecological design and planning. John Wiley and Sons, New York, New York, USA.

Tjallingii, S. P. 1995. Ecopolis: strategies for ecologically sound urban development. Backhuys Publishers, Leiden, Netherlands.

Turner, M. G., and R. H. Gardner, editors. 1991. Quantitative methods in landscape ecology: the analysis and interpretation of landscape heterogeneity. Springer-Verlag, New York, New York, USA.

Van der Ryn, S., and S. Cowan. 1996. Ecological design. Island Press, Washington, D.C., USA.

Wheeler, R. R. 1967. Concord: climate for freedom. Concord Antiquarian Society, Concord, Massachusetts, USA.

Wilson, E. O. 1984. Biophilia. Harvard University Press, Cambridge, Massachusetts, USA.

Zonneveld, I. S., and R. T. T. Forman, editors. 1990. Changing landscapes: an ecological perspective. Springer-Verlag, New York, New York, USA.

PART II

Perspectives on Theory and Practice

What do designers and planners need to know to address current trends in natural and cultural environments? The authors in this section emphasize divergent themes, including the need to increase the relevance of design and planning to public policy, the difference between ecological science and ecological understanding, and the need for ecologists, designers, and planners to recognize the relevance of their fields to public health concerns. Their essays are written from a range of professional, theoretical, practical, and philosophical perspectives: a landscape architect in public policy as the head of a federal institute, a practicing urban designer, an ecologist who offers a biological view of design and planning, and an ecologist and a designer who jointly focus on issues of human health.

Planner Carolyn A. Adams explores the strategic knowledge of ecology and natural processes that a landscape architect needs to take a leadership role in contemporary public planning in Chapter 5, "Lead or Fade into Obscurity: Can Landscape Educators Ask and Answer Useful Questions about Ecology?" Ecologist James R. Karr considers the nature of ecological knowledge and how we can apply it to reduce the biological impacts of design and planning interventions over time in Chapter 6, "What from Ecology Is Relevant to Design and Planning?" In Chapter 7, "Toward an Inclusive Concept of Infrastructure," designer and practitioner William E. Wenk examines the historic split between the science and aesthetics of urban infrastructure design and describes the practical knowledge needed to fulfill its ecological potential. Chapter 8, "Human Health and Design: An Essay in Two Parts," consists of paired essays about human health, ecology, and land-use practices. Ecologist and poet Sandra Steingraber writes about the consequences of human-generated toxic materials for both human health and that of other species, in "Exquisite Communion: The Body, Landscape, and Toxic Exposures." Designer Kristina Hill examines the potential for design and planning to address a broad range of human health issues in "Design and Planning as Healing Arts: The Broader Context of Health and Environment."

CHAPTER 5

Lead or Fade into Obscurity: Can Landscape Educators Ask and Answer Useful Questions about Ecology?

Carolyn A. Adams

Reinvention, redesign, and reform are all words describing the incredible metamorphosis of organizations associated with landscape planning and design. Nowhere is this more obvious than in the public service sector where shrinking budgets are driving dramatic changes in structure, responsibilities, and staffing levels. For many natural resource agencies, a more compelling catalyst for change is the shift from managing a single resource or single species to managing entire ecosystems. While a focus on ecosystems makes intuitive sense to natural resource professionals, moving in this new direction creates great conflicts within public institutions as they attempt to integrate ever-evolving ecological sciences into the more slowly changing arenas of policy and legislation.

The primary message of this commentary is that landscape design and planning program educators should step forward now to shepherd this shift in partnership with public agencies that increasingly rely on ecological sciences. If they do not, the professions will lose their ability to influence the landscape either directly through public policy or indirectly through their graduates who may work for public organizations. Program educators must also aggressively control the destiny of landscape professions in the climate of substantial change within universities. If they do not, our future professionals will lose the opportunity to lead and guide progress toward a sustainable and culturally responsible future for our continent's natural and tended environments. If programs do not adapt and produce leaders who are well versed in ecological principles and who can "design ecologically" at large scales, the public sector of landscape practice could well diminish into obscurity. This chapter discusses how program educators might better prepare future practitioners for leadership in and consultation to evolving public agencies. It does so by examining current challenges in public practice, anticipating future challenges, reflecting on the past, and posing some strategic concepts for programs to consider.

The Contemporary Balancing Act of a Public Practitioner

Contemporary public work in ecological planning is best characterized as requiring a balancing act, not unlike that used by the fiddler in Broadway's *Fiddler on the Roof.* Public professionals are always precariously perched, trying to operate in the mèlange of mandated requirements, institutional protocol, constituent desires, political realities, and personal ethics. Attempting to be successful balancers, public practitioners try to juggle disparate activities such as the following:

- Determining who is a credible expert for the natural resource issue *du jour*
- Gathering technical facts from credible expert sources
- Keeping communication channels open so that information flows freely
- Synthesizing information for institutional managers who can budget only a miniscule amount of time to absorb and then act only on snippets of knowledge
- Interpreting and conforming to policies and regulations
- Finding and encouraging interested stakeholders to participate and actively lead local planning efforts
- Ensuring broad access to decision-making processes
- Preparing defensible documentation sufficient to respond to legal challenges

It is overwhelming to consider these activities from the vantage of our safe, historically derived landscape professions. Add to this the need to be conversant in the still evolving ecological sciences, and it becomes easy to understand why opportunities for conflict, disaster, and burnout abound.

A Viewpoint for the Future

An investment in change is necessary for landscape design and planning programs to prepare for future challenges and issues of ecological planning and ecosystem management. Changes external to programs will continue to be rapid, influential, largely unpredictable, and widespread. Citizen expectations about safe food, clean water, breathable air, and an environment that can be proudly passed on to future generations will only increase. There is clear evidence that environmentalism is a core American value on par with parental responsibility and religious convictions (Kempton, Boster, and Hartley 1995). Public agency attention to ecosystem management is not a fad. It will continue as an operational concept for the foreseeable future.

> We are now in the early destabilized phase of a process of co-evolution between humans and natural systems on a planetary scale. The future is not just uncertain; it is inherently unpredictable. (Holling 1994, p. 100)

Although noted ecologist C. S. Holling states emphatically that the future is unpredictable, I believe there are some tea leaves that can be read and should be considered in recasting some landscape design/planning programs. Academic institutions must be prudent and persistent in creating inventive, adaptive, and evolutionary models of learning that address, to some level, the following fifteen "tea leaf" predictions.

1. Narrowly defined problems will not be the order of the day; problems will be more and more complex. If you think they are complex now, just wait.
2. Interdisciplinary work will be absolutely essential, and more "nontraditional" and arcane disciplines will be needed to address increasingly complex issues.
3. Effective ecosystem management will require that scientists agree on working definitions for "success" and "failure" and establish thresholds for tolerance. Measurement tools and techniques will be needed to assess trending directions prior to reaching the thresholds. Monitoring and evaluation will become essential components of all public projects that involve ecosystem protection, modification, or restoration.
4. Sciences, especially the biological sciences, will have to find familiar ways to explain their work so that they limit confusion for the public and decision makers. Scientific work must lead to *greater* societal learning and understanding.
5. Ecological changes caused by cumulative, seemingly insignificant human actions will continue to cause surprises and occasional disasters.
6. Increased attention will be directed toward those ecosystems, or remnants of ecosystems, most directly impacted by human actions—the agricultural-based countryside and urban areas.
7. More resources will be directed toward community-led processes that empower citizens to take ownership for identifying and achieving their own goals for ecological improvements in nearby landscapes and watersheds.
8. Analysis tools such as geographic information systems and remote sens-

ing and digital data will become more affordable and commonly used to help decision makers understand the effects of their decisions on ecosystem functions. At the same time, analysis tools will become increasingly mobile and data will be more accessible to the public as the information highway (Internet) grows.

9. Citizens and politicians will demand place-specific, consistent answers about expected ecological outcomes, especially cause and effect. For example, they will want to know that "If Action X occurs in Watershed A," what will happen in downstream receiving waters, when will it happen, who and what will be impacted, and with what level of certainty can we make such predictions?

10. Individual and community counterfactual beliefs will always influence natural resource decisions to varying degrees.

11. Legislative bodies will demand accountability (using performance measures) for resources spent annually for remedial and restorative efforts in spite of mounting evidence that biological reactions to interventions may take decades.

12. Ecological sciences will continue to evolve at the same time as public agencies struggle with adaptive actions/reactions. Agencies will likely remain bogged down in inflexible policies and regulations—generally several steps behind the evolution of ecological sciences.

13. Ecological problems will increasingly be viewed in the context of multiple scales, that is, problems that have traditionally been viewed as local or regional (e.g., acid precipitation) will also be considered at global scales.

14. Local ecological problems will continue to proliferate, but these problems will be created more often by geographically distant organizations. The global economy and actions of multinational companies will increasingly influence local natural resource issues. (Gould, Schnaiberg, and Weinberg 1996)

15. We live in an ever-changing ecological, economic, and social environment. The organizations that survive, including landscape design/planning programs, will be those that attract visionary leaders, are highly adaptive, encourage shared collaboration, are constantly open to new uses and sources of information, and strategically concentrate on mental processes that foster innovation and never-ending learning.

Reflections on the Past

Twenty-five years ago, most landscape design and planning programs were on the "knife edge" of learning about and working with the environment. Those programs encouraged collaboration and were among the first to use interdisciplinary efforts. Many landscape architecture programs taught fundamentals about relationships between landscape elements using Ian McHarg's *Design with Nature* (1969) principles. They recognized humans as important components of the environment and emphasized design processes and principles that sought a best fit between human needs and environmental conditions. Twenty-five years ago, these pursuits represented progressive thinking. Even though programs were advanced in these respects, hindsight reveals certain inadequacies in their chain of learning. For example, design itself was taught as an iterative process, but the design function was embedded in a linear series of activities that went something like this: plan, design, construct, take photographs of the project, and walk away. While that might work for inert architectural elements, it ignored the absolute necessity for monitoring and evaluating ecological changes in the biota affected by design implementation.

But while McHarg's work and writing focused on a range of spatial scales, the scale of choice in many studio planning and design exercises was "the site." Little emphasis was given to larger landscape contexts for understanding formative processes, spatial and temporal variation, and ecological capacities or vulnerabilities. There were some notable exemptions, such as the teaching and studies of Carl Steinitz, Julius Fabos, Erv Zube, David Streatfield, Tito Petri, and Phil Lewis, but many curricula ignored the contributions of basins, landscapes, or ecological regions in determining the behavior of natural systems on sites being planned or designed. Environmental impacts of development schemes were viewed as confined to the site of consideration and usually mitigated in isolation from other landscape patterns and conditions. The cumulative effects of many small-site "impacts" to a region were rarely addressed. The interactions, complexities, and interdependencies of biotic components were not emphasized, though the relationships of landscape elements were.

Possibly the greatest shortcoming is that only a few landscape educators assumed intellectual leadership to understand ecology. Those early years of ecological inquiry were based on a "science of parts" (Holling 1994) that focused on single species, populations, or communities. In hindsight, it seems reasonable to expect that the landscape professions (that had already embraced interdisciplinary work) would foresee the need to collaborate with others to better understand the integration of biotic parts, that is, ecosystem interactions. Landscape professionals did not learn enough about ecology to be able to ask the "hard" questions that would have positioned them as ecological design leaders. Rather, we comfortably retained our preoccupation with Frederick Law Olmsted, the English garden tradition, or simplistic overlays.

Fortunately, many program educators are now teaching more "holistically" and using ecological principles in planning, analysis, and design. Woefully, some programs have not or are only giving lip service to ecology in the curriculum. In my opinion, graduates emerging from programs tied to formerly useful paradigms that focus mainly on site scales will not be well prepared to work in the public sector. Successful public practitioners in natural resource management agencies must not only have a good grasp of ecological concepts, they must also be strong leaders, facilitators, motivators, problem solvers, mediators, knowledgeable in policy and legislative process, and expert in strategic and process thinking. They need to be *conversant* in a wide range of science and technology including, but not limited to, engineering, forestry, terrestrial ecology, landscape ecology, soils, wildlife biology, air quality, aquatic ecology, and planning. These practitioners must possess the critical skills to balance information and resources, and use leadership to guide present actions and anticipate critical future directions.

In light of reflections on the contemporary setting and the past, it is clear that landscape design/planning professionals in the public sector have major challenges. It is also clear that academic programs cannot, and perhaps should not, try to provide all of the skills and learning opportunities to prepare public practitioners totally to right the wrongs of the past and successfully deal with all of the contemporary conflicts. Further, after looking at the pace of change that is affecting economies, technology, institutions, cultural diversity, and ecology, I can safely predict that challenges of preparing for tomorrow will be much greater than simply addressing the shortcomings of the past and the issues of the present. What then, should programs do? What are the critical issues? Is a strategy possible? I believe so, but it will not be simple or easy.

An Adaptive Strategy: Distinctive Concepts for Putting a Program's Whole Brain to Work

> Disciplinary knowledge is not static. It progresses in lurches of expanded understanding as theory and practice confront reality and as expanded debate with other disciplines widens comprehension. (Holling 1994, p. 79)

Landscape design and planning programs will remain relevant only if they aggressively pursue and document "lurches" in their own knowledge base as well as keep their finger on the pulse of the progress of other disciplines. Program educators must lead and participate in forums for expanded debate. Landscape practitioners must not simply repackage old concepts; rather, they must search for new truths. We must ask ourselves not only how can we best apply new learning but what must we unlearn? We should not do this in isolation, but rather in an intellectual dialogue with others.

I propose that this can best be accomplished through the development and use of strategic intellectual frameworks. As a first step, educational units must define a philosophical base for their particular approach to ecology and share it with regional practitioners, colleagues, and prospective students. Perhaps equally important is that program educators develop an adaptive strategy that reflects their unique philosophy and identify concepts that would make their approach distinct from others. Once articulated, these should form the structure or framework for any compelling strategy.

No single strategic framework can be devised that would best accommodate learning and unlearning in all landscape programs as they attempt to apply ecological knowledge in teaching. Each program must customize its point of view. There are numerous frameworks that would work equally well, depending on the desired outcomes or emphasis areas of a particular program. For example, the "integration-focused" intellectual framework developed at the University of Vermont's School of Natural Resources for their undergraduate natural resources management program may not be appropriate elsewhere, but has proven a unifying factor for their administration, faculty, and students (Ginger, Wang, and Tritton 1999).

In the remainder of this chapter, I suggest concepts worthy of consideration for any strategy. One concept is especially pervasive: Programs should base any framework on a committed, long-term investment in anticipating changes that might benefit or erode the successful delivery of landscape-based learning experiences. Further, the framework should embrace key characteristics such as flexibility and perpetual learning. Such a strategic framework, if managed well, could provide an exploratory learning atmosphere within which ideas can be nourished and new technologies explored as a basis for improved landscape design and planning theory and practice.

Emphasize What Landscape Professions Do Best, and Do It with Increasing Excellence and Innovation

Some specific aspects make the landscape professions different from all others: synthesis, spatial understanding, visualization of ideas, and a command of landscape aesthetics. These professions, especially landscape architecture, will do best by recasting and emphasizing their essential distinguishing characteristics in terms of ecological principles.

Synthesis is defined as "the process or result of building up separate elements, especially ideas, into a connected whole, especially into a theory or system" (*Oxford Dictionary and Thesaurus* 1996). At the heart of practicing landscape architecture is the art and skill of synthesis. Some landscape educators, such as those at the University of Pennsylvania, cast a wide net to include the knowledge and skills of other disciplines in teaching synthesizing processes. Many programs did not follow this model. Tomorrow's program must reach beyond this example. Ecological and social sciences are multiplying and now include such specialties as ethnobotany, environmental psychology, fluvial geomorphology, and ecological anthropology, just to name a few. The astute and successful landscape professional needs to learn who knows what, who does what, and when to ask for input. Synthesizing and facilitation skills are also needed to clarify issues and differences between disciplines with diverse opinions. It is critically important to have professionals who add perspective and insight from a broad view of landscape systems in addition to those who are experts in specialized areas.

Landscape programs excel in teaching *spatial understanding*, or the design of volume to accommodate functional human uses in land planning. There are many design expressions for spatial organizations meaningful to human interpretations, such as enclosing, directional, static, stimulating, and so on (Simonds 1961). On the other hand, scientists are only beginning to explore the metrics of spatial composition as they relate to ecosystems. Landscape architects should be leaders in that exploration. We have the language and concepts needed to design and organize spaces for human occupation, and must work harder to assist biologists and ecologists in documenting characteristics and devising language helpful to plan and design ecological restorations.

Creative planning and design ideas are enhanced by the use of *visualization*. Ideas communicated through verbal processes alone often sound overly complex, can be misinterpreted, or are simply misunderstood. No matter how clearly ecology is explained in words, the average person cannot visualize future ecological design outcomes on the landscape. Landscape design training includes skills to articulate ideas, concepts, and design notions using visual tools. Expressing design ideas visually is a clear asset, but it may be equally important to use these skills helping visualize the ideas of others. Many scientists have a difficult time expressing their thoughts and concepts in three dimensions. There will be a special need for certain landscape professionals to interpret ecological systems and visual changes.

Landscape architecture is the primary profession studying the *aesthetics* of landscape. Landscape architects must continue to be the champions for landscape aesthetics and must continue to develop usable theories and concepts of aesthetics that highlight all of the characteristics of functioning ecosystems: all the tensions, cycles, complexity, stability, equilibrium, and fluxes. The profession's past reliance on mainly an artful understanding of aesthetics is inadequate for incorporating ecological principles into design. We must develop concepts to express not only ecological designs, but also their values. For example, the environmental philosopher Robert Elliot (1997) argues that nature has both aesthetic and moral value as compared to human-produced works of art that have aesthetic value but questionable intrinsic moral value. New theories are welcomed, such as making ecology visible (Thayer 1994), creating landscapes that are "rough and refined" (Quayle 1993), combining culture with ecology (Nassauer 1997a), and creating theoretical concepts that are hybrids (Meyer 1997).

At the same time, we must also move toward sustainable landscape systems. Recently, the federal executive branch issued orders to "require the use of environmentally beneficial landscaping techniques, including increased use of native species and reduced use of water and chemicals, at federal facilities and federally funded projects" (Gore 1993, p. 143). The highly controlled and manicured garden-style landscapes dependent on ornamental plants and high nutrient inputs for survivability are outmoded for most public work.

Provide Skill-Building in Critical Nontraditional Areas

There is clear evidence that adept facilitation can help to resolve conflicts among contentious constituents. Understanding group dynamics, improving listening techniques, using confirming mechanisms, giving power, and recognizing the interpersonal forces at play are essential and are teachable/learnable skills. Landscape professionals as facilitators could be particularly effective since we are already trained in transforming verbal thoughts/ideas into graphic representation for clarifying, illuminating, and validating.

Closely linked to, but not synonymous with, facilitation is the increasing need for public practitioners to possess excellent skills in mediation and conflict resolution. Ecosystem management issues are teeming with emotions, beliefs, passions, fears, and high-pressure tactics all mixed up with and complicated by scientific and technical information. Groups enmeshed in natural resource situations can learn incrementally, but groups are too often dominated by behaviors such as poor listening and intimidation tactics. Skills are critically needed in understanding how to achieve effective communication, create a vision of desired outcomes, develop goals/objectives, foster empowerment, and develop effective teams and teamwork. Landscape professionals can be both educators and leaders in these settings.

Teach Students the Difference between "Science, Nonscience, and Nonsense"

Michael Zimmerman, in *Science, Nonscience, and Nonsense: Approaching Environmental Literacy* (1995), argues that there is substantial evidence of misunderstandings about science. This leads to misinterpretations of scientific studies, more reliance on pseudoscience or anecdotal examples, and in general, according to Zimmerman, a significant growing illiteracy of the real nature of science.

We must become conversant in the language, process, and progress of nearby science so that we can represent the interests and characteristics of regional landscapes. Examples of technological fads abound, but one that is particularly distasteful is the recent dissemination of information about and the desire to construct "lunkers" or "skyhook" structures for use in Pacific Northwest stream restoration projects. Lunkers and skyhooks are wooden structures designed for use in stream channels that are wide and shallow and have erosion-resistant bottom materials with low streambanks (Hunt 1993). They are used extensively in the upper Midwest to provide hiding covers for adult trout. These structures are often used in active agricultural areas where riparian vegetation was historically removed to accommodate tillage very close to the edge of the streambank. Although successful in the upper Midwest, these structures are inappropriate as quick fixes in the forested landscapes of western Oregon and Washington.

Regional practitioners should not fall victim to using inappropriate technology based on science not appropriate to nearby landscapes. Programs have a major role in transferring innovative science for bioregional practice. In the case of lunkers and skyhooks, researchers at the University of Washington are rejecting their use in favor of engineered log jams, an environmentally derived technology based on science specific to resource issues in the regional landscape.

Embrace New Modes of Learning

> There are two kinds of learning: one for a stable world and one for a world of uncertainty and change. Learning appropriate for the former world has to do with learning the right answers and learning how to adapt and settle into another mode of being and doing. Learning appropriate for our world has to do with learning what are the useful questions to ask and learning how to keep on learning since the questions keep changing.
>
> There will be no place to settle down, no time to stop asking. It is this kind of learning that befits ecological management, the kind that befits our human ecology as it struggles to sustain the natural one. (Michael 1995, p. 484)

Above all else, learning involving ecological sciences should be progressive, adaptive, evolutionary, and experimental. It should be progressive in the sense that one should learn to build on old projects; we should reject the notion that every project is new and different. The world of public practice involving ecosystems is a constant work in progress; fewer and fewer projects start with no previous intervention. More commonly, they are revamps, redesign, rehabilitation, or restoration. Learning should be adaptive so that there is always room for new ideas and learning from mistakes. The ecological sciences are evolving; thus design using ecological principles must also evolve. Learning should be evolutionary: a gradual development that moves from the simple to the more complex. Principles of ecology note that evolution is constrained by what went on before; that is, what we study in the contemporary setting has an inescapable history (Allen and Hoekstra 1992). Thus, our learning cannot be fruitful if we fail to consider the point of view that captured our predecessors' perceptions. And learning should be *experimental*. Working with ecosystems is at best uncertain, so incorporating ecological principles into landscape programs should also be viewed with a measure of uncertainty. Surprises, errors, and misunderstandings should not only be accommodated, but encouraged.

> Perhaps catastrophe is the natural human environment, and even though we spend a good deal of energy trying to get away from it, we are programmed for survival amid catastrophe. (Greer 1984)

Program educators should use *disasters as occasions for learning*. Disasters and ecological crises are a necessary part of the disturbance regime necessary for ecosystems to function. It seems that landscape based disasters, both human-induced and natural, are commonplace today. Federal agencies plan "emergency funds" into their budgets, even though Congress does not appropriate money until the president declares a disaster. Agencies are not hoping for disasters, they are simply certain that they will occur. While we are saddened by personal losses resulting from disasters, these remain clear opportunities for learning. Certain questions should be posed:

- What could have been done differently to prevent the disaster from occurring in the first place?
- Could things be done differently to minimize losses in the future?
- Did the disaster create opportunities for individual or institutional education about how responses could have been different during the emergency?

Further, disasters do not have to be real to be used as learning opportunities. Modes of learning such as simulation and scenario building/evaluation should be used in a similarly constructive mode as actual catastrophes.

Focus Research on Evolving Ecological Areas with Strong Connections to Landscape Design and Planning

Academic landscape professionals must engage in scholarly production. Not only is it necessary for survival in institutional frameworks, but it is an imperative for credibility from peers in engineering and the ecological and social sciences. Having stated that, I am not in any way encouraging landscape professionals to desert our existing body of knowledge. I am, however, suggesting that we apply social and natural scientific methods and procedures to the study of areas that are both important to ecological design and have strong connections to our professional knowledge base. A few candidate areas for future research are human attitudes and behavior toward landscape, landscape as a necessity for human health and welfare, countryside landscapes, and planning and designing at large scales (the watershed revolution).

Human Attitudes and Behavior toward Landscape

A great deal remains to be learned about why humans act the way they do toward landscapes. This knowledge is essential before we can make significant strides in ecological improvement and progress toward sustainability. Psychologist Deborah Winter (1996) is exploring what she terms "social psychology" or the scientific study of social influence:

> Although we like to think that our attitudes and behaviors are based on rational and logical assessment of facts, a brief glimpse at social psychology reveals the enormously powerful (although usually unconscious) influences that other people have on us, our reasoning, our beliefs, and our behavior. The main point is that our understandings and actions about environmental issues are largely social phenomena. (p. 63)

As the design profession committed to designing human infrastructures that fit natural systems, it seems ironic that landscape architectural principles and concepts originate primarily from art and engineering and have little to do with an understanding of why humans behave the way they do toward spaces and environments we create. We take great pleasure in our accomplishments, such as the public using and enjoying downtown plazas that we have designed. But we do not seem intellectually puzzled by the destructive behavior of nearby residents who apply contact herbicides to the riparian plants we specified. Programs must intensify their efforts to understand the feedback mechanisms between self and the environment (Anderson 1996), the moral language and images of our culture (Bellah et al. 1985), and the intriguing possibilities of a "woman-centered" approach to environmental interactions (Silliman 1995).

Landscape as a Necessity for Human Health and Welfare

The bookshelves of the extensive "Ecology" section at the University Bookstore in Seattle are filled with titles suggesting a multitude of human impacts and sins that have caused havoc in the natural landscape. It is an irrefutable fact that humans extract, use, and pollute tremendous quantities of the earth's physical resources: air, water, soil, and rock. Of equal or greater importance, however, may be what the earth's resources give us that cannot be readily consumed—its restorative and healing powers. On this subject, the literature is silent for the most part. Some inquiry has begun (Kaplan 1995), yet it remains an area of great potential for landscape professionals.

Countryside Landscapes

With a few exceptions (Nassauer 1997b; Schauman 1986), landscape professionals have largely ignored agricultural landscapes as an area for study or potential practice. The land used for food and fiber production constitutes over two-fifths (about 930 million acres) of all land in the United States and has far-reaching characteristics that touch everyone, including incredibly rich ecosystems. It is privately owned for the most part and was historically managed on individual values set to achieve a certain livelihood, independence, cultural identity, and quality of life. This situation is rapidly changing. No continental landscapes are more representative of endemic changes than those dominated by agricultural production. Technology is producing bigger machines that use on-board geographic information systems to "precision farm," farms continue to be consolidated under corporate ownership, environmental challenges increase, conflicts abound in some regions over share of water consumption, and the list goes on. Two big issues of the 1990s were the safety of the food supply and agriculture's contribution to the pollution of our soil, water, and air. The direct connections are not clear, but it is well documented that agriculture bears substantial responsibility (Steingraber 1997; Watzin and McIntosh 1998).

These are perceived by many landscape professionals to be "on farm" issues best handled by others. The reality is that there are few places in the United States where the issues of agriculture do not affect urban and suburban areas—the landscapes of choice for many practitioners. As noted on many occasions in this commentary, ecological systems are interrelated and interact. Landscape professionals would be well advised to begin understanding the complex countryside.

The Watershed Revolution—Planning and Design at Large Scales

Watersheds have recently become the operational scale of choice for public agencies, especially the U.S. Environmental Protection Agency's Office of Water in Washington, D.C. Watersheds will be used as the common framework for federal agencies in implementing the new Clean Water Action Plan

(Browner and Glickman 1998). Several states have established statewide systems of watershed councils and watershed priority programs, and some are considering reorganizing their state agencies along watershed, instead of geopolitical, boundaries. This trend will likely continue. Landscape programs should reexamine and refit the work of McHarg, Steinitz, and Lewis as frameworks for watersheds and future involvement in large planning efforts. Equally important is the need for programs and practitioners to become familiar with inquiry and research related to other broad spatial scales. Of particular interest is the work of the nation's preeminent macrogeographers (Omernik and Bailey 1997) and bioregionalists (Berg 1995; McCloskey 1995).

Represent Human Ecology—The Culture, Gender, and Ethnicity of Our North American, Hemispheric, and Global Society

Landscape design and planning programs must make heroic efforts to be representative of our continent's pluralistic society. It is simply unacceptable for landscape design and planning professions to continue to be strikingly unrepresentative of our changing cultural diversity. Students must be exposed to and receive guidance from role models that bring different perspectives, experiences, and backgrounds into the classrooms and studios. These role models should represent varying practice types, ethnic backgrounds, and gender, as a minimum. No matter what culture, gender, or ethnic background, we all owe our existence to ecological systems. This common thread could be used as a powerful unifying force for promoting ecosystem management, but landscape educators could use the differences as a basis for creative interpretations in design and planning.

Programs must provide experiences in working with nontraditional constituents such as Native Americans or people of limited resources who are living in or managing a piece of the landscape. Our culturally derived collective differences and shared history should be the basis for new design theories and ecosystem management models. For example, public agencies are increasingly relying on ethnobotanists to understand environmental management techniques used by indigenous groups, such as the sowing or broadcasting of seeds and pruning or coppicing of plants to encourage particular patterns of growth (Blackburn and Anderson 1993).

Bring Ecological Ethics Out of the Closet

The ASLA Code of Professional Conduct contains a detailed set of "rules" about how one should act responsibly and conduct business as a professional

(American Society of Landscape Architects 1997). This is essential knowledge, but it is equally, and perhaps more, essential that we understand environmental ethics. Consider, for example, an accepted professional strategy stated in the ASLA's Declaration on Environment and Development: "Avoid the use of plants that are known to be invasive to indigenous ecosystems." In the Northwest, not only is English ivy used, but wetland restoration projects designed by landscape architects continue to use yellow iris, another exotic plant that, under many conditions, outcompetes natives. These practices continue, in part, due to a lack of agreement within landscape professions about an environmentally based belief system. Some practicing landscape architects do not consult plant ecologists or wetland plant specialists before preparing plant palettes for restoration projects. Arguably, it may be unethical to design ecological restoration projects without consulting experts with specialized knowledge. Programs should assist the profession in defining appropriate ethical behavior for ecological work and prepare students to recognize appropriate thresholds for this behavior.

Adapt to a Changing World

Taking risks should be assumed in work related to ecosystems. In the adaptive management process being attempted by public agencies and businesses, there is an effort to redefine success and failure in terms of experience. As Lee (1993) states, "The behavior of natural systems is incompletely understood. Predictions of behavior are accordingly incomplete and often incorrect" (p. 61). Some of the fundamental precepts of adaptive management should be evaluated for use in teaching programs—for example, flexibility; emphasis on exploration, experimentation, and discovery; acceptance of conflict as a learning mechanism; finding new ways to conceptualize and frame problems; achieving shared meaning; and being alert to surprises.

Another adaptation that programs should consider is the use of technological innovations to inspire and achieve collaboration. Landscape architects have historically relied on "trace" and colored felt-tip markers as collaborative media, but the media of the future will be digital. Emerging media will require new levels of conversational literacy: The spoken word will be secondary to visuals that are accessible on the World Wide Web and transmitted in real time. New models for collaboration through computer media already exist and are being used to assist planning on U.S. National Forests. A networked system of computers allows participants to enter data and effect changes interactively with other participants and instantly see the results of their collaborative actions on a centrally viewed screen. All of the information generated is stored in digital form and available for repetitive use. Computer-enhanced or computer-augmented meetings can achieve collaborative outcomes through electronic data collection, connection, and correction. Collaboration using digital information is the future.

> The slick presentation that has all the facts and has touched all the bases and is bulletproof has become unsatisfying. Why? Because it doesn't invite collaboration. It doesn't encourage others to make a contribution, to add value. Collaborative tools, on the other hand, will encourage people to design presentations that invite further input and ideas, that beg improvement and closure from other parties. (Schrage 1989, p. 193)

Embrace the Entire Landscape Intellectually

Landscape professionals, except for those in the public arenas, have carved out a relatively small piece of the landscape on which to work. The majority of private practitioners still work on a few particular problem types in urban and urbanizing areas. Perhaps that speaks to one of the primary reasons why programs have approached the integration of ecological sciences with different degrees of urgency; that is, many ecological systems in urban areas ceased to function in any substantial way decades ago. Perhaps public practitioners feel so passionate about the need for increased attention to ecology because our practice primarily occurs in the rest of the landscape where more functioning ecosystems remain. Of the 1.89 billion acres of land in the United States, less than 5 percent (92 million acres) is developed; 1.39 billion are rural land and 408 million acres are in federal ownership (USDA, NRCS 1996).

To be relevant in the next several decades, landscape design and planning programs and professionals must take intellectual ownership of the entire landscape. This should be accomplished in two critical ways: looking at developed urban lands through a new set of ecological lenses and taking on emerging issues in the countryside—an area of rapidly increasing environmental challenges.

No matter how skillfully addressed, ecological planning often leads to turbulent disputes, conflicts, and decisions that almost always alienate some constituency and rarely put ecological considerations first. What has become increasingly clear is that humans act toward the environment and natural resources based on their own versions of the truth and their uniquely derived values (Anderson 1996). These systems of values, if they change, will change slowly. More likely, they will persist and continue to influence whether landscape design professionals and ecologists will be successful as they struggle with sustaining our continent's ability to repair itself and provide for the needs of living organisms, both human and nonhuman.

It's All about Balance

Many will point out that certain academic realities may deter a program's success in achieving resiliency and flexibility. I acknowledge that there are barriers, but I contend that academic barriers are no more difficult to overcome than those facing any contemporary organization. Some will debate this statement; in fact, a great and good friend who works at a public university once told me that compared to her academic institution, the federal government runs like a well-oiled machine. In response to those who say that academic barriers cannot be overcome, I reiterate that contemporary public work is a balancing act. Incorporating ecological principles into landscape planning and design curricula will also require a great deal of balance.

Will something have to be dropped from existing curricula in order to add ecological principles? I don't think so. I believe that the notion of "adding ecology" is no different than previous challenges of "adding computers" or future challenges of "adding something else" that we don't yet know about. Besides, it really isn't about "adding"; it's about "incorporating." The challenge of incorporating new or different information can be met using three fundamental principles: emphasis, integration, and collaboration. Individually, we have choices about how we allocate time. I advocate that we don't drop core content from existing curricula, but rather we de-emphasize some subject areas and emphasize ecological principles. Further, I am not suggesting that ecology, facilitation, or conflict management be taught as separate instructional units; these and other topics should be integrated into existing courses and studio exercises. I am also not suggesting that landscape professionals assume total responsibility for presenting and teaching the "new" topics suggested. Scientists and specialists should be brought into the classroom and studio so that they can teach and learn in surroundings dominated by design principles and processes.

If we are going to teach students to adapt to a changing world, then there is no better way to teach them than by our own actions. The content of landscape planning and design programs will continue to be challenged and expanded in the future. It could be extremely beneficial to devise a flexible approach or framework that uses ecology as a pilot endeavor now. This could only help in preparing for the next challenge.

Conclusions and Prospects

There is no single prescription for landscape planning and design programs to use as they consider how to incorporate ecological sciences into teaching. After all, the workings of academic institutions are no less complex than those of public agencies. Each program must struggle through its own context of confusion and change.

Two things, however, are fundamentally clear to me. First, landscape professionals must assume leadership in ecological planning and design. It is critically important—the landscape will be better because of it. As a profession, we are now simply toying with ecological design. We must quickly become serious about it. If we do not, this entire area of inquiry and implementation will belong to engineers, biologists, and restoration specialists. Second, we must use the tension this will create in our professions in a positive way. Many landscape professionals will counter this argument by stating that we should focus more on plants, our name isn't right, nobody understands us, we are really artists, we are relying too much on computers, and so on. Others will bemoan the notion of adding yet something else to professions already struggling for identity. Landscape educators may still discredit these notions because of "academic realities"—there is simply not enough time to teach ecological concepts, facilitation, or conflict resolution in programs that are already overtaxed with too many other new things to teach. I contend that there is little choice if we want to remain relevant in public arenas. These challenges simply must be met.

Above all, we must understand that the contemporary tensions in the landscape professions can be used positively or destructively. It is not healthy or productive for all of us to think alike, the key is that we all must "think." Dorothy Leonard of the Harvard Business School reminds us, "Conflict is essential to innovation. The key is to make the abrasion creative" (Leonard and Straus 1997, p. 111). Landscape professionals are trained to be creative and use innovative thinking. If we do not use these skills to bring ecological principles into our work, the landscape will be the victim.

Citations

Allen, T. F. H. and T. W. Hoekstra. 1992. Toward a unified ecology. Columbia University Press, New York, New York, USA.

American Society of Landscape Architects. 1997. Member handbook 97. Atwood Convention Publishing, Overland Park, Kansas, USA.

Anderson, E. N. 1996. Ecologies of the heart. Oxford University Press, New York, New York, USA.

Bellah, R. N., R. Madsen, W. M. Sullivan, A. Swidler, and S. M. Tipton. 1985. Habits of the heart. University of California Press, Berkeley, California, USA.

Berg, P. 1995. Putting bio in front of regional. <http://www.tnews.com/text/berg.html>.

Blackburn, T. C., and K. Anderson. 1993. Before the wilderness: environmental management by native Californians. Ballena Press, Menlo Park, California, USA.

Browner, C., and D. Glickman. 1998. Clean water action plan: restoring and protecting America's waters. USEPA National Center for Environmental Publications and Information, Cincinnati, Ohio, USA.

Elliot, R. 1997. Faking nature. Routledge, London, UK.

Ginger, C., D. Wang, and L. Tritton. 1999. Integrating disciplines in an undergraduate curriculum. Journal of Forestry 97(1): 17–21.

Gore, A. 1993. The Gore report on reinventing government: creating a government that works better and costs less. Time Books, New York, New York, USA.

Gould, K. A., A. Schnaiberg, and A. S. Weinberg. 1996. Local environmental struggles. Cambridge University Press, Cambridge, UK.

Greer, G. 1984. "Sex and destiny" in The Columbia dictionary of quotes. 1985. Columbia University Press, New York, New York, USA.

Holling, C. S. 1994. An ecologist's view of the Malthusian conflict. Pages 79–104 in K. Lendahl-Kiessling and H. Landberg, editors. Population, economic development and the environment. Oxford University Press, New York, New York, USA.

Hunt, R. L. 1993. Trout stream therapy. University of Wisconsin Press, Madison, Wisconsin, USA.

Kaplan, S. 1995. The restorative benefits of nature: toward an integrative framework. Special Issue: Green Psychology. Journal of Environmental Psychology 15(3): 169–182.

Kempton, W., J. S. Boster, and J. S. Hartley. 1995. Environmental values in American culture. MIT Press, Cambridge, Massachusetts, USA.

Lee, K. N. 1993. Compass and gyroscope: integrating science and politics for the environment. Island Press, Washington, D.C., USA.

Leonard, D., and S. Straus. 1997. Putting your company's whole brain to work. Harvard Business Review, July–August 1997, pages 111–121.

McCloskey, D. 1995. Ecology and community: the bioregional vision. <http://www.tnews.com./text/mccloskey2.html>.

McHarg, I. 1969. Design with nature. Doubleday, Garden City, New York, USA.

Meyer, E. 1997. The expanded field of landscape architecture. Pages 45–79 in G. F. Thompson and F. R. Steiner, editors. Ecological design and planning. John Wiley & Sons, New York, New York, USA.

Michael, D. N. 1995. Barriers and bridges to learning in a turbulent human ecology. Pages 461–485 in L. H. Gunderson, C. S. Holling, and S. S. Light, editors. Barriers and bridges to the renewal of ecosystems and institutions. Columbia University Press, New York, New York, USA.

Nassauer, J. I. 1997a. Cultural sustainability: aligning aesthetics and ecology. Pages 67–83 in J. Nassauer, editor. Placing nature. Island Press, Washington, D.C., USA.

———. 1997b. Placing nature. Island Press, Washington, D.C., USA.

Omernik, J. M., and R. G. Bailey. 1997. Distinguishing between watersheds and ecoregions. Journal of the American Water Resources Association 33(5): 1–15.

Oxford dictionary and thesaurus: American edition. 1996. Oxford University Press, New York, New York, USA.

Quayle, M. 1993. The rough and the refined: expanding design vocabulary in the public urban landscape. Pages 30–42 in Robert Ribe, editor. CELA 1993: Public landscapes. Proceedings of the 1993 CELA Conference. University of Oregon, Eugene, Oregon, USA.

Schauman, S. 1986. Countryside landscape visual assessment. Pages 103–114 in Richard C. Smardon, J. F. Palmer and J. P. Felleman, editors. Foundations for visual project analysis. John Wiley & Sons, New York, New York, USA.

Schrage, M. 1989. No more teams: mastering the dynamics of creative collaboration. Doubleday, New York, New York, USA.

Silliman, J. M. 1995. Ethics, family planning, status of women, and the environment. Pages 251–261 in H. Coward, editor. Population, consumption and the environment. State University of New York Press, Albany, New York, USA.

Simonds, J. O. 1961. Landscape architecture: the shaping of man's natural environment. McGraw-Hill, New York, New York, USA.

Steingraber, S. 1997. Living downstream. Addison-Wesley, Reading, Massachusetts, USA.

Thayer, R. L. 1994. Gray world, green heart. John Wiley & Sons, New York, New York, USA.

U.S. Department of Agriculture, Natural Resources Conservation Service. 1996. A geography of hope: America's private land. USDA, Washington, D.C., USA.

Watzin, M. C., and A. W. McIntosh. 1998. Aquatic ecosystems in agricultural landscapes: achievable ecological outcomes and targeted indicators of ecological health. An unpublished paper. University of Vermont, Burlington, Vermont, USA.

Winter, D. D. 1996. Ecological psychology. HarperCollins, New York, New York, USA.

Zimmerman, M. 1995. Science, nonscience, and nonsense: approaching environmental literacy. Johns Hopkins University Press, Baltimore, Maryland, USA.

CHAPTER 6

What from Ecology Is Relevant to Design and Planning?

James R. Karr

Education is the instruction of the intellect in the laws of Nature, under which name I include not merely things and their forces but men and their ways

—T. H. Huxley, 1868

Ecology (from the Greek *oikos,* or house) began as a scientific discipline in 1866, when Ernst Haeckel first defined it as "the body of knowledge concerning the economy of nature" (Allee et al. 1949, frontispiece). Ecological thinking began much earlier. Nineteenth-century contemporaries of Haeckel, from Henry David Thoreau (1854) and Frederick Law Olmsted (Olmsted and Kimball 1970; Sutton 1971) to George Perkins Marsh (1865), wove ecological thinking into their lives and their writings. Yet, as ecology the discipline moved ahead in the twentieth century, world-altering technologies led society away from ecological thinking and toward a different "science of house"—economy. Limited ecological thinking in broad public debates has left declining living systems, and this decline threatens human well-being. European settlement of North America, guided by the dominant American myth of "the inexhaustibility of resources" (Nash 1989, p. 35), embodied this new antithesis of ecological thinking. It fueled a hungry human economy with capital plundered from the natural economy. The failure to understand or to work within the economy of nature spawned the environmental challenges we face today.

Many of those challenges are important to the actions of designers and planners. Infusing the design and planning disciplines with ecological thinking is crucial because designers and planners, as disciplinary generalists and integrators, work at the interface of many disciplines. From the design of buildings to growth management and landscape design, ecological considerations impose constraints. But they also offer inspiration. The efficiency of ecological systems is unparalleled. Recycling is standard in those closed systems; waste does not occur. Air and water are not fouled in ways that cause biological collapse. The beauty of plants and animals and the views of sweeping landscapes also inspire humans. But those systems are collapsing, largely because of humans' own actions, the most important force for change on the surface of Earth today.

Early in the twenty-first century, we need a new science and art of home maintenance (the original meaning of economy), one that helps us understand and interpret the consequences of human-driven change. We need a new view of design and planning's purpose. We must seek a balance between our modern industrial economy—and the designs that support it—and our homelands' natural economies. As individuals and as societies, we need to understand the consequences of our actions for the present and the future; ecological thinking is and will be central to attaining that goal.

Designers and planners will play pivotal roles in a new vision of home maintenance. Their artistic tradition—finding unique and esthetically pleasing solutions to design problems—has always been shaped by functional and economic constraints. Now they must add a third dimension to their thinking and learning: ecological thinking, both its constraints and its inspiration. Although their relative importance may vary somewhat among projects, the three Es—esthetics, economics, and ecology—are crucial in all planning and design.

Ecological thinking can always be found in the work of visionary designers and planners (e.g., McHarg 1969; McHarg and Steiner 1998) as they, to paraphrase Huxley, "strive to combine the laws of nature and the ways of men." Good design requires ecological intelligence, an intimate familiarity with how nature works. But how can those charged with defining the appropriate curriculum in design and planning programs foster ecological thinking? Who will make the decisions? How do they decide which among the many facts and concepts in ecology are crucial in a design curriculum? Is it enough to familiarize design and planning students with what they need to know to design buildings or landscapes that are in compliance with environmental regulations? Or should the goal be more than legal compliance? How can designers and planners be trained so that they provide solutions to modern environmental challenges rather than contribute to them?

Those questions cannot be addressed until one understands the connections between those questions, ecology the science, and living systems. Many people express concern about the effects of environmental degradation, but many others consider environmental concerns unwarranted, even silly (Karr

2000). The first section of this chapter defines what I see as the most serious dimension of the current ecological crisis: biotic impoverishment. I then turn to the evolution of ecology (the discipline) over the past century. While ecology the discipline has advanced rapidly, ecologists themselves, whether in their ivory towers or maximizing harvest of some commodity, have often been as isolated from ecological thinking as have many designers and planners.

I then review several generations of ecologists' views on important ecological concepts. My goal is to challenge you to think deeply, to distinguish the key concepts in the science of ecology from the concepts that are key to ecological thinking. Finally, I illustrate my choices for the keys to ecological thinking in a discussion of a regional Pacific Northwest crisis—the decline of Pacific salmon.

What Ecological Crisis?

The crisis is that Earth's living systems are threatened by the actions of human society. This biotic impoverishment (Woodwell 1990) is visible today in three major forms (Table 6 1; Karr and Chu 1995; Chu and Karr 2001) The loss of species, the destruction of agricultural lands, the depletion of forests and fisheries, the loss of human cultural diversity, declining urban cores, political instability, and the exposure of all humans to increased environmental hazards—especially the economically disadvantaged and minorities everywhere—degrade the quality of human life. Collectively, the many faces of biotic impoverishment illustrate the challenges that living systems face in the twenty-first century. It also shows the close connections between environmental and social concerns. Local, regional, and global biological systems are no longer what they were three hundred years ago, and the change threatens the life-support systems of human society (Hannah et al. 1994; Costanza et al. 1997; Daily 1997; Pimentel et al. 1997, 2000). Recognizing this reality can be depressing: "One of the penalties of an ecological education is that one lives alone in a world of wounds" (Leopold 1966, p. 197). But recognition is also essential to reverse the trend.

The root cause of ecological decline is a society behaving as if no risks followed from degrading its living systems. Neoclassical economics reinforced humans' self-appointed dominion over nature's free wealth and brought unparalleled gains in societal welfare in some places, but it also seemed to divorce the human economy from the natural one on which it stands. Environmental challenges faced by modern society are a direct result of widespread human failure to understand the risks associated with degradation in living systems. Perhaps worse is the failure to act even when the risks were understood.

TABLE 6-1. Biotic impoverishment, or losing life, the systematic reduction in Earth's ability to support healthy living systems, can be grouped in three major types with associated examples (modified from Karr and Chu 1995; Chu and Karr 2001)

A. Indirect Depletion of Living Systems

1. Soil depletion and degradation (erosion, degradation of soil structure, salinization, desertification, acidification, nutrient leaching, destruction/alteration of soil biota)
2. Degradation of water (nutrient enrichment, surface-water and groundwater depletion, extinction, redirected flows, homogenization of aquatic biota, wetland drainage)
3. Alteration of global biogeochemical cycles (alteration of water cycle, nutrient enrichment, acid rain, fossil fuel emissions, particulate pollution)
4. Chemical contamination (land, air, and water pollution from pesticides, herbicides, heavy metals, and toxic synthetic chemicals; bioaccumulation, cancer, immunological deficiencies, hormone disruption, developmental anomalies, intergenerational effects)
5. Global atmospheric and climate change (rising greenhouse gases, altered precipitation and airflow patterns, rising temperatures, ozone depletion)

B. Direct Depletion of Nonhuman Living Systems

1. Renewable resource depletion (overfishing, excessive timber harvest, altered food webs)
2. Biotic homogenization (extinctions, spread of alien taxa; homogenization of crops, including loss of ancestral stocks and traditional cultivars, and reductions in genetic diversity within remaining varieties)
3. Habitat destruction and fragmentation (homogenization of biota, destruction of landscape mosaics and connectivity, red tides and pest outbreaks tied to habitat destruction, crop and livestock pests, pesticide resistance, spread of diseases)
4. Genetic engineering (antibiotic resistance, potential extinctions and invasions if genes escape, other unknown ecological effects)

C. Direct Depletion of Human Systems

1. Epidemics; emerging and reemerging diseases (occupational hazards; asthma and respiratory ills; pandemics; Ebola, AIDS, Hanta virus, and many others; antibiotic resistance; diseases of overnutrition; higher human death rates)
2. Loss of human cultural diversity (genocide, loss of knowledge, loss of cultural and linguistic diversity)
3. Reduced quality of life (malnourishment and starvation, overnutrition, failure to thrive, poverty)
4. Environmental injustice (environmental discrimination and racism; economic exploitation; intragenerational inequity; growing gaps between rich and poor individuals, segments of society, and nations; trampling of the environmental and economic rights of future generations)
5. Political instability (resource wars, increased numbers of environmental refugees, international terrorism)
6. Cumulative effects (environmental surprises, increased frequency of "natural" catastrophes, boom-and-bust cycles in marginal regions, collapse of civilizations)

History and Critique of Ecology

For at least fifty years, textbooks have defined ecology as the "study of the interactions of organisms with one another and with their physical and chemical environment." Ecologists identify patterns in living systems and connect those patterns to the processes generating and maintaining those patterns in space and time. Although the definition remains unchanged, concepts are in constant flux. Concepts such as succession and physiological ecology, intrinsic rate of increase and population regulation, competition and niche, community and ecosystem, logistic growth and carrying capacity, and equilibrium and the balance of nature have long made up the fabric of ecology and ecology books. They were defined and debated as ecology evolved from a mostly observational science (natural history) to a modern theoretical discipline with subdisciplines such as population, community, and ecosystem ecology. New subdisciplines such as conservation and restoration ecology, watershed management and landscape ecology, and urban and industrial ecology at the frontiers of ecology are blurring the old boundaries and yielding new concepts (and jargon).

Most of this is the work of basic ecologists studying or attempting to model natural systems subject to relatively little influence from human activities. To them, "Ecology is neither an emotional state of mind nor a political point of view. . . . Ecology is a science" (Dasmann, Milton, and Freeman 1973, p. 20). In most universities, basic ecologists are housed in colleges of arts and sciences. They ask questions such as, Why does the number of species vary from place to place on the surface of the earth? What regulates the size of animal and plant populations? How do global biogeochemical cycles regulate ecosystem structure and function? What is responsible for the spatial properties of natural systems? Unfortunately, in trying to understand nature, too many basic ecologists isolated themselves from real-world problems faced by society. One distinguished ecologist recently argued that the premier North American ecological journal, *Ecology*, is "packed with papers describing more and more sophisticated analyses applied to more and more trivial problems" (Ehrlich 1997, p. 111).

The British Ecological Society and the Ecological Society of America initiated new applied journals, *Journal of Applied Ecology* (1964) and *Ecological Applications* (1991), respectively. An analysis of 60 papers published over thirty years in the *Journal of Applied Ecology* (Pienkowski and Watkinson 1996), however, showed that most articles lacked practical applications or management recommendations. About half the papers (31) indicated an applied purpose in the paper's introduction, but only 8 actually provided management recommendations in their summaries; 16 gave recommendations in the discussion or conclusions sections. Pienkowski and Watkinson recognized that failure to incorporate explicit recommendations made it unlikely the research would influence management situations.

But not all scientists who use ecological understanding have avoided applied, or "solution," sciences. A variety of disciplines (agriculture, forestry, fisheries, wildlife ecology) that depend on the application of ecological principles have long manipulated living systems to achieve a goal—generally to maximize production of a specific commodity. These "applied ecologists" are typically trained in professional schools (forestry, fisheries, agriculture). Supplies of grains, fruits, livestock, wood fiber, and fish are a testament to the success of humans behaving as applied ecologists. Today, despite public awareness and legislation prompted by visibly degraded biological systems, the disciplines dependent on applying ecological concepts and principles still pursue their commodity goals. Specialization created focused disciplines and related professional schools. Numerous government agencies with specific mandates likewise worked to maximize "their" individual commodities with minimal interest in the effects of their actions on the commodity mandate of other agencies, or on broader questions of ecological health. Biological targets have been too narrow (more wood, ducks, or salmon; hardier corn and potatoes); ecological thinking has been rare. This problem persists despite clear mandates such as the U.S. Clean Water Act's call for protecting biological integrity, the Endangered Species Act's call to protect species and their habitats, and the past decade's rhetoric of "ecosystem management."

In short, the problem of biotic impoverishment derives from the failure of ecologists who focus on understanding the natural world to connect with ecologists who strive to maximize harvests of narrowly defined commodities. And neither group is accurately communicating how the world works to other scholarly disciplines or to the general public. Even the recent emergence of conservation biologists and ecosystem managers leaves gaps. Many conservation biologists emphasize protecting endangered species as their primary goal instead of protecting life-support systems more broadly. The single-mindedness of this wing of conservation biology differs little from that of other special-interest commodity groups. Many ecosystem managers still place emphasis on the utilitarian value of living systems in the drive to supply consumers and their ever-expanding human economy.

Few ecology graduate students today will accept the intellectual elitism that was pervasive in basic ecology when I was a student and young professor in the 1960s and 1970s. Students arriving at universities and becoming members of professional societies no longer accept the isolation that characterized academic ivory towers through much of the twentieth century. Fortunately, many senior ecologists are climbing down from those towers (Lubchenco et al. 1991; Ehrlich 1997; Lubchenco 1998). They are less willing to ignore the unraveling of the fabric of life as they pursue their scientific interests. In her presidential address to the American Association for the

Advancement of Science, Jane Lubchenco (1998) called for a more effective interdisciplinary effort on the environment from all the sciences, nothing less than a new "social contract," a commitment by scientists to "devote their energies and talents to the most pressing problems of the day . . . in exchange for public funding" (p. 491). Increasingly, ecologists are recognizing that science is not limited to the production of new knowledge; a critical role of the scientist (Castillo and Toledo 2000) and the university (Karr and Chu, in review) involves the transmission, exchange, and use of that knowledge. Perhaps we can expect more papers by ecologists to include management recommendations and more university curricula to include ecological thinking in this new millennium.

The historical dichotomy between basic ecology and applied ecology needs to give rise to a seamless "new ecology" (Karr and Chu 1999). Whereas basic ecology has tried to understand the natural world and applied ecology has largely concentrated on extracting commodities from that natural world for consumption by humans, a new ecology will protect local, regional, and global life-support systems. This more integrative ecology shares its emphasis on human activities with the commodity branches of applied ecology. But where commodity ecology sought to increase human influence and to use that influence to maximize harvests of wild and cultivated species, a seamless ecology would seek to understand the biological consequences of human activity and to minimize the harmful ones. Understanding and communicating those consequences to all members of the human community is perhaps the greatest challenge of modern ecology—a challenge that landscape designers and planners can help to address because of their broad connections to other dimensions of human knowledge.

A broader applied ecology should, for example, seek to discover the consequences of activities such as grazing, logging, and urbanization on particular places as well as the influence of those activities as they spread across the landscape (Wackernagel and Rees 1996). All branches of ecology should ask: How can we better understand and thereby predict biological responses to human activities? What methods and measurements best isolate the signal produced by human impact from background noise? How do we interpret the results? What are the likely consequences of changes we see? How do we apply those lessons to improve design and planning? How do we design and plan to avoid unacceptable consequences? How do we tell citizens, policy makers, and political leaders what is happening and how to fix it? These questions form the basis of an "integrative" (Holling 1996) or seamless ecology, one more directly useful to designers and planners. Effective solutions must come from applying ecology to find better, broader indicators of biological condition and using those indicators alongside the economic and esthetic guidelines that have long shaped design and planning.

Ecology Is Adopted by the Public: Ecologists and Designers Should Take Note

Just a few decades ago, *ecology* and *environment*, and the rich meaning they now convey, were not on the public radar. Environment was not included, for example, in the 1955 *New York Times Index* (Thiele 1999). Only a single citation for environmental sciences appeared in the 1960 index, and neither environmentalist nor environmentalism was included in the 1971 *American Heritage Dictionary* (Thiele 1999). Since 1970, increased concern about environmental trends injected *environment* and *ecology* into the thoughts of an increasing number of citizens. Voices from all corners of society now draw attention to the severity of present ecological crises (see Appendix for this chapter).

Ecology thereby escaped from the confines of academia and a few government agencies and took on very different meanings. Rachel Carson's *Silent Spring* was a crucial catalyst in this transition. A recent survey in Sweden demonstrated that a belief in the balance of nature is widespread, as is the feeling that the balance should not be interfered with by humans (Westoby 1997). Perhaps a more important lesson of the Swedish study was the public's understanding of ecology as a life philosophy, a source of guidance or a link to morality. The Swedish public clearly recognizes the relevance and importance of ecological thinking beyond the dimensions of ecology as science. These insights stand in stark contrast to the scientist's view of ecology. Yet there are and must be connections between ecology and people.

Biophilia, peoples' innate tendency to focus on life and lifelike processes (Wilson 1984; Kellert and Wilson 1993; Kellert 1996), is an affinity shaped during the long course of evolutionary time. Its functional significance is to connect humans to their life-support systems; truth be told, it is the inspiration that led most ecologists to become ecologists.

Ecological Thinking and Design Inspiration

Ecology the discipline should be more effective at communicating connections between the needs of human society and the understanding of ecological principles, patterns, and processes. I suggest that it should also underpin the creative inspiration that is central to design and planning, from landscape architecture to engineering. As David Orr (1998, p. 334) notes, "Industrial civilization, of course, wasn't designed at all. It was mostly imposed by single-minded individuals armed with one doctrine of human progress or another, each requiring a homogenization of nature and society." Ecological design arts, especially design with an understanding of living systems, have not been key components of the conceptual framework of the design professions.

Good ecological design should focus on correct spatial and temporal scales, simplicity, efficient use of resources, a close fit between means and ends, durability, redundancy, and resilience. Design, too, must be place specific. Design should focus on more than one problem at a time to avoid unwanted side effects. Orr (1998) gives three reasons for poor design: (1) As long as land and energy were cheap and the world was relatively empty, we did not need to master the discipline of good design. (2) Design intelligence fails when greed, narrow self-interest, and individualism take over. Good design is a cooperative community process requiring people who share common values, common understanding of consequences, and goals that bring them together and hold them together. (3) Poor design results from poorly equipped minds. Good design requires ecological thinking, that is, a conscious effort to avoid unexpected biological consequences (Karr 1993; Chu and Karr 2001). Without ecological thinking, our technological society will be seen in hindsight to have collapsed because of the biological disintegration we caused ourselves.

What to Do: Lessons and Directions

The propensity to ignore, overlook, even deny ecological risks goes back at least two centuries. It originated in a battle between scientific ideas (Sachs 1995). The eighteenth-century biogeographer Baron Alexander von Humboldt wanted to catalog the natural world in detail and to formulate a grand theory that would unify and link natural phenomena, including humanity's place within these interdependent relationships. He essentially created and popularized a new profession, that of "scientist," a word that entered the English language in 1830.

Von Humboldt's view ("integration") was soon supplanted by reductionism ("specialization"), by the breaking down of phenomena into their supposed component parts—a theme that has dominated science for the last two centuries. Science was taken over by narrow specializations, and, in the rush to gain in-depth knowledge, science and society lost sight of the need to tie knowledge together. Worse, society began to deny the reality that humans are tied to the complex interrelationships that fascinated von Humboldt, Darwin, and other critical thinkers, such as the landscape architect Ian McHarg (McHarg and Steiner 1998). The struggle between specialization and integration continues today as optimists (cornucopians) and pessimists (environmental doomsayers) vie for dominance.

Pessimists contend that ecological studies call into question some of the cultural and economic premises widely accepted by Western societies (Sears 1964). Optimists lead society to behave as if no negative ecological consequences derived from our actions—as if there were no ecological risks. Optimists suggest that advanced technological cultures release human society from the limitations, or "laws," of nature (Dunlap 1980). Repair and replacement of lost or broken parts is easy. We buy new houses. We get new jobs. We replace teeth and severed limbs, even hearts. We replace natural biological systems with corn, cows, cars, cancer, carbon dioxide, and computers. But we cannot replace places, whether *place* means King County (Washington), the Pacific Northwest, the western United States, or planet Earth. We cannot replace food and fiber, water and air with the virtual worlds generated by computers.

We know much about human health risk and short-term economic risks, and do much to plan for them. Why don't we give as much attention to ecological risks and the environmental deficits that result from our failure to anticipate ecological risks? Because we organize our knowledge in pieces; we choose to ignore connections. We have lost von Humboldt's profound insight. But we can regain it. Landscape designers and planners, even more than most members of society, make decisions each day that can worsen the damage to Earth's living systems, or that can minimize that damage. They can be important in restoring the connections.

Key Concepts in Ecology

The definition of key concepts and principles within a discipline is important for several reasons. First, self-assessment is a stimulus for clear thinking and disciplinary advance. Second, self-assessment organizes knowledge and shows the relationships among the facts and principles in a discipline. Recent efforts to integrate knowledge within ecology is overcoming a legacy of fragmentation and fostering synthesis through development of new subdisciplines (e.g., landscape ecology capturing and integrating key concepts from population, community, and ecosystem ecology). Third, it frames our understanding in a way that captures current knowledge without losing the insights of earlier generations. Fourth, it facilitates transfer of knowledge from other disciplines into ecology and from ecology to other disciplines.

Since the dawn of the twentieth century, ecologists have attempted to define the discipline's general principles or key concepts (e.g., Cowles 1904; McIntosh 1985; Cherrett 1988, 1989; Pool 1991; Klemow 1991; Odum 1992). In perhaps the most extensive survey (Cherrett 1988, 1989), the British Ecological Society asked all society members to rank their top ten from a list of fifty "critical concepts" (Table 6-2); they received 645 responses. The breadth of concepts in the list reflects a broad range of perspectives and approaches in ecology. I think the list reflects the hot concepts in 1986, especially as influenced by ecological science in Britain; ranking would likely be very different now. Concepts such as stochastic processes, natural distur-

TABLE 6-2. A 1986 list by British ecologists ranking in order "the most important concepts in ecology" (from Cherrett 1989)

1. The ecosystem	26. Natural disturbance
2. Succession	27. Habitat restoration
3. Energy flow	28. The managed reserve
4. Conservation of resources	29. Indicator organisms
5. Competition	30. Competition and species exclusion
6. Niche	31. Trophic level
7. Materials cycling	32. Pattern
8. The community	33. r and K selection
9. Life-history strategies	34. Plant-animal coevolution
10. Ecosystem fragility	35. The diversity stability hypothesis
11. Food webs	36. Socioecology
12. Ecological adaptation	37. Optimal foraging
13. Environmental heterogeneity	38. Parasite-host interactions
14. Species diversity	39. Species-area relationships
15. Density-dependent regulation	40. The ecotype
16. Limiting factors	41. Climax
17. Carrying capacity	42. Territoriality
18. Maximum sustainable yield	43. Allocation theory
19. Population cycles	44. Intrinsic regulation
20. Predator-prey interactions	45. Pyramid of numbers
21. Plant-herbivore interactions	46. Keystone species
22. Island biogeographic theory	47. The biome
23. Bioaccumulation in food chains	48. Species packing
24. Coevolution	49. The 3/2 thinning law
25. Stochastic processes	50. The guild

bance, and restoration would clearly move higher on the list today. Conservation biology, landscapes, metapopulations, sources and sinks, ecological health, and population genetics did not even appear on the 1986 list. Biodiversity was not listed, but species diversity was. Concepts from applied ecology, a minor component in the 1986 list, would surely occupy more prominent positions in a current compilation.

Another approach (Odum 1992; Klemow 1991) involves fewer concepts listed as short statements (Table 6-3). These lists rely heavily on ecological theory and jargon that is not agreed upon or even understood by many ecologists, let alone those outside the field. "Theory" in ecology expanded rapidly in the 1960s. When published in the 1960s (MacArthur and Wilson 1967), island biogeography, for example, brought together disparate lines of evidence, made testable predictions, addressed important questions, applied

TABLE 6-3. "Twenty great ideas in ecology for the 1990s" developed by a leading American ecosystem ecologist (from Odum 1992)

1. An ecosystem is a thermodynamically open, far from equilibrium, system.
2. The source-sink concept: One area (the source) exports to another area (the sink).
3. Species interactions that tend to be unstable, nonequilibrium, or even chaotic are constrained by the slower interactions that characterize large systems.
4. The first signs of environmental stress usually occur at the population level, affecting especially sensitive species.
5. Feedback in an ecosystem is internal and has no fixed goal.
6. Natural selection may occur at more than one level.
7. There are two kinds of natural selection, or two aspects of the struggle for existence.
8. Competition may lead to diversity rather than extinction.
9. Evolution of mutualism increases when resources become scarce.
10. Indirect effects may be as important as direct interactions in a food web and may contribute to network mutualism.
11. Since the beginning of life on Earth, organisms have not only adapted to physical conditions but have modified the environment.
12. Heterotrophs may control energy flow in food webs.
13. Biodiversity should include genetic and landscape diversity.
14. Ecosystem development or autogenic ecological succession is a two-phase process.
15. Carrying capacity is a two-dimensional concept (number of users and per capita use).
16. Input management is the only way to deal with nonpoint pollution.
17. Expenditure of energy is required to produce (maintain) energy flow or material cycle.
18. Need to connect human-made and natural capital and sustainability.
19. Transition costs are always associated with major changes in nature and in human affairs.
20. A parasite-host model for man and the biosphere is a basis for turning exploiting the Earth to taking care of it.

widely, and was supported by some empirical evidence. But not all ecologists then or now would consider island biogeographic theory an important advance. "Although theoretical ecology is one of the more highly touted aspects of recent ecology, it is not easy to find consensus among ecologists about established theories, their basic postulates, sources or even their names or pseudonyms" (McIntosh 1985, p. 257).

Philosopher Mark Sagoff (1997) suggests that the proliferation of theory in ecology—much of which, he contends, borrows from other sciences—fogs the recent ecological literature with scores of inadequately tested conceptual frameworks.

Smith (1976, p. 546) noted a "tendency in ecology to call a hypothesis a theory or to give credence to concepts through repetition," leading to "confusion between what has been verified and what has been postulated." This confusion between ideas and reality has resulted in hypotheses being advanced as solutions to real-world problems before the hypotheses are adequately tested.

Because the road to ecology as a discipline is littered with the carcasses of discarded theories, premature advocacy of theory is at best foolish and at worse dangerous. In many respects, the situation is similar to the position of some in economics who advance economic theory as a guide to public policy and decision making in the face of evidence that key components of economic theory do not apply in the real world (Norgaard 1994; Prugh et al. 1995; Hawken, Lovins, and Lovins 1999; Jacobs 2000).

Unfortunately, the language of ecology the science—and ecologists' efforts to define key concepts focuses narrowly on ecological detail, often with limited relevance to public policy challenges. Moreover, that language permits us to forget the biological core of ecology. Watershed analysis and ecosystem management—mantras for scientists, politicians, and environmentalists in the U.S. Pacific Northwest and beyond—are often narrowly framed. Ecosystem management as interpreted by state and federal agencies in the United States has a "distinctly utilitarian" emphasis (Noss et al. 1999, p. 104), one that all too often focuses on select commodities. Protocols for watershed analysis include physical models (e.g., landslides, erosion, hydrology, bank stability), but they almost never include any direct biological measures (e.g., WFPB 1993, p. xii). The closest these analyses come to biological factors is the presence of physical habitats thought to be required by fish. Monitoring to determine if fish or other components of the biota are present (often referred to as "validation monitoring" because it is meant to validate the dominant physical models) is rarely attempted, usually because funds are exhausted before validation is accomplished. Policies based on relationships that are assumed before they are empirically validated are common.

An experience with a federal land management agency is illustrative. The U.S. Forest Service convened a meeting (Eugene, Oregon, November 1994) to discuss the components of a watershed analysis protocol. Its protocol, like the one advanced by the Washington Forest Practices Board (WFPB 1993), included no measures of biological condition. When I asked about the biological context that I assumed would be central to this new vision for management of Northwest landscapes, I was told that the Forest Service hoped to be able to implement watershed analysis without collecting any new data. Watershed management is explicitly designed to prevent recurrence of the failures of the past (e.g., declining spotted owl and salmon populations), yet its implementation is devoid of the biological focus that is central to its stated goal. Surely we can do better. Definitions of key components of watershed analysis or ecosystem management that exclude the biota and biology are simply too narrow to be effective.

Schneider (1997) takes a very different approach to defining what is important from ecology. Rather than asking what the key concepts are, he asks about the key components of education. He emphasizes the importance of "illuminating the policy process as it deals with science and societal decision-making" (p. 457). He argues that the goal depends less on being well versed in knowledge or methods than on an interdisciplinary process of integration. For Schneider, a critical component of literacy is learning how assessment teams must integrate content (traditional disciplinary knowledge) with context in real places (addressing real-world problems with interdisciplinary synthesis). Interdisciplinary means a combination of knowledge and methods or paradigms; interdisciplinary teams construct an original synthesis that would probably not emerge from the collection of multidisciplinary subcomponents. Schneider's top priority is for individuals to ask three questions of experts inside and outside their discipline: What can happen? What are the odds? How do you know?

The truth no doubt lies in the middle ground between facts and integration. Our society must be rich in critical and creative scientific thinkers, individuals with scientific literacy, with its emphasis on scientific ways of knowing and the process of thinking critically and creatively in the long-term about the natural world (Maienschein 1998). It must also be rich in knowledgeable experts, individuals with detailed technical and scientific knowledge (science literacy). Knowing how to ask the "And then what?" question is critical, but so is enough disciplinary knowledge to recognize that the answer to that question fails the laugh test.

In my experience, the most successful interdisciplinary team members have in-depth knowledge of their own field and are open to learning from others. They are willing to establish a basis for communication across disciplines. They are willing to ask for explanations when they do not understand. They are willing to challenge with respect and an open mind the dogma of their discipline as well as the dogma of others on the team. The most important attribute of all members of a truly interdisciplinary team is recognition that no individual is knowledgeable enough and no discipline is broad enough to grasp the many dimensions of the complex ecological issues faced by modern society. Multidisciplinary teams that lack these abilities are not likely to become truly interdisciplinary. Planners and designers, who pride themselves on being generalists, can and should play a key role in bringing together the diverse disciplines. But their success depends on more than a superficial familiarity with the three critical *Es*—esthetics, economics, and ecology.

As implied in Schneider's comments, it is no longer enough for ecologists to think in terms of what ecologists should know and understand. Ecology's escape from the confines of its academic home (see earlier section) requires an effort to define concepts key to ecological thinking. Success in protecting the environment and the quality of life for people requires disciplinary synthesis and clear communication that makes the real challenges clear and the solutions tractable. How does one do that?

David Orr, one of the nation's most respected environmental educators, asks not what from ecology is relevant but what knowledge is needed to produce whole persons with intellectual breadth to "join the struggle to build a humane and sustainable world" (Orr 1992, p. 108). His approach is less about ecology the science and its concepts and more about creating a syllabus for ecological literacy, an effort to foster ecological thinking. His "Syllabus for Ecological Literacy" (pp. 109–124), meant as an antidote to Bloom's *The Closing of the American Mind* (1987), includes books grouped in sets under such headings as "How the World Works," "Trends, Forecasts, Probabilities, Possibilities, Uncertainties," "Sources of Environmental Problems," "The Question of Scientific Knowledge," "Ideas of Nature," "And Human Nature," "The Concept of Sustainability," "Tools of Analysis," and "Resources/Perspectives." Orr's goal is to show that the crisis of sustainability has roots that extend from public policies and technology to assumptions about science, nature, culture, and human nature.

Key Ecological Concepts for Designers and Planners

For several years, I have been compiling a list of key words for ecological thinking (Table 6-4), words whose meaning and content are crucial to understanding the workings of living systems. The following few paragraphs weave those key words (in italics) together in a brief synopsis of ecological thinking.

Plants, animals, and microorganisms interact as interdependent parts in complex *systems*. The primary *parts* of those systems (the species) operate over *spatial scales* from a few millimeters to thousands of kilometers (e.g., migratory birds and fish); they may live for only a few minutes or hundreds of years (*temporal scales*). Mosaics of these ecosystems form complex *landscapes* that are in turn influenced by the *context* of the surrounding landscape.

Each species has a *natural history* and *life cycle* shaped by its evolutionary history in the local landscape. Unique natural histories allow each species to exist in specific spatial and temporal contexts. The parts of living systems are

TABLE 6-4. Key concepts for ecological thinking

1. INTEGRITY and HEALTH

Integrity: unimpaired condition, quality, or state of being complete or undivided; equivalence with some original condition; natural; a living system's capacity to organize, regenerate, reproduce, sustain, adapt, develop, and evolve; a local or regional biota that is the product of evolutionary and biogeographic processes with minimal human influence

Health: flourishing condition, well-being, vitality; culturally preferred state that is sustainable; the goal or condition of a site cultivated for crops or otherwise used by people

2. SYSTEM and SCALE

System: a regularly interacting or interdependent group of items forming a unified whole, which is in, or tends to be in, equilibrium

Scale: spatial and temporal dimensions of a system, including its parts and processes

3. LANDSCAPE and CONTEXT

Landscape: mosaic of natural ecosystems and land uses across a region; the aggregate landforms of a region

Context: interrelated conditions in which something exists or occurs; setting or environment

4. PARTS and PROCESSES

Parts: elements, entities, or constituents that make up a system; measured as number of kinds of things, such as the number of species or alleles

Processes: natural phenomena that define system condition by defining functional relationships; measured as rates, such as birth, death, or metabolic rates

5. NATURAL HISTORY and LIFE CYCLE

Natural history: the behavior, ecology, distribution, physiology, systematics, and anatomy of organisms

Life cycle: the series of stages (both form and function) through which an individual passes during its lifetime

6. RESILIENCE and RESISTANCE

Resilience: ability to recover from or adjust easily to misfortune or change; ability to return to previous state following disturbance

Resistance: inherent capacity to withstand untoward circumstances; ability to remain unchanged under the influence of outside events

7. DISTURBANCE and EQUILIBRIUM

Disturbance: an event or action that alters something

Equilibrium: balance between opposing forces or actions that is either static (unchanging) or dynamic (varying narrowly around a central tendency)

8. CHANCE and CHANGE

Chance: something that happens unpredictably without discernible human intention or observable cause; an element even of natural events that are contingent on historical context and current condition

Change: transformation, transition, or substitution; a loss of original identity or a substitution of one thing for another; fundamental fact of everything; nature's constant

9. TRAJECTORY and CYCLES

Trajectory: path, progression, or line of development

Cycles: tendency of system to experience recurrent patterns, usually leading back to the same starting point

10. CONNECTIONS, LIMITS, and COLLAPSE

Connections: links among system parts or elements; everything linked to everything else

Limits: something that bounds, restrains, confines, or imposes constraints, such as limited food or space

Collapse: complete breakdown; sudden failure

11. ROOT CAUSES and PATTERNS

Root causes: ultimate reasons for an event or sequence of events

Patterns: natural configurations; traits, tendencies, or other observable characteristics; detectable conditions of a system in space, time, or both

12. EFFECT, CONSEQUENCE, and AFTERMATH

Effect: something that inevitably and directly follows from a cause; the increased demand for metals and fuel caused by invention of the internal combustion engine

Consequence: complex medium term results; more remote or less obvious connections linked to a cause; the fragmentation of natural environments and human communities by roads; deaths (human and nonhuman) from cars, trucks, trains, and planes; increased carbon dioxide emissions

Aftermath: belated, often complex, consequences; global climate change that follows from increased carbon dioxide emissions caused by internal combustion engine

13. SIMPLIFICATION, COMPLEXITY, and DIVERSITY

Simplification: reduction in scope or complexity

Complexity: something made up of complicated or interrelated parts that are likely hard to separate, analyze, or solve

Diversity: the variety and variability in living systems and the environments that they occupy; genetic, life-history, species, and ecosystem diversity

14. UNCERTAINTY and SURPRISE

Uncertainty: lack of knowledge about an outcome or result; lack of sureness about something

Surprise: something that is unexpected at a particular time or place

connected to one another and to their physical and chemical environments through a variety of *processes*—biological (metabolism, genetics, natural selection, demographics, nutrient cycling) as well as chemical and physical (hydrology and the water cycle, weathering, biogeochemical cycles)—that regulate species abundances and distributions. The *connections* among the parts and processes of these systems are convoluted and rarely understood in great detail; nonetheless, they are pivotal.

Although natural systems are never in *equilibrium,* organisms have evolved to survive and reproduce in a narrow range of physical and chemical environments (that is, in a *dynamic but bounded equilibrium*). Pushing systems beyond those *limits* may result in species extinction or even system *collapse* (complete system breakdown). *Change* is normal, because environments fluctuate naturally. Although natural systems can withstand diverse *natural disturbances* within the range of their evolutionary experience, *human-induced disturbances* that are beyond that experience in type, magnitude, or frequency may radically shift system configuration. Natural systems are especially vulnerable when human-induced disturbance is combined with rare natural events; the *chance* of system collapse in such situations is high. The ability of living systems to *resist* change and to adapt to change is vital.

When human actions alter systems, it is often difficult to define the *root cause* of the changes, or even to discern patterns that may suggest collapse is near. The *trajectories* of natural systems often include cycles of abundance or other aspects of system condition. Those trajectories often change drastically under the influence of human actions. Often we know much about the short-term, direct, and immediate *effects,* or results, of our actions. Direct but looser or more remote *consequences* need to be better understood, as do belated, often complex, consequences that may spell disaster (*aftermath*). In the end, failure to understand the natural course of events in living systems, and how they are affected by and affect human systems (their *resilience* and *resistance*), can lead to collapse with potentially ominous boomerang effects for humans.

Human actions are often directed toward simplifying systems as we strive to concentrate production in those parts and processes we value most. *Simplification* often produces unexpected consequences. The resultant loss of *complexity* and *diversity* may threaten critical processes that provide utilitarian or functional value to humans. Loss of natural system *integrity* is the all-to-common result. Some divergence from integrity may be culturally acceptable, even necessary, in some areas dedicated to support the needs of humans. Care is necessary to maintain the condition or *health* of such areas to ensure that their use by human society does not alter their ability to provide the needs of both human and nonhuman living systems over the long term (sustainability). Because systems are so complex and human actions influence numerous nonlinear dynamics, the precision of human predictions about system responses is limited. *Surprise* is inevitable and *uncertainty* is high because we are relatively ignorant of the many parts of living systems, the processes that generate and maintain those parts, and their interactions.

Individuals (human and nonhuman) and societies are embedded in the cyclical processes of nature, in a "globally interconnected world, in which biological, psychological, social, and environmental phenomenon are all interdependent. To describe this world appropriately, we need an ecological perspective" (Capra 1982, p. 16).

As important as the concepts themselves is the framework in which they are used and learned. Designers, planners, and other professionals are more likely to understand and use key concepts if they are learned in a context that includes the following four components:

1. *Basic familiarity with the language of ecology.* Because the number of concepts, words, and definitions in ecology is so great and changes so rapidly, I am uncomfortable prescribing a list. As noted earlier, I place emphasis on those that are more relevant to ecological thinking than to the details of ecological science. But, inevitably, students and practitioners will be exposed to an expanding array of ecological concepts. They should be ever vigilant, taking personal responsibility to understand the concepts they encounter during a project or a career.

2. *Exposure to and familiarity with organisms, their natural histories, and their environments.* It is especially important that students become familiar with the many kinds of organisms and the way they make a living, with the major environment types in different regions, and with the way human actions influence these organisms and ecosystems. Although they can't learn the life histories of all organisms, or even all major taxa, students should be aware of the spatial and temporal scales of life and life histories across organisms from algae to trees and from protozoa to large predators. They should understand that designs that fail to factor in organisms' life histories may have unexpected consequences for local and regional living systems.

3. *Supplement learning of abstract concepts on global scale with walks in the real (local) world to examine the biology of real places* (see Chapter 11). These places should include both natural systems—places minimally disturbed by human activity—and built systems highly modified by humans. Too often ecologists and ecology courses do not expose students to both, or discuss how the systems are similar and different. Courses rarely discuss the efficiencies and inefficiencies of natural versus built systems or the broad array of goods and services gained or lost with each type. Students should think about, describe, and discuss the differences between these

two kinds of places. They should thoughtfully explore the consequences of those differences and for whom.

4. *Practicums, problem sets, or studios to require individuals to work in interdisciplinary teams on specific problems such as they are likely to face upon leaving the academic environment* (see Chapter 16). A balance of learning, discussion, and real-world application will serve students long past the memorization of current ecological buzzwords. Knowledge and experience are a crucial counterweight to "pure" artistry inspired only by aesthetics and incompatible with ecological reality. Aesthetically appealing designs should be grounded in both economic and ecological reality.

Designing and Planning with Living Systems in Mind

Whether large or small, virtually all design and planning efforts should evaluate likely project effects on living systems. Failing to routinely do so risks damage to those systems. Considering living systems might even prevent excess expense by taking advantage of living systems' natural benefits. Broad planning should explicitly incorporate living systems, from the fact-finding at the beginning of a design study to collecting and analyzing data, and, ultimately, to project planning and implementation. Perhaps most important, planning should be evaluated to determine if stated goals are accomplished as well as if the aftermath of projects includes unexpected consequences.

Yet designers and planners know that completed projects are rarely evaluated after completion in any but the most perfunctory way. As we learn more about the effects of humans and see their consequences on living systems with greater clarity, responsible professionals will encourage such assessments. Opportunities for this kind of adaptive management (Walters 1986; Lee 1994) are legion; done correctly, they will improve the performance of the design and planning professions.

Fortunately, new biological monitoring tools (multimetric biological indexes such as the index of biological integrity, or IBI) are available to designers and planners to evaluate current biological condition and to track site condition after project completion (Davis and Simon 1995; Barbour et al. 1999; Karr and Chu 1999; Simon 1999). These biological measures are akin to commonplace economic indexes such as the index of leading economic indicators or the consumer price index used by investors to track the health of the U.S. economy. Economic indexes integrate indicators of economic health, such as housing starts and sales of durable goods, while biological indexes integrate indicators of biological condition, covering levels of biological organization from the health of individuals to taxa richness and trophic organization (Figure 6-1). Much as a physician relies on a collection

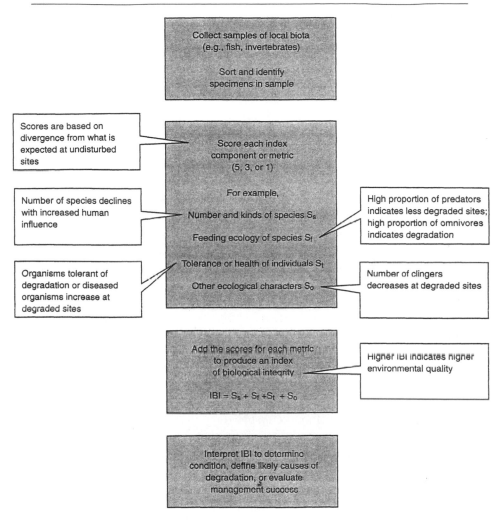

FIGURE 6-1.
Steps or components of a biological assessment using an index of biological integrity (IBI) (after Karr et al. 1986 and Molles 1999).

of medical tests, not just one, to diagnose illness, anyone can use an IBI to diagnose the condition of a landscape.

Scientifically rigorous yet easily developed and adapted for particular locales, the IBI is a straightforward, biological means of measuring the health of places. It is founded on common sense and observations from the real world as well as on established ecological theory. The IBI is simple to craft and to communicate, adaptable for any geographical location, and applicable in a wide variety of contexts to meet human and other ecological needs. It can be used with diverse plant and animal groups in diverse management settings. In addition, the IBI is objective, scientifically rigorous, and easy for scientists and planners to communicate to nontechnical audiences.

The biological attributes incorporated into an IBI are called metrics. They are chosen because they reflect specific, predictable responses of organisms to changes in landscape condition. They are sensitive to a range of physical, chemical, and biological factors that alter biological systems, and they are relatively easy to measure and interpret. They are not simply narrow indicators of commodity production or threatened and endangered status. Moreover, because the IBI's foundation is heavily empirical, its use does not require resolving of all the higher-order theoretical debates in contemporary ecology or solving the formal mathematical models of ecological functions.

The IBI and analogous measures are gradually replacing a variety of bureaucratic benchmarks (numbers of permits, fines levied, pollutants removed). Unlike conventional benchmarks, biological measures are one step closer to the factors that constitute environmental quality (Keeler and McLemore 1996; Yoder and Rankin 1998), the core goals in most environmental legislation (Karr 1991). Biological monitoring has been used most to evaluate the impacts of projects on water quality, but terrestrial applications are now available (Kimberling, Karr, and Fore, in review), and the concept has been introduced in the planning literature (Karr 1998).

Streams in nearly pristine areas of Grand Teton National Park, Wyoming, for example, had near maximum IBIs (44 out of a possible 45). Streams with light recreational use in their watersheds (hiking, backpacking) had IBIs (41) that did not differ significantly from those in pristine areas. Places where recreation was heavy were clearly damaged (28). Urban streams in the town of Jackson, Wyoming, had the lowest indexes in the region (21) but not as low as urban streams in Seattle (9) or Japan (9–11) (Karr 1997).

IBIs have detected the effects of effluents from a bauxite plant in Guinea (Hugueny et al. 1996) and from salmonid aquaculture on small streams in France (Oberdorff and Porcher 1994); the cumulative effects of channelization, agricultural runoff, and urbanization in France's Seine River basin; and the impact of diverse land uses on streams in dry west-central Mexico (Lyons et al. 1995). Within stream channels, an IBI fully reflects chemical contamination in diverse geographical settings: a single pollutant, chlorine, in U.S. midwestern streams (Karr, Heidinger, and Helmer 1985); more complex pollutant mixtures, such as metal and organic pollution in central Indian rivers (Ganasan and Hughes 1998); and urban point and nonpoint pollution gradients from "pristine" to "grossly polluted" in Thailand, Ghana, and Brazil (Thorne and Williams 1997). IBIs also reflect impacts of human activities on a watershed scale, no matter which taxonomic group was used to build the index. For example, an IBI based on fishes declined as forest cover declined in Wisconsin (Wang et al. 1997). An IBI based on benthic invertebrates declined with increasing urbanization in the Puget Sound, Washington, area (Karr and Chu 2000) and did not rise after restoration projects in the Pacific Northwest (Larsen et al., in press).

When biological monitoring and assessment are integrated with knowledge of regional human activities, managers, policy makers, and citizens can use this information to decide if measured alterations in biological condition are acceptable and set policies accordingly. By enabling us to identify the biological and ecological consequences of human actions, biological monitoring provides an essential foundation for assessing the likely consequences of specific design or planning activities.

CASE STUDY: SALMON IN THE PACIFIC NORTHWEST

I have been fortunate over the past three decades to work on projects in the agricultural lands of the U.S. Midwest, the wetlands of south Florida and Venezuela, the forested landscapes of Panama and the U.S. Pacific Northwest, and urbanizing Puget Sound, Washington. Each is a real-world example of the relevance of ecological thinking. The story of salmon in the Pacific Northwest illustrates the consequences of ignoring ecological thinking.

By the middle of the nineteenth century, citizens in the Pacific Northwest recognized that local and regional salmon populations were threatened by modern human society (Lichatowich 1999; Taylor 1999). The first territorial legislature (1848) in Washington established regulations to protect salmon and their habitat. Since then, each generation of citizens watched as many salmon populations went extinct and most others declined (Nehlsen, Williams, and Lichatowich 1991). Each generation has called for protection of salmon and supported those calls with laws and regulations. But the best of intentions are hollow if the institutions and agencies responsible for carrying out the many mandates are ineffective, or their efforts are thwarted by special interests that benefit when the mandates are ignored (Karr 1995; Chasan 2000).

Waves of settlers to the Pacific Northwest divided the landscape according to their economic interests (Lichatowich 1999): Fur trappers took beavers; miners extracted gold; loggers took trees; ranchers took grasslands and riparian zones; irrigators took water; hydroelectric dams took the rivers' energy; urbanites used it all. All worked to feed a human economy that rapidly outgrew the natural economy of local and regional ecosystems. With little or no government regulation, these special interest groups "pulled apart, fragmented, and destroyed the salmon's home" (Lichatowich 1999, p. 78).

All the special economic interests saw pieces; none saw a connected system of parts and processes that were intimately tied to and dependent upon each other. Few saw, or were willing to acknowledge, that clear-cut logging in the highlands benefited loggers but damaged both salmon and the livelihoods of lowland fishing communities with debris flows, sedimentation, and downstream flooding. When it was acknowledged that dams blocked fish migration, fish ladders that aided adults heading upstream were installed, but no plans were made for juvenile fish swimming downstream to the ocean. The lack of baby fish led to construction of hatcheries in the misplaced belief that we could have "salmon without rivers" (Lichatowich 1999). Those responsible for protecting salmon populations failed to consider salmon's natural history, the complex life cycles that link them to a landscape that stretches from the highest reaches of mountain streams to the North Pacific Ocean.

Ecological thinking that encompasses understanding systems (whole things), limits, consequences, uncertainty, simplification, disturbance, and trajectories has been lacking. Solutions often made matters worse. Narrow solutions focused on narrow symptoms rather than root causes. For example, the use of hatcheries to produce salmon for release assumed that the places into which salmon were released could support those hatchery fish and that those hatchery fish were suited to life in those rivers. It assumed that laws and harvest policies would not need to be changed when they should have been; the presence of hatchery fish, for example, allowed harvests to continue that further damaged wild stocks in the same regions (Lichatowich 1999).

During the nineteenth century, salmon—like all of North America's vast resources—seemed so abundant that no one could imagine a human impact large enough to diminish them. Repeatedly through the twentieth century, optimists behaved as if there were no limit to their resilience. Human actions simplified the diverse and complex salmon landscape; the prognosis for salmon became more and more uncertain. As we enter the new millennium, many see certainty returning, but in this case, that certainty may well be the regional extinction of the once vast runs.

The listing of a dozen or more regional salmon stocks under the Endangered Species Act (ESA) in the last decade has awakened many forces in society. Early listings emphasized the role of dams and logging. Recent listings, especially the inclusion of Puget Sound chinook, emphasize the effects of urbanization on salmon numbers; collectively, the listings demonstrate that all northwesterners contribute to the crisis. Is there any evidence that the present round of planning to protect salmon will be different from those of earlier generations? At the very least, it seems prudent to examine carefully any and all efforts to produce a new generation of "fixes."

Historically, a number of hypotheses have guided salmon protection and restoration. In the early days, it seemed there were so many salmon in the Pacific Northwest that human actions could not influence their numbers. That optimistic view was soon discarded, replaced by a century of narrowly conceived technofixes—fish ladders, fish hatcheries, cleaner water, less woody debris, barging smolts (baby salmon) downstream, spawning channels, riparian corridors, bounties on predators, more woody debris, fewer dams. Inevitably, the problem was defined narrowly, symptoms were treated, and the disease (rapidly degrading salmon landscapes) continued unabated. The current push to protect salmon habitat is, in my view, just as shortsighted.

A recent symposium, Salmon in the City (Mt. Vernon, Washington, 20–21 May 1998), carried the subtitle "Can habitat in the path of development be saved?" Conversations at the symposium often reverted to analyses of physical habitat structure, to the even more narrow topic of woody debris or sediment in the channel. This narrow focus will inevitably lead to a new generation of narrow habitat fixes rather than analyses designed to sustain the complex landscapes and associated living systems on which salmon and other fish depend. No matter how important for commerce, sport, or even as a regional icon a particular species is to humans, it cannot persist outside the biological context that sustains it. Failing to protect plankton, insects, bacteria, higher plants, or other fish species ignores the key contributions of these taxa to healthy living landscapes. Salmon depend on those landscapes. Designers and planners should understand the many dimensions of landscapes important to sustaining regional aquatic systems, including salmon; they should not fall into the trap of planning to attain some narrowly conceived and defined physical (habitat) goal.

A focus on processes presumed to create habitat, or worse, efforts to manufacture selected components of river habitat as by placing large woody debris, cannot substitute for understanding the presence and condition of living systems occupying those places. The habitat approach reminds me of the philosophy that made the film *Field of Dreams* so popular: "Build it and they will come." We do know how to build baseball diamonds, but I am less confident in our ability to define and then construct the complex landscapes that salmon require to complete their life cycles.

What precisely is the goal? Should we focus on saving salmon, all fish, or the complex living systems that salmon and other fish depend on? Or should we think more comprehensively about the landscapes that salmon depend on for their survival? That landscape clearly includes rivers, their watersheds, and the ocean environments where salmon reside through much of their life cycle. But there is not a single salmon landscape. Each salmon stock is uniquely adapted from the timing of its migrations to its adult size to occupy and successfully reproduce in its area of the region. That diversity of salmon species— and stocks within each species—is central to the success of salmon in the region.

But the salmon landscape also includes the administrative, institutional, and political landscapes that salmon must traverse in their migrations. Those landscapes are created by the attitudes and philosophies of the people of the Pacific Northwest. If those attitudes and philosophies do not include ecological thinking, there can be little hope for salmon. Failure to understand and focus on all dimensions of the salmon landscape is likely to limit the success of policies or the plans of ecologists or designers and planners to protect and restore salmon. Clearly, saving habitat in the path of development will not be enough, just as making baby fish or clean water has not been enough.

Efforts to protect and restore the salmon landscape, and thus protect salmon populations, should keep several lessons in mind. First, we must overcome narrow conceptions of the challenge of salmon conservation and the legacies that come from those conceptions. Second, we must be realistic. We cannot and should not expect to restore all rivers to the level that they will support healthy salmon populations. But we can and should strive for better rivers everywhere. Third, we must also recognize that we can't repair more than a hundred years of damage in five to ten years. Fourth, we should recognize that knowledge is, and always will be, limited. At the same time, we know enough today to do a better job of protecting salmon landscapes and thus salmon. Fifth, we must begin to establish priorities in a more comprehensive and incisive way. Decisions to protect (conservation), develop intelligently (planning and development), and restore (design and restoration) must be guided by thoughtful integration of scientific information, economic consequences, and the values and attitudes of people.

Ecological thinking does make a difference, as demonstrated on the Fraser River in Canada (Lichatowich 1999). Pressure to undertake activities that would modify spawning areas, pollute the river, or dam or divert water was resisted. Instead, research on the life histories of Fraser River salmon stocks was initiated, and new regulations were delayed until those life histories were better understood. Emphasis in salmon management was placed on natural production and the protection of river reaches critical to early stages in the salmon life cycle (eggs, fry, juveniles). Regional scientists, political leaders, and the public heeded the warnings that dams, hatcheries, and decisions based on short-term economic gains from development and excessive harvest were not a substitute for careful management of salmon and salmon landscapes. Ecological thinking prevailed in British Columbia, and Fraser River stocks remained healthy as Columbia River stocks declined.

Decisions based on integrative ecological thinking to protect and restore salmon should focus on three questions: How do we decide what to do? How do we decide where to do it? How do we decide if it worked? Landscape designers and planners have much to contribute to providing answers to all three questions.

Throughout North America and the world, decision making in the absence of ecological thinking has impoverished human and nonhuman living systems. Rational people cannot and would not strive to restore the planet to its condition a few centuries ago. At the same time, rational people cannot hope to maintain viable human cultures in a world devoid of the complex life-support systems that make human life possible. Design and planning professionals will play a critical role as society knits its future, a future that must seek a balance between our modern industrial economy—and the designs that support it—and our homelands' natural economies.

Acknowledgments

This paper was prepared with support from the Consortium for Risk Evaluation with Stakeholder Participation (CRESP) by Department of Energy Cooperative Agreement #DE-FC01-95EW55084.S. My research over the past twenty-five years has been supported by a variety of agencies and organizations, including USEPA, NSF, Forest Service, Fish and Wildlife Service, Smithsonian Institution, National Geographic Society, and others. Richard T. T. Forman, Ellen W. Chu, Elena S. Karr, and one anonymous reviewer provided invaluable advice that substantially improved the paper.

Citations

Allee, W. C., A. E. Emerson, O. Park, T. Park, and K. P. Schmidt. 1949. Principles of animal wcology. Saunders, Philadelphia, Pennsylvania, USA.

Barbour, M. T., J. Gerritsen, B. D. Snyder, and J. B. Stribling. 1999. Rapid bioassessment protocols for use in streams and wadeable rivers: periphyton, benthic macroinvertebrates and fish, second edition. EPA 841-B-99-002. U.S. Environmental Protection Agency, Office of Water, Washington, D.C., USA.

Capra, R. 1982. The turning point. Simon and Schuster, New York, New York, USA.

Castillo, A., and V. M. Toledo. 2000. Applying ecology in the third world: the case of Mexico. BioScience 50: 66–76.

Chasan, D. J. 2000. The rusted shield: government's failure to enforce—or obey—our system of environmental law threatens the recovery of Puget Sound's wild salmon. The Bullitt Foundation, Seattle, Washington, USA.

Cherrett, J. M. 1988. Ecological concepts: a survey of the views of the members of the British Ecological Society. Biologist 35: 64–66.

————. 1989. Key concepts: the results of a survey of our members' opinions. Pages 1–16 in J. M. Cherrett, A. D. Bradshaw, F. B. Goldsmith, P. J. Grubb, and J. R. Krebs, editors. Ecological concepts: the contribution of ecology to an understanding of the natural world. Blackwell Scientific Publications, Oxford, UK.

Chu, E. W., and J. R. Karr. 2001. Environmental impact: concept and measurement of. Pages 557–577 in S. A. Levin, editor. Encyclopedia of biodiversity, vol. 2. Academic Press, Orlando, Florida, USA.

Costanza, R., R. d'Arge, R. de Groot, R. Farber, M. Grasso, B. Hannon, K. Limburg, S. Naeem, R. V. O'Neill, J. Paruelo, R. G. Raskin, P. Sutton, and M. Belt. 1997. The value of the world's ecosystem services and natural capital. Nature 387: 253–260.

Cowles, H. C. 1904. The work of the year 1903 in ecology. Science 19: 879–885.

Daily, G. C., ed. 1997. Nature's services: societal dependence on natural ecosystems. Island Press, Washington, D.C., USA.

Dasmann, R. F., J. P. Milton, and P. H. Freeman. 1973. Ecological principles for economic development. John Wiley and Sons, New York, New York, USA.

Davis, W. S., and T. P. Simon, eds. 1995. Biological assessment and criteria: tools for water resource planning and decision Making. Lewis, Boca Raton, Forida, USA.

Dunlap, R. E. 1980. Paradigmatic change in social sciences. American Behavioral Scientist 24: 5–14.

Ehrlich, P. R. 1997. A world of wounds: ecologists and the human dilemma. Ecology Institute, Oldendorf/Luhe, Germany.

Ganasan, V., and R. M. Hughes. 1998. Application of an index of biological integrity (IBI) to fish assemblages of the rivers Khan and Kshipra (Madhya Pradesh), India. Freshwater Biology. 40: 367–383.

Hannah, L., D. Lohse, C. Hutchinson, J. L. Carr, and A. Lankerani. 1994. A preliminary inventory of human disturbance of world ecosystems. Ambio 23: 246–250.

Hawken, P., A. Lovins, and L. H. Lovins. 1999. Natural capitalism: creating the next industrial revolution. Little, Brown, Boston, Massachusetts, USA.

Holling, C. S. 1996. Two cultures of ecology. Conservation Ecology <www.consecol.org/Journal/editorial/editorial/html.>

Hugueny, B., S. Camara, B. Samoura, and M. Magassouba. 1996. Applying an index of biotic integrity based on fish communities in a west African river. Hydrobiologia **331**: 71–78.

Huxley, T. H. 1868. Science and education. Reprinted 1900. P. F. Collier, New York, New York, USA.

Jacobs, J. 2000. The nature of economies. The Modern Library, New York, New York, USA.

Karr, J. R. 1991. Biological integrity: a long-neglected aspect of water resource management. Ecological Applications **1**: 66–84.

————. 1993. Protecting ecological integrity: an urgent societal goal. Yale Journal of International Law **18**: 297–306.

————. 1995. Clean water is not enough. Illahee **11**: 51–59.

————. 1997. The future is now: biological monitoring to ensure healthy waters. Pages 31–36 in Streamkeepers: aquatic insects as biomonitors. Xerces Society, Portland, Oregon, USA.

————. 1998. Biological integrity: a long-neglected aspect of environmental program evaluation. Pages 148–175 in G. J. Knaap and T. J Kim, editors. Environmental program evaluation: a primer. University of Illinois Press, Urbana, Illinois, USA.

————. In press. Protecting life: weaving together environment, people, and law. In R. G. Stahl, Jr., A. Barton, J. R. Clark, P. deFur, S. Ells, C. A. Pittinger, M. W. Slimak, and R. S. Wentsel, editors. Risk management: ecological risk-based decision making. SETAC Press.

Karr, J. R., and E. W. Chu. 1995. Ecological integrity: reclaiming lost connections. Pages 34–48 in L. Westra and J. Lemons, editors. Perspectives on ecological integrity. Kluwer Academic Publishers, Dordrecht, The Netherlands.

————. 1999. Restoring life in running waters: better biological monitoring. Island Press, Washington, D.C., USA.

————. 2000. Sustaining living rivers. Hydrobiologia **422/433**: 1–14.

————. In review. Academia and our water resources. BioScience.

Karr, J. R., K. D. Fausch, P. L. Angermeier, P. R. Yant, and I. J. Schlosser. 1986. Assessment of biological integrity in running waters: a method and its rationale. Special Publication No. 5. Illinois Natural History Survey, Champaign, Illinois, USA.

Karr, J. R., R. C. Heidinger, and E. H. Helmer. 1985. Effects of chlorine and ammonia from wastewater treatment facilities on biotic integrity. Journal of the Water Pollution Control Federation **57**: 912–915.

Keeler, A. G., and D. McLemore. 1996. The value of incorporating bioindicators in economic approaches to water pollution control. Ecological Economics **19**: 237–245.

Kellert, S. R. 1996. The value of life: biological diversity and human society. Island Press, Washington D.C., USA.

Kellert, S. R., and E. O. Wilson, eds. 1993. Biophilia. Island Press, Washington, D.C., USA.

Kimberling, D. N., J. R. Karr, and L. S. Fore. In review. Measuring human disturbance using terrestrial invertebrates in shrub-steppe of eastern Washington (USA).
Ecological Indicators.

Klemow, K. M. 1991. Basic ecological literacy: a first cut. ESA Education Section
Newsletter 2: 4–5. Reprinted: ESA Education Newsletter. Fall 1997, pages 2–3.

Larson, M. G., D. B. Booth, S. M. Morley. In press. Effectiveness of large woody
debris in stream rehabilitation projects in urban basins. Ecological Engineering.

Lee, K. N. 1994. Compass and gyroscope. Island Press, Washington, D.C., USA.

Leopold, A. 1966. A Sand County almanac, with essays from Round River. Ballantine Books, New York, New York, USA.

Lichatowich, J. 1999. Salmon without rivers: a history of the Pacific salmon crisis.
Island Press, Washington, D.C., USA.

Lubchenco, J. 1998. Entering the century of the environment: a new social contract
for science. Science 279: 491–497.

Lubchenco, J., A. M. Olson, L. B. Brubaker, S. R. Carpenter, M. M. Holland, S. P.
Hubbell, S. A. Levin, J. A. MacMahon, P. A. Matson, J. M. Mellilo, H. A.
Mooney, C. H. Peterson, H. R. Pulliam, L. A. Real, P. J. Regal, and P. G. Risser.
1991. The sustainable biosphere initiative: an ecological research agenda. Ecology
72: 371–412.

Lyons, J., S. Navarro-Perez, P. A. Cochran, C. E. Santana, and M. Guzman-Arroyo.
1995. Index of biotic integrity based on fish assemblages for the conservation of
streams and rivers in west-central Mexico. Conservation Biology 9: 569–584.

MacArthur, R. H., and E. O. Wilson. 1967. The theory of island biogeography.
Princeton University Press, Princeton, New Jersey, USA.

Maienschein, J. 1998. Scientific literacy. Science 281: 917.

Marsh, G. P. 1865. Man and nature; or, physical geography as modified by human
nature. Charles Scribner, New York, New York, USA.

McHarg, I. 1969. Design with nature. Doubleday, Garden City, New York. 1992,
Second edition. Wiley Press, New York, New York, USA.

McHarg, I. L., and F. R. Steiner, eds. 1998. To heal the earth: selected writings of Ian
L. McHarg. Island Press, Washington, D.C., USA.

McIntosh, R. P. 1985. The background of ecology: concept and theory. Cambridge
University Press, Cambridge, UK.

Molles, M. C., Jr. 1999. Ecology: concepts and applications. McGraw-Hill, Boston,
Massachusetts, USA.

Nash, R. 1989. The rights of nature: a history of environmental ethics. University of
Wisconsin Press, Madison, Wisconsin, USA.

Nehlsen, W., J. E. Williams, and J. A. Lichatowich. 1991. Pacific salmon at the cross-
roads: stocks at risk from California, Oregon, Idaho, and Washington. Fisheries
16: 4–21.

Norgaard, R. B. 1994. Development betrayed: the end of progress and a coevolu-
tionary revisioning of the future. Routledge Press, London, UK.

Noss, R. F., E. Dinerstein, B. Gilbert, M. Gilpin, B. J. Miller, J. Terborgh, and S.
Trombulak. 1999. Core areas: where nature begins. Pages 99–128 in M. E. Soulé
and J. Terborgh, editors. Continental conservation: scientific foundations of
regional reserve networks. Island Press, Washington, D.C., USA.

Oberdorff, T., and J. P. Porcher. 1994. An index of biotic integrity to assess biological impacts of salmonid farm effluents on receiving waters. Aquaculture **119**: 219–235.

Odum, E. P. 1992. Great ideas in ecology for the 1990s. BioScience **42**: 542–545.

Olmsted, F. L., Jr., and T. Kimball. 1970. Frederick Law Olmsted, landscape architect, 1822–1903. B. Blom, New York, New York, USA.

Orr, D. W. 1992. Ecological literacy: education and the transition to a postmodern world. State University of New York Press, Albany, New York, USA.

———. 1998. The ecological design arts. Pages 334–335 in G. T. Miller Jr., editor. Living in the environment: principles, connections, and solutions. Tenth edition. Wadsworth, Belmont, California, USA.

Pienkowski, M. W., and A. R. Watkinson. 1996. The application of ecology. Journal of Applied Ecology **33**: 1–4.

Pimentel, D., L. Lach, R. Zuniga, and D. Morrison. 2000. Environmental and economic costs of nonindigenous species in the United States. BioScience **50**: 53–65.

Pimentel, D., C. Wilson, C. McCullum, R. Huang, P. Dwen, J. Flack, Q. Tran, T. Saltman, and B. Cliff. 1997. Economic and environmental benefits of biodiversity. BioScience **47**: 747–757.

Pool, R. 1991. Science literacy: the enemy is us. Science **251**: 266–267.

Prugh, T., R. Costanza, J. H. Cumberland, H. Daly, R. Goodland, and R. B. Norgaard. 1995. Natural capital and human economic survival. ISEE Press, Solomons, Maryland, USA.

Sachs, A. 1995. Humboldt's legacy and the restoration of science. World Watch **8**: 28–38.

Sagoff, M. 1997. Muddle or muddle through? Takings jurisprudence meets the Endangered Species Act. College of William and Mary Law Review **38**: 825–993.

Schneider, S. H. 1997. Defining and teaching environmental literacy. Trends in Ecology and Evolution **12**: 457.

Sears, P. B. 1964. Ecology: a subversive subject. BioScience **14**: 11–13.

Simon, T. P., ed. 1999. Assessing the sustainability and biological integrity of water resource quality using fish communities. CRC Press, Boca Raton, Florida, USA.

Smith, F. E. 1976. Ecology: progress and self-criticism. Science **192**: 546.

Sutton, S. B., ed. 1971. Civilizing American cities: a selection of Frederick Law Olmsted's writings on city landscapes. MIT Press, Cambridge, Massachusetts, USA.

Taylor, J. E., III. 1999. Making salmon: an environmental history of the Northwest fisheries crisis. University of Washington Press, Seattle, Washington, USA.

Thiele, L. P. 1999. Environmentalism for a new millennium: the challenge of coevolution. Oxford University Press, New York, New York, USA.

Thoreau, H. D. 1854. Walden. Reprinted 1950. Harper's Modern Classics, New York, New York, USA.

Thorne, R. St. J., and W. P. Williams. 1997. The response of benthic invertebrates to pollution in developing countries: a multimetric system of bioassessment. Freshwater Biology **37**: 671–686.

Wackernagel, M., and W. E. Rees, 1996. Our ecological footprint: reducing human impact on the earth. New Society Press, Gabriola Island, British Columbia, Canada.

Walters, C. F. 1986. Adaptive management of renewable resources. Macmillan, New York, New York, USA.

Wang, L., J. Lyons, P. Kanehl, and R. Gatti. 1997. Influences of watershed land use on habitat quality and biotic integrity in Wisconsin streams. Fisheries 22: 6–12.

Washington Forest Practices Board (WFPB). 1993. Board manual: standard methodology for conducting watershed analysis. Washington Forest Practices Board, Olympia, Washington, USA.

Westoby, M. 1997. What does "ecology" mean? Trends in Ecology and Evolution 12: 166.

Wilson, E. O. 1984. Biophilia: the human bond with other species. Harvard University Press, Cambridge, Massachusetts, USA.

Woodwell, G. M., ed. 1990. The Earth in transition: patterns and process of biotic impoverishment. Cambridge University Press, Cambridge, UK.

Yoder, C. O., and E. T. Rankin. 1998. The role of biological indicators in a state water quality management process. Environmental Monitoring and Assessment 51: 61–88.

Appendix 6A

Statements of concern about global environmental trends from citizens, scholars, and labor, business, and religious leaders demonstrate that the "environmental movement" comprises not only "environmentalists," but all segments of society.

Citizens

C-1. Gallup Health of the Planet Survey (November 1993). Scientific survey of over 28,000 individuals in twenty-four countries (including industrialized and developing nations) shows "strong public concern for environmental protection throughout the world, including regions where it was assumed to be absent" (Dunlap et al. 1993).

Scholars

S-1. Union of Concerned Scientists, World Scientists' Warning to Humanity (December 1992). "Human beings and the natural world are on a collision course . . . A great change in our stewardship of the earth, and life on it, is required if vast human misery is to be avoided and our global home on this planet is not to be irretrievably mutilated." Signed by a worldwide collection of 1,575 scientists, including 99 Nobel Prize winners (UCS 1992).

S-2. U.S. National Academy of Sciences and Royal Society of London (1992). The first joint statement ever by these two organizations recognizes that, if industrial countries do not make major changes, irreversible damage to the earth's capacity to sustain life is likely to result. "[F]uture of our planet is in the balance. Sustainable development can be achieved but only if irreversible degradation can be halted in time. The next 30 years could be crucial" (Press et al. 1992; Perrin 1992).

S-3. Joint Statement on Population by the World's Scientific Academies (October 1993). Fifty-eight national academies participated in the 1993 Population Summit in New Delhi. Summit convened to explore issues of population growth, resource consumption, socioeconomic development, and environmental protection. Natural and social sciences are crucial for understanding and developing new options for protecting the environment and improving the quality of human life today and for future generations. "Let 1994 be remembered as the year when the people of the world decided to act together for the benefit of future generations" (WSA 1993).

S-4. Ecological Society of America and International Association of Ecology (1991). Sustainable Biosphere Initiative statement by these two organizations notes wide agreement among scientists that the future of planet Earth is at risk. They call for research initiatives to move society toward sustainable use of ecological resources (Lubchenco et al. 1991; Huntley et al. 1991).

S-5. Sigma Xi Conference, "Global Change and the Human Prospect: Issues in Population, Science, Technology and Equity" (November 1991). Focused on the driving forces behind problems such as ozone depletion, global warming, and biotic impoverishment in order to address their root causes (Sigma Xi 1992a).

S-6. Carnegie Commission. Four major reports deal largely or exclusively with environmental issues (Carnegie Commission 1992a, b, c; 1993).

S-7. Committee for the National Institute for the Environment. A grassroots initiative of more than 5,000 scientists and educators and scores of organizations, professional societies, and universities with the mission to "improve the scientific basis for making decisions on environmental issues." Goals include environmental research, assessment of environmental knowledge and its implications, expanded access to environmental information, and a strengthened capacity to address environmental issues by sponsoring higher education and training. The committee was formed in 1990 and seeks to form a national program modeled after the National Institutes of Health (CNIE 1993). The U.S. National Science Foundation announced a new biocomplexity initiative in 1999 that includes many of the components originally advocated by CNIE.

S-8. International Geosphere-Biosphere Programs: A Study of Global Change (June 1992). An international effort to explore the causes and impacts of global climate change with a focus on reducing uncertainties, both in our knowledge and with respect to the consequences of human actions. The June 1992 report synthesizes some ten years of research (IGBP 1992).

Universities

U-1. Talloires Declaration (October 1990). "We, the presidents, rectors, and vice chancellors of universities from all regions of the world are deeply concerned about the unprecedented scale and speed of environmental pollution and degradation, and the depletion of natural resources." They agree on twelve actions to be taken by their institutions to respond to this urgent challenge. Originally signed by the presidents of 20 world universities, the declaration was signed by 125 university presidents from 32 countries by 1992, including all the state universities in Virginia. In addition, the Conference of European Rectors (representing 490 university heads) endorsed the declarations principles (Talloires Declaration 1990; see Cortese 1993).

U-2. New Perspectives on Environmental Research and Education (September 1992). A report on the University Colloquium on Environmental Research and Education, a gathering of academic program directors, administrators, and faculty from fifty-five universities plus business and government leaders. Their interest is the future of environmental studies at the nation's universities. The colloquium's agenda concentrated on the need to nurture interdisciplinary environmental research and education within the traditional departmental structure of those institutions (Sigma Xi 1992b).

U-3. U.S. Universities. A number of major universities in the United States, including California (Berkeley, Davis), Colorado State, Duke, Florida, Illinois, Michigan, Tufts, Washington, Wisconsin, and Yale, have recently completed or are working on major initiatives to reframe their approaches to environmental issues in teaching, research, and outreach programs. More needs to be done (Cortese 1991; Orr 1992; Smith 1992; Bowers 1993).

U-4. University Responsibility. Despite major initiatives on a number of campuses, many ask why so little is done "compared with what environmental-studies faculty members think needs to be done" (Perrin 1992). Also see Eagan and Orr (1992).

U-5. Magna Carta of European Universities. (September 1998). "[U]niversities must give future generations education and training that will teach them, and through them, others to respect the great harmonies of their natural environment and of life itself" (Leal Filho 2000; van Weenen 2000).

Governments

G-1. U.N. Conference on Environment and Development: The Earth Summit (June 1992). An unprecedented gathering of 170 nations (the official conference and largest meeting ever of world leaders) and grassroots organizations (the unofficial conference) to explore international dimensions of environmental issues and to define steps necessary to run our economies and secure our future (Centre for Our Common Future 1993).

G-2. Committee on Life Sciences, National Research Council. A recent report, *Research to Protect, Restore, and Manage the Environment,* issued by the National Research Council (June 1993) concludes that science and engineering provide many "tools to address environmental problems of enormous consequence to our social and economic well-being. But we are not using those tools most effectively" (NRC 1993).

Business

B-1. Business Council for Sustainable Development (1992). Forty-eight international industrialists and business leaders from more than twenty-five countries call for renewed efforts by business and government to make ecological imperatives part of the market forces that govern production, investment, and trade (Schmidheiny 1992).

B-2. *Sunday Times* (London) (October 21, 1990). "Sir James Goldsmith—corporate predator extraordinaire, scourge of board rooms, one of the most feared men on Wall Street—[is] retiring from business. From now on, he said, he would devote his energies and much of his fortune of more than $1 billion to ecological and environmental causes." Great wealth, the fifty-seven-year-old billionaire argued, was of no value in a crumbling world (Fallow 1997).

B-3. *Science* (September 27, 1992). Column in the weekly journal of the American Association for the Advancement of Science entitled "The Greening of 'Green Science' Means New Jobs" notes an increase in jobs and business commitment to respond to the environmental perils facing the earth, but that universities are still behind the times because change comes slowly in academia (Holden 1992).

B-4. Business Environmental Leadership Council (1999). Formed by twenty-one major corporations to address climate change issues. "Businesses can and should take concrete steps now in the U.S. and abroad to assess opportunities for emissions redcutions, . . . and invest in new more efficient products, practices, and technologies." William Clay Ford Jr., chairman of Ford, says, "I want Ford Motor Co. to be a leader in the second industrial revolution—the clean revolution" (Knickerbocker 2000; <www.pewclimate.org/belc/index.html>).

Labor

L-1. United Steelworkers of America (1990). The report *Our Children's World: Steelworkers and the Environment* was overwhelmingly endorsed by

2,163 delegates to the annual USWA convention. The report notes, "We cannot protect steelworker jobs by ignoring environmental problems." Further, the "greatest threat to our children's future may lie in the destruction of their environment" and "the environment outside the workplace is only an extension of the environment inside" (USWA 1990).

Religions

R-1. Parliament of World's Religions (August 1993). Leaders of Christianity, Buddhism, Islam, Judaism, Hinduism, and other faiths have drawn up a global ethic. Among other things, it condemns environmental abuses, lamenting that in an age of unparalleled technological progress, poverty, hunger, the death of children "and the destruction of nature have not diminished but rather have increased" (Briggs 1993).

R-2. His Holiness Bartholomew I (November 1997). Spiritual leader of the world's 300 million Orthodox Christians, speaking at a symposium on religion, science, and environment in Santa Barbara, California, said, "For humans to cause species to become extinct and to destroy the biological diversity of God's creation, for humans to degrade the integrity of the Earth by causing changes in climate, stripping the Earth of its natural forests, or destroying its wetlands . . . these are sins." (Stammer 1997, text of speech at <http://home.goarch.org/patriarchate/us-visit/speeches/Address_at_Environmenta.htm>

R-3. Pope John Paul II (1989). "The seriousness of ecological degradation lays bare the depth of man's moral crisis" (Wilkinson 1999). "An education in ecological responsibility is urgent. . . . [T]he ecological crisis is a moral issue" (Pope John Paul II 1990).

R-4. National Religious Partnership for the Environment (1998). The 271 signatories represent 116 religious leaders from North America, 93 from the Soviet Union, 27 from Europe, and 35 from Africa, Latin America, India, and the Far East. Of the 271 signatories, 181 attended the Global Forum in Moscow. "Many in the religious community have followed, with growing alarm, reports of threats to the well-being of our planet's environment. . . . We believe the environmental crisis is intrinsically religious. All faith traditions and teachings firmly instruct us to revere and care for the natural world. Yet sacred Creation is being violated and is in ultimate jeopardy as a result of long-standing human behavior. A religious response is essential to reverse such long-standing patterns of neglect and exploitation" (New World Dialogue 1991).

Other

O-1. The Morelia Declaration (1991). An international group of 100 scientists, writers, representatives of native tribes, and political activists express "concern that life on our planet is in grave danger." They urged the leaders of the world to commit themselves to ending ecocide and ethnocide and to create an International Court of the Environment modeled on the International Court of Justice at The Hague (Group of 100 1991).

O-2. Compact for a New World (October 1991). An open letter to heads of state and government and legislators of the Americas. More than two dozen political leaders, businesspersons, and scholars from throughout the Americas are "alarmed by a decade of accelerating environmental damage and rising poverty." They appeal to political leaders to "forge the international initiatives and agreements necessary for lasting prosperity and environmental protection in the Americas." Signatories included U.S. Vice President Al Gore, Senator John H. Chaffee, and a number of university presidents (New World Dialogue 1991).

O-3. Earth Rights and Responsibilities: Human Rights and Environmental Protection (June 1992). A special conference held at Yale Law School and sponsored by the Yale Law School, American Association for the Advancement of Science, and other organizations. A multidisciplinary discussion on the future of the planet (YJIL 1993; Yale Law School 1992).

Citations for Appendix 6A

Bowers, C. A. 1993. Education, cultural myths, and the ecological crisis. State University of New York Press, Albany, New York, USA. [U-3]

Briggs, D. 1993. World's clerics draft global ethic: violence, sexism, environmental abuse are all targeted. Seattle Times, September 1, 1993. [R-1]

Carnegie Commission. 1992a. International environmental research and assessment: proposals for better organization and decision making. Report, Carnegie Commission of Science, Technology, and Government, New York, New York, USA. 82 pp. [S-6]

———. 1992b. Partnerships for global development: the clearing horizon. Report, Carnegie Commission of Science, Technology, and Government, New York, 129 pp. [S-6]

———. 1992c. Environmental research and development: strengthening the federal infrastructure. Report, Carnegie Commission of Science, Technology, and Government, New York, 143 pp. [S-6]

———. 1993. Science, technology, and government for a changing world. Concluding Report, Carnegie Commission of Science, Technology, and Government, New York, 94 pp. [S-6]

Centre for Our Common Future. 1993. The Earth Summit's agenda for change: a plain language version of agenda 21 and the other Rio agreements. Centre for Our Common Future, Geneva, Switzerland. 70 pp. [G-1]

Chu, E. W., and J. R. Karr. 2001. Environmental impact: concept and measurement of. Pages 557–577 in S. A. Levin, editor. Encyclopedia of biodiversity, vol. 2. Academic Press, Orlando, Florida, USA.

Committee for the National Institute for the Environment (CNIE). 1993. A Proposal for a National Institute for the Environment: need, rationale, and structure. CNIE, Washington, D.C., USA. 99 pp. [S-7]

Cortese, A. D. 1991. Training professionals: toward environmental responsibility. EPA Journal. Sept./Oct. 1991: 31–34. [U-3]

———. 1993. Building the intellectual capacity for a sustainable future: talloires and beyond. Pages 1–9 in B. Wallace, J. Cairns Jr., and P. A. Distler, editors. Environmental Literacy and Beyond. President's Symposium, Vol. V. Virginia Polytechnic and State University, Blacksburg, Virginia, USA. [U-1]

Dunlap, R. E., G. H. Gallup, Jr., and A. M. Gallup. 1993. Of global concern: results of the Health of the Planet Survey. Environment 35: 7–15, 33–39. [C-1]

Eagan, D. J. and D. W. Orr, editors. 1992. The campus and environmental responsibility. Josey-Bass, San Francisco, California, USA. [U-4]

Fallon, I. 1990. The jolly green giant. Sunday Times (London). October 21, 1990, pp. 1–15. [B-2]

Group of 100. 1991. The Morelia Declaration. The Morelia Symposium: approaching the year 2000. New York Times. October 10, 1991, p. A19. [O-1]

Holden, C. 1992. The greening of "green science" means new jobs. Science 257: 1730–1731, 1766. [B-3]

Huntley, B. J., et al. 1991. A sustainable biosphere: the global imperative. Ecology International 1991: 20. 14 pp. [S-4]

International Geosphere-Biosphere Program (IGBP). 1992. Global change: reducing uncertainties. International Geosphere-Biosphere Program, Royal Swedish Academy of Sciences, Stockholm, 40 pp. [S-8]

Knickerbocker, B. 2000. Businesses take "greener" stand on global warming. Christian Science Monitor. January 24, 2000, p. 2. [B-4]

Leal Filho, W. 2000. Dealing with misconceptions on the concept of sustainability. International Journal of Sustainability in Higher Education 1: 9–19. [U-5]

Lubchenco, J., et al. 1991. The sustainable biosphere initiative: an ecological research agenda. Ecology 72: 371–412. [S-4]

Lubchenco, J., A. M. Olson, L. B. Brubaker, S. R. Carpenter, M. M. Holland, S. P. Hubbell, S. A. Levin, J. A. MacMahon, P. A. Matson, J. M. Mellilo, H. A. Mooney, C. H. Peterson, H. R. Pulliam, L. A. Real, P. J. Regal, and P. G. Risser, 1991. The sustainable biosphere initiative: an ecological research agenda. Ecology 72: 371–412. [S-4]

National Research Council NRC. 1993. Research to protect, restore, and manage the environment. Committee on Environmental Research, Commission on Life Sciences. National Academy Press, Washington, D.C., 242 pp. [G-2]

New World Dialogue on Environment and Development in the Western Hemisphere. 1991. Compact for a new world: an open letter to the heads of state and government and legislators of the Americas. World Resources Institute, Washington, D.C., USA. 26 pp. <www.nrpe.org/openletter.html>. [O-2]

Orr, D. W. 1992. Ecological literacy: education and the transition to a post-modern world. State University of New York Press, Albany New York, USA. [U-3]

Perrin, N. 1992. Colleges are doing pitifully little to protect the environment. The Chronicle of Higher Education. October 28, 1992, pp. B3–B4. [S-2, U-4]

Pope John Paul II. 1990. The ecological crisis: a common responsibility. Message of His Holiness Pope John Paul II for the Celebration of the World Day of Peace. 1 January 1990. <http://www.nccbuscc.org/opps/johnpaulii.htm> [R-1]

Press, F., et al. 1992. Population growth, resource consumption and sustainable world. The Royal Society of London and U.S. National Academy of Sciences, London, UK. [S-2]

Schmidheiny, S. 1992. Changing course: a global business perspective on development and the environment. MIT Press, Cambridge, Massachusetts, USA. 374 pp. [B-1]

Sigma Xi. 1992a. Global change and the human prospect: issues in population, science, technology and equity. Sigma Xi Forum, November 16–18, 1991. Sigma Xi, Research Triangle Park, North Carolina, USA. 294 pp. [S-5]

———. 1992b. New perspectives on environmental education and research: a report on the University Colloquium on Environmental Education and Research. September 24–26, 1992. North Carolina State University, Raleigh, North Carolina. Sigma Xi, Research Triangle Park, North Carolina, USA. 58 pp. [U-2]

Smith, G. A. 1992. Education and the environment: learning to live with limits. State University of New York Press, Albany, New York, USA. [U-3]

Stammer, L. B. 1997. Harming the environment is sinful, prelate says. Los Angeles Times, November 9, 1997, p. A-3. [R-2]

Union of Concerned Scientists (UCS). 1992. World scientists' warning to humanity. UCS, Cambridge, Massachusetts, USA. [S-1]

United Steelworkers of America (USWA). 1990. Our children's world: steelworkers and the environment. Report of the USWA Task Force on Environment, United Steelworkers of America, Pittsburgh, Pennsylvania, USA. 34 pp. [L-1]

van Weenen, H. 2000. Towards a vision of a sustainable university. International Journal of Sustainability in Higher Education 1: 20–34. [U-5]

Wilkinson, T. 1999. Recruiting in pews to save planet. Christian Science Monitor. (December 23, 1999), pp. 1, 4. [R-3]

World's Scientific Academies (WSA). 1993. Population summit of the world's scientific academies. National Academy of Sciences, National Research Council, Washington, D.C., USA. 16 pp. [S-3]

Yale Journal of International Law (YJIL). 1993. Symposium. Earth rights and responsibilities: human rights and environmental protection. YJIL 18: 213–411. [O-3]

Yale Law School. 1992. Earth rights and responsibilities: human rights and environmental protection. Conference Report. April. Yale Law School, New Haven, Connecticut, USA. [O-3]

CHAPTER 7

Toward an Inclusive
Concept of Infrastructure

William E. Wenk

McMillan Reservoir Park is a nearly forgotten 25-acre spot in the Northeast
quadrant of Washington, D.C. Its reservoir once supplied much of the city's
drinking water. Before delivery, water was fed by gravity into twenty sand-
filled underground vaults that filtered out impurities. A deck above the vaults
featured a formal park, green with plants, designed in 1904 by Frederick Law
Olmsted Jr. Even the aboveground sand-storage towers at this reservoir were
stunning examples of industrial design.

Until 1941, the reservoir was used for recreation that included boating.
With the advent of World War II, officials became concerned that saboteurs
would poison the water. The park and reservoir were closed and fenced off.
In the 1980s, the sand filtration plant itself was retired. Various redevelopment schemes have failed. Today, the site remains empty and fenced-off. Yet
even as an urban ruin, the McMillan Reservoir remains an excellent example
of a cultivated public landscape that served as vital public infrastructure.

At one time, architects, landscape architects, engineers, and artists worked
in concert to fashion the "public realm," including streets, parks, riverfronts,
water-delivery systems, and drainage ways. When these collaborations were
successful (most frequently during the turn-of-the-century City Beautiful
movement), they created new types of landscapes: places that combined high
civic function with recreation, high art, and in some cases, natural areas with
wildlife habitat.

These were stunning landscapes of every scale that are not being replicated today—and not for reasons of budget alone. Rather, these civic landscapes were born from a collaborative professional mind-set that no longer
exists. As a result of these collaborations, these landscapes were multifunc-

tional, created through the application of nineteenth-century technologies that worked with natural processes.

Natural Technologies

In the twentieth century, we—meaning the average person as well as the designer and ecologist—tended to deride any landscape touched by humans as "despoiled." Indeed, in our current model for development we scrape and regrade the landscape to prepare for large-lot suburbs, overly wide roads, and commercial buildings surrounded by asphalt. This model creates an unflattering public image for developers, designers, and planners. It also suggests strongly that "stewardship" is not the foundation of our professional ethics, in spite of claims made by our professional societies. No wonder so many people mistrust human alteration of the land. Our most ubiquitous model of development really does represent a despoiled landscape, designed with no recognition of ongoing natural processes.

Yet there are many historical precedents for a more benevolent shaping of the land that use "natural technologies," such as the combination of gravity and the natural material of sand to cleanse water stored at the McMillan Reservoir. Natural technology is a concept that has deep historical roots, much deeper than nineteenth-century engineering. In what is now metropolitan Phoenix, the Hohokam built 1,250 miles of gravity-fed irrigation canals across the Salt River Valley. They recycled water in and out of the river, rather than damming and depleting this basic resource. Their systems lasted more than 1,500 years (from about 100 B.C. to A.D. 1450) and later became the basis for the ditch irrigation systems still used in Arizona. Gravity-fed systems of irrigation ditches and canals are still prevalent throughout the arid American West.

The key to defining natural technologies is in their ability to make use of natural processes, such as water flow or sand filtration, in such a way that the resource they manipulate is both conserved and used to create multiple benefits. Resources such as water or wildlife can flow through both built and natural systems; the hard part is understanding enough about those flows to work with them instead of against them in our built infrastructure. Building *with* natural flows and processes often has the added benefit of allowing built systems to make more efficient use of energy and materials. Properly applied, natural technologies can create a cultivated ecological landscape that provides high-quality human habitat while minimizing impacts on natural ecosystems. Anyone who walks or bikes along an intact section of Colorado's historic Highline Canal, for example, will find a ribbon of green among the sere prairies. Nourished by water, the canal area teems with wildlife and is shaded by huge cottonwoods. The infrastructure of a water-delivery system becomes an integrated landscape with places for habitat and recreation.

Today, public utilities are less likely to invite the public in to enjoy multiple uses of an infrastructure system or facility. Because of concerns about safety, liability, or vandalism, we see fenced-off reservoirs and ditchways. Sewage plants and garbage dumps are often off-limits. We move our water (including our creeks and the stormwater runoff that feeds them) through underground culverts. We rarely hear design professionals or members of the public express the idea of using natural technology to design our infrastructure. Our modern drinking water is treated with chemicals, not filtered through sand. Waters like those the Hohokam treasured are hidden from view, buried beneath the ground in pipes or concealed in concrete channels where we are less likely to notice their condition or understand their potential.

The Impact of Divisions among the Design Professions

Just as we have segregated ourselves from beneficial natural forces, so we have segregated our disciplines. It is now rare to find productive collaborations among the engineering and design and planning professions.

This essay proposes that both the disciplines that are involved in landscape development and the functions of landscapes can be reunited, but only when we begin to view public infrastructure and the infrastructure of nature as one. Until then, we will be restricted by our own incomplete vision of the metropolis. We will continue to import small, isolated (and basically inconsequential) "slices" of nature into the city. Conversely, we will conduct our planning at an overly broad, macroregional scale that misses opportunities to provide urban services, wildlife habitat, and recreation in the same setting because it ignores materiality and the specificity of place.

In order to understand the dynamics of this separation, it is helpful to explore the split between planning and landscape architecture that occurred earlier in the century. Frederick Law Olmsted Sr. and his contemporaries worked closely with other disciplines to shape American cities as they grew rapidly after the Civil War. All the professions involved were broadly educated in the arts and sciences, and were committed to improving quality of life in cities. The scale and scope of influence of our urban bureaucracies was more limited. Perhaps because the engineering and design professions were relatively new, there was freedom to explore new models for addressing sanitation, stormwater control, transportation, and public open space. Teams of designers, planners, and engineers developed integrated solutions.

In those solutions, they expressed an antiurban bias (which we share today) that resulted in the design of pastoral-looking open spaces and natural areas, such as the meadows, glades, and woodlands of Central Park. But they also created urbane spaces, such as City Beautiful street plans and plazas that enrich the social interactions and beauty of many cities. Parks were often part of a larger system that integrated infrastructure to encourage human use.

In the 1920s, city planning evolved as a separate academic discipline that had once been part of landscape architecture. The increasing tendency in the past century to value specialization in both the academies and in private practice contributed to the split between planning and landscape architecture. Planners such as Ian McHarg and Phil Lewis expressed a regional view, concerning themselves with nature and socioeconomic patterns on a broad scale. They collaborated with sociologists, economists, and natural scientists—not with designers. From this context, planners proposed systems that integrated human and natural landscape patterns on a regional scale.

Although these concepts were sound, they failed to address such realities as fragmented land use and ownership, the limited scope of most design projects, or other political realities. In at least one instance, at The Woodlands, even some significant natural patterns were not recognizable at the site scale, and therefore were impossible to use in guiding site-scale design decisions (Claus 1994). Ecologically ambitious projects that were actually built, such as The Woodlands, remain anomalies that have not been widely replicated.

The specialization we see in the planning, engineering, and design disciplines is also reflected in city bureaucracies, where responsibilities are frequently split among departments that rarely communicate. Public works and parks departments often compete for public attention and dollars. This limits or precludes the collaboration needed to develop multifunctional, environmentally sound infrastructure.

Designers must become reinvolved in the design of urban infrastructure at the scale of the site and the system. The design of systems requires a pragmatic approach to the development of components that when linked together create systems that address engineering, environmental, and civic design issues. When other design professions participate knowledgeably, they can help develop a palette of design responses and types. For example, we can rethink all components of urban stormwater systems, from individual storm drain inlets to trunk storm sewers, to create surface stormwater systems that are functional and beautiful. We can accommodate both naturalistic and formal expressions and the use of native and nonnative plant species. We can enrich leftover spaces such as the edges of parking lots, which can become wonderful wetland or conventionally planted environments. We can build on the scientific research and engineering talents of related professionals to create landscapes that function in specific, quantifiable ways, and that are integral with the fabric of the city.

In recent decades, a number of exemplary designers and planners have tried to close the distance between the natural and urban worlds. But we lack the proper tools. We become passionate about preserving wilderness while writing off urban landscapes because they are "defiled" and no longer "pristine." This is a prudish and even Victorian attitude toward nature that needs to evolve. William Cronon (1995) has described our culture's conflicted views of wilderness and our desire to return to a natural ideal that has never existed. Indeed, the concepts of wilderness and nature are cultural constructs that affect how we evaluate and design landscapes (see Chapter 2).

The Problem of Scale in Design with Natural Systems

The impulse to seek out wild nature also leads us to import nature into the city in "boutique" quantities. The result is small, decorative, native-plant or xeric gardens that may enrich some small corner of a city or suburb. But this approach fails to integrate nature and culture in a more meaningful way.

Such miniaturized landscapes have little to do with the land, ecosystems, or for that matter, the infrastructure around them. In these instances, we have become too focused on site design while ignoring the full picture of the urban environment and the water flows, wildlife patterns, and plantings that sustain it.

In the macrocosm, we have made major advances in the study of regional ecosystems. In the 1960s, landscape architects Ian McHarg and Phil Lewis put forth pioneering theories of ecological planning on a regional scale. In his landmark book, *Design with Nature*, McHarg (1969) described how a more environmentally responsible practice of planning and design could address many of the challenges facing the industrialized world. He described a bold vision that could transform the landscape, and envisioned profound changes in how we use the land at a regional scale.

Unfortunately, most change occurs at a much smaller scale, incrementally over time, not in the sweeping changes of a planned development. Although changes in policy and practice at a regional scale are essential to address many environmental problems, it is equally important to practice environmentally responsible design at a site scale, where changes can accumulate to produce landscape-scale patterns over time. This is the scale at which most landscape architects, engineers, and other designers practice— and it is also the scale that must respond to and implement new environmental policies. McHarg offered little direction on how to effectively integrate the cultural and the natural within a single, specific site where property boundaries have already been drawn without regard for larger-scale landscape patterns.

There are problems with the approaches favored by McHarg and Lewis in application as well as in theory. Both landscape architects proposed an approach in which a designer or planner would study a region's ecological attributes (including the spatial patterns of waterways, wetlands, wildlife habitat, and important landforms) and then build in places that would do the least harm to these resources. These powerful and original ideas continue to inspire much regional planning, as well as the design of large-scale development.

But these projects can exhibit major failings in site design, in exactly the reverse situation of the "boutique" native gardens that are too narrowly focused on their sites. In one example, I am aware of a neighborhood creek at a McHarg-planned project that has been engineered to have an 8-foot vertical drop. (This example can be found at The Woodlands in Houston, Texas, frequently cited as one of the most influential ecological planning projects of the 1970s. See Chapter 4.) This may provide aeration that improves water quality, but it is unsightly, visually boring, and dangerous. Similarly, open drainage swales (considered an ecological "improvement" over buried drainage culverts) contribute little to local biodiversity when they are simply ditches planted with bluegrass. If planted with native species, they might at least add to the wildlife habitat available in the development.

In *The Granite Garden*, Anne Whiston Spirn (1984), a student and protégé of McHarg, described cultured landscapes that were both environmentally responsible and integral to our daily lives. Unfortunately, the gap between cultured landscapes and environmentally responsible "natural" landscapes remains. With a few exceptions, the imperatives of landscape maintenance practices, regulatory agencies, and client and public expectations for cultured urban landscapes have discouraged innovations.

We can fault both practitioners and educators for failing to deal with the reality of implementing McHarg's vision at a site scale in a way that incorporates formal design traditions. In thirty years since the publication of *Design with Nature*, our professions have still largely failed to integrate environmental concerns into the mainstream of practice. We need a new approach that combines ecological regional planning with ecological site design, or, at the least, design for public use. For this to occur, landscape architects, urban designers, and planners must reclaim their historic roles as shapers of urban infrastructure, a role we have abdicated to the engineering disciplines.

Design and planning professions have largely ignored the possibility of creating landscapes as living instruments that address urban environmental issues. Landscape architects have failed to integrate their professional legacy of creating artful, cultured landscapes with environmental awareness and expertise. Engineers, in turn, have limited their roles to an excessively narrow focus (for example, flood control or water quality) that excludes design sensibilities. This mentality also often excludes the public from using public landscapes, since these landscapes are often defined as serving only one functional purpose. In focusing on narrow measures of "success" such as containment of a 2- or 100-year flood event, or quantifiable reduction in pollutants, engineers have missed the opportunity to rethink the design of infrastructure systems in light of new urban patterns. In particular, they have ignored the transformation of the city from a dense urban fabric into more loosely structured suburban and ex-urban patterns. Their work often focuses on compliance with federal, state, and local environmental regulations, resulting in clumsy attempts at integrating environmentally sound projects humanely into the urban fabric. A project such as The Woodlands may function well at a regional or project scale, but fail at a site scale because the design was left to the devices of design engineers without the involvement of landscape architects or ecologists. The result frequently is a single-use project that adds little value to the community other than to minimize engineering costs. The ecological value, other than to improve water quality, is certainly questionable.

In reaction to this, I propose an approach to addressing environmental issues that embraces current approaches to public works, draws from native and cultured landscapes for inspiration, and creates humane, artful landscapes that are environmentally sound and functional. To affect the form of our cities, landscape architects, planners, engineers, and allied designers must grasp the intrinsic value of infrastructure and the theories and history of public works. By understanding and appreciating the functional aspects of these systems, and by contributing a knowledge of natural system functions, designers can again give form to urban infrastructure. This will allow us to become involved at both a site and a system scale.

Steps Designers Can Take to Recognize Opportunities in Their Projects

Urban systems in the public realm are primarily controlled by public works departments, and secondarily by the citizens that they serve. In cities, design with nature requires design within the context of urban systems. A pragmatic and eclectic approach to ecologically responsible design could provide new models that are both functional and beautiful, using the following strategies.

Look for Design Opportunities within Urban Infrastructure

Most "natural technology" design opportunities originate in problems with urban infrastructure. Much of this work has been focused on urban rivers and streams, urban flooding, federal regulations limiting impacts on urban wetlands, and more recently, regulations requiring the cleanup of non-point-source pollutants. These have become the basis for a significant body of work, including waterfront parks, greenways, promenades, and recreation-based parks. Often, the work designers currently do in this area quickly becomes more about public works and urban systems than about natural systems. To do differently requires us to remember to speak the language of ecologists during our projects, as well as at the outset.

To restore ecological health to natural systems in the city, we must speak the language of engineers as well, by addressing the requirements of those engineers who control urban systems. To be successful, the project must solve the problem or comply with regulations as well as fulfill its primary function as a public works facility. Within those limitations there remains tremendous opportunity to address ecological issues, and to enhance the livability of the city.

In our professional zeal to restore the natural environment, we often ignore the expressive potential of structures that are sometimes required to stabilize urban streams or the road crossings that punctuate them. When designed sensitively with concern for human interaction, engineering structures can enhance the recreational value of a stream corridor as well as provide necessary protection and stability to a stream whose hydrology has changed as a result of urbanization.

In the past century, landscape architectural designers have left the design of these structures to engineers, and in the process have missed an opportunity that has significant creative potential, and that contributes to livability.

Engage Allies and Capitalize on the Needs and Desires of Citizens

Citizens can be a powerful force for change and strong advocates for the natural realm, as can individuals within public agencies. Some of the most interesting, creative, and environmentally responsible work has been realized only because a citizen or public servant has championed a position that has forced changes in engineering standards. Citizens and elected officials are often looking for alternative solutions to problems with urban systems that are more environmentally responsible and integral with urban and natural systems. It is our responsibility as designers to seek out precedents and to demonstrate how they address engineering concerns.

Don't Ignore Political or Administrative Realities

We must understand how political and administrative issues affect the design of urban systems. Most public works departments are administered by engineers who are driven by maintenance concerns. Often there are federal or state mandates, such as floodplain regulations, that must be accommodated. Alternatives to current engineering practices must be made within the context of regulatory mandates and through known solutions that are maintainable. They must protect the public's investment. We must mediate between conflicting interests, and propose solutions in a manner that addresses a broad range of engineering, community design, and environmental issues.

Use Natural Systems in the Context of Solving Urban Problems

The key to this issue may be to bring a landscape architect's perspective to the solution of engineering problems. For example, simple, engineered stormwater channels and retention ponds can be modified to create visually and ecologically rich landscapes. We should think in terms of multifunctional landscapes that meet specific environmental goals. For example, my own design work often includes functional elements that improve water quality and are integral to the urban parks. These functional elements incorporate native and horticultural plant species to address both water quality and aesthetic issues. They are at the same scale as what has been called "boutique" or "pocket" wetlands, yet they genuinely serve the functional purpose of cleaning the first flush of stormwater runoff. They are formal in that they are obviously designed. Their formality welcomes public access, and reinterprets past traditions of park design as part of addressing contemporary ecological issues.

For example, the closure of Lowry Air Force Base in Denver allowed the city to redevelop the 1,900-acre site as a mixed-use community, which is organized around a 300-acre parks and open space network. The spine of the park system will be Westerly Creek, which has two regional stormwater detention ponds on the base to protect downstream areas from flooding. The ponds will become part of the system's open space network, and the smallest of the two will receive the majority of the urban runoff from the new community. Figure 7-1 illustrates the integration of stormwater treatment areas

FIGURE 7-1.
Lowry Park design proposal. A regional detention pond will be modified to incorporate treatment areas for non-point-source pollutants, a more diverse range of native plant communities, and recreational use. The linear path and landform left of center in the illustration are an emergency spillway that must be maintained as part of the detention pond's function. The crescents provide upland prairie to enhance diversity, to provide separation between treatment wetlands and natural areas along the creek, and to provide controlled access to the open space. (Drawing by Bill Wenk.)

and a series of formal park elements into the detention pond open space area in a manner that maintains its original function. Detention, water treatment, and recreational use each read as distinct "layers" of function and use.

Find the Inherent Beauty in the Ordinary

J. B. Jackson taught us to see the intrinsic qualities of the vernacular landscape around us, including the ever-present elements of the city. In *Discovering the Vernacular Landscape* he writes, "Over and over again I have said that the commonplace aspects of the contemporary landscape, the streets and houses and fields and places of work, could teach a great deal not only about American history and American society but about ourselves and how we relate to the world. It is a matter of learning how to see" (Jackson 1984, p. ix). Viewing the world with a critical eye is a skill that all landscape architects should develop.

I would take that idea a step further. The vernacular landscape can be an inspiration for our professional work. As designers, we must train ourselves to see urban infrastructure as an important part of the vernacular landscape. Even the simplest element, such as a storm drain, should not escape our curiosity and attention. It is our responsibility to exert influence on the design of those systems to make them humane and to promote ecological health. We can go even further.

Jackson (1984, p. xii) goes on to say, "For far too long we have told ourselves that the beauty of a landscape was the expression of some transcendent law: the conformity to certain universal esthetic principles or the conformity to certain biological or ecological laws. But this is true only of formal or planned political landscapes. The beauty that we see in the vernacular landscape is the image of our common humanity: hard work, stubborn hope, and mutual forbearance striving to be love. I believe that a landscape which makes these qualities manifest is one that can be called beautiful." The challenge is to integrate the qualities of formal, planned landscapes into a highly functional and ecologically responsible infrastructure.

Recent Projects That Can Be Studied as Successful Cases

In recent years, there has been an encouraging yet limited trend toward genuine collaboration among the design professions. The result is a small number of public landscapes where ecological goals have been integrated with infrastructure needs. Ironically, these projects are often driven by public discontent or concern with aspects of new development, such as loss of open space or odors from an expanded sewage plant. In dealing with empowered citizens, public officials are left with two choices: Move the proposed new development or infrastructure expansion elsewhere or "mitigate" side effects of the project to make it palatable.

In Seattle, the mitigation scenario was chosen in the expansion of the West Point Wastewater Treatment Plant. Built in 1952 on a point jutting into Puget Sound, the plant required upgrade by the 1980s to meet new treatment standards for water pollution. Residents in Seattle's well-to-do Magnolia neighborhood protested loudly. They argued that the plant should be moved elsewhere and that the site should be reclaimed to expand adjacent Discovery Park, a 535-acre natural preserve with forested bluffs overlooking the water.

Faced with the enormous costs and logistics of moving a sewage plant that serves 1.1 million people, public officials devised an alternative that would keep the plant in situ. They proposed depressing the wastewater treatment plant into the earth as much as possible, enveloping it within massive, sculptural concrete walls and berms, and wrapping it on three sides with waterfront trails, dunes, reclaimed marshes and beaches, and a mile-long trail connecting to hiking trails in Discovery Park.

A team of 200 consultants worked on this $578 million project. Along with the requisite engineers, this team included landscape architects, urban designers, plant ecologists, and nursery growers. Through dramatic models and beautiful, easily understood drawings, the public was encouraged to accept this concept. More than 150 mitigating factors were included in the final permit authorizing reconstruction of the plant. Perhaps the most persuasive single one was the $25 million allocated for landscape reconstruction to create an elaborate native-plant garden. Among 2 miles of retaining walls, some 13,000 trees, 51,000 shrubs, 133,000 ground-cover plants, and 100,000 plugs of beach grass were planted according to the instructions of landscape architects Luanne Smith and Ann Bettman. Once inaccessible to the public, the point and its beaches are now fully accessible by trails. Thanks to the sinuous nature of the concrete retaining walls (designed by Angela Danadjieva of Danadjieva & Koenig), the entire landscape is clearly fabricated. With its swooping curves and dramatic landforms, it succeeds as landscape art. Yet it is a self-sustaining ecological landscape (the extensive irrigation system will be disconnected once plantings are established) that provides habitat and recreation. In some ways, it represents the late-twentieth-century realization of McMillan Reservoir.

On the south side of Seattle, the Waterworks Gardens represents another collaboration to create a public landscape on the edge of a sewage treatment plant. Located in Renton, Washington, the 8-acre Waterworks cleanses runoff from 50 acres of paving at the adjacent plant. Using native plants almost exclusively, it achieves this in a highly artistic and symbolic fashion. Stormwater is pumped up to a promontory, where it filters through scalloped ponds set into terraces on a slope. The cleansed water is then released through a marsh for polishing before it reaches Springbrook Creek. A pathway of crushed stone connects a series of "rooms" that include a folk-art-inspired grotto that doubles as a hanging garden of ferns and mosses. The site plan resembles a leaf, symbolically expressing the cleansing power of plants.

The project was conceived by installation artist Lorna Jordan, who had never implement a landscape or garden project. She collaborated with landscape architects Jones & Jones and engineers Brown and Caldwell. The project has been extensively covered in the local press as well as design magazines. In 1996 it received an "outstanding local project" award from the Seattle section of the American Society of Civil Engineers. The plantings at Waterworks have matured beautifully, and the public has discovered that a park next to a sewage plant is a wonderful place to have a wedding ceremony.

In Portland, Oregon, the landscape architecture firm Murase Associates recently collaborated with architects and engineers to create the new Water Pollution Control Laboratory. Seamlessly merged with a neighborhood park, the 6-acre facility conducts research on how contaminants affect water quality. Left to themselves, the engineers may have simply replaced the failing underground storm sewer with a utilitarian pond. The Murase team created sculptural forms. Within a pond shaped like the figure 8, a partial circle and an arc of stone rise from the water's surface to almost meet. The longer arc is a stone-lined flume that filters water drained from a 50-acre urban area. The semicircular stone wall serves as a marker for water level. The pond banks were gently graded and planted with native wetland species along with ornamental grasses (Figure 7-2). Some 900 acres of the Willamette riverfront also has been regraded and reclaimed. The site is not natural, but it employs natural technologies, such as the abilities of plants to remove pollution from water through transpiration. It is clearly engineered, but it also is clearly designed, with smooth symmetrical forms and perfectly laid stone walls. It embraces the interplay of culture while honoring nature. Daniel Winterbottom and William McElroy invented the term "infra-garden" to describe this form of infrastructure that supports ecological and horticultural values (Winterbottom and McElroy 1993).

Arid regions also have abundant opportunities to create a new hybrid infrastructure. In Phoenix, people are looking for ways to revive the ecologi-

FIGURE 7-2.
Stormwater pond and garden. In Portland, Oregon, collaboration among landscape architects, architects, and engineers led to an artistic and functional pond for cleansing stormwater at the new Water Pollution Control Laboratory. (Photo by Chrissie Rowe.)

cal and recreational aspects of irrigation and drainage ditches. As the city grew after 1903, citizens enjoyed ditchways as oases lined by huge cottonwoods. In the 1950s, Phoenix started to view the canals and "lateral" ditches as a safety threat and usurper of precious water. Trees were cut down so workers could maneuver large maintenance machinery to muck out canals. Originally clay-lined and concave, the canals were sprayed with gunnite and made into trapezoidal channels. Many laterals were buried in underground pipes. The Salt River Project (SRP), a local utility that manages the canals for the federal Bureau of Reclamation, began spraying the banks with herbicides to suppress plant growth. Although they remain accessible to cyclists and joggers, the canals today are bleak places that some residents compare to abandoned railroad tracks. The irrigation networks are highly refined systems for collecting, transporting, and distributing water. Urban stormwater systems are efficient collectors, but distribution has only recently been of concern because of the concentration of pollutants in urban stormwater. By bringing the distribution networks of irrigation systems to the city, stormwater can be used to irrigate urban landscapes, and pollutants removed before they enter urban rivers and streams.

In the 1980s, a group of citizens, prodded by the local Junior League of Phoenix, began promoting reclamation of canal rights-of-way for new public uses. Some 65 percent of the canal water is now used within the city of 2.2 million, and not in citrus groves as it was historically. As balance of power in water politics shifts from rural to urban, Phoenix is seriously considering the idea the canals could augment the city's 2,220 acres of parks, or create an off-road network for bicycle transportation. In addition, in the 1990s, the city funded a 1.5-mile demonstration project to replant the edge of the Sunnyslope Canal with desert vegetation and to create "rooms" where visitors can cool off as misters fed by canal water spray sandstone paving, creating a form of natural, evaporative cooling within the canal-side rooms.

At Shop Creek wetlands near Denver, my colleagues and I created drop structures inspired by the geology of the American Southwest. Designed to slow and cleanse floodwaters before they reach a reservoir, these structures resemble crescent-shaped "outcrops" made by layering slabs of native sand combined with Portland cement. They look like natural structures, and are designed to erode like the sandstone from Utah's Canyonlands. Stormwater has added texture and patina to the surfaces. These structures are expected to last 500 years, gaining a sculptural quality over time that represents the geomorphological evolution of the prairie (Figure 7-3).

Only 10 miles away at Wallace Park, we employed stormwater control and cleansing structures that use a completely different aesthetic. Inspired by the work of geometrical painter Piet Mondrian, drop structures are composed of hard-edged concrete surfaces that include rectangles, squares, and notches

FIGURE 7-3.
Shop Creek wetlands. Soil cement structures are hydraulically efficient energy dissipaters and are formed to allow safe access to the water's edge. Wetlands created have contributed to removing 50 percent of the stream's phosphorous load, one of the project's primary goals.

protruding from a basin in a skewed checkerboard pattern. In low water flow, these structures create outdoor seating and tables. While they provide little ecological value, they celebrate the value of an urban stream as a neighborhood resource, and the potential of drainage structures to be both a recreation resource and a functional part of the city's stormwater system (Figure 7-4).

These examples suggest a broad range of approaches to environmentally responsible design. They are excellent models for practice, whether one considers the landscape a functional machine or a sacred body that must be restored. But the first step will still be for designers and planners to engage project opportunities in ways that combine ecological, functional, and aesthetic goals. This approach allows us to become involved in a broader range of urban issues and to more effectively address environmental concerns.

For the design and planning professions to regain their voice and prosper in the urban realm, it is imperative that we look beyond old paradigms. We must reach past the immediate site or policy objective to the system scale, and begin to synthesize and integrate the work of related design and engineering disciplines into new design solutions.

(a)

(b)

FIGURE 7-4.
Wallace Park stormwater control structures. (a) In low water flow, the
structures create outdoor seating and tables used by people. (Photo by
Thorny Lieberman.) (b) During storms, the "dragon's teeth" are engulfed
by water as they dissipate the energy of storm flows. (Photo by Wenk Associates.)

To have an inclusive concept of infrastructure means that urban, natural, and cultural systems become an integrated whole that serves to sustain society physically and spiritually. As good stewards of the land, we can assert our humanity to create landscapes of a cultured naturalness that are ecologically responsible in that they promote a richness and diversity of the natural realm. It is our collective responsibility to define what richness and diversity mean to us all as educators and practitioners, and to articulate an inclusive vision that links policy system and site, and that welcomes a rich and diverse range of expressions on the land.

Citations

Claus, R. 1994. The Woodlands, Texas—a retrospective critique of the principles and implementation of an ecologically planned development. Master's Thesis, Department of Urban Studies and Planning, Massachusetts Institute of Technology, Cambridge, Massachusetts, USA.

Cronon, W. 1995. The trouble with wilderness, or, getting back to the wrong nature. Pages 69–90 in W. Cronon, editor. Uncommon ground: toward reinventing nature. W.W. Norton & Co., New York, New York, USA.

Jackson, J. 1984. Discovering the vernacular landscape. Yale University Press, New Haven, Connecticut, USA, and London, UK.

McHarg, I. 1969. Design with nature. Doubleday, Garden City, New York. Second Edition (1992), Wiley Press, New York, New York, USA.

Spirn, A. W. 1984. The granite garden: urban nature and human design. Basic Books, New York, New York, USA.

Winterbottom, D., and W. McElroy, 1993. Toward a new garden: a model for an emerging 21st century middle landscape. Critiques of Built Works of Landscape Architecture, Louisiana State University School of Landscape Architecture, Volume 7, Fall, 1997.

CHAPTER 8

Human Health and Design:
An Essay in Two Parts

Sandra Steingraber and Kristina Hill

Editors' Note: When we organized the Shire Conference, we felt it was important to include environmental influences on human health as a complement to ecological concepts related to biodiversity. We were fortunate to receive an invited essay from Dr. Sandra Steingraber, poet, author, and ecologist. The birth of her daughter and the success of her book, *Living Downstream*, prevented Dr. Steingraber from joining us at The Shire that year and made it difficult for her to update her essay based on the meeting, as other authors have done. To this end, Kristina Hill has added a second piece that establishes a context for Steingraber's essay within design and planning. In it, she proposes a broad definition of "health" and uses examples from the past, current, and future practice of design and planning to identify the significance of health issues in these fields. Although contemporary medical models often ignore the ecological dimensions of human health and disease, we believe that this is an issue of global importance.

One

Exquisite Communion:
The Body, Landscape, and Toxic Exposures

Sandra Steingraber

At the time of the 1998 Shire Conference, I was thirty-eight years old and pregnant with my first child—a daughter. I am a biologist, an adoptee, and a former cancer patient. All of these parts of my identity persuaded me to undergo amniocentesis—a procedure that, among other things, assesses the chromosomal structure of a developing fetus.

Knowing my professional interest in the subject, the obstetrical technician allowed me to hold and examine the vial of amniotic fluid that she had just withdrawn from my uterus. It was still warm to the touch and radiated a pleasing golden hue—the color of a fine chardonnay. Although I had just finished a book on the topic of human health and the environment, nothing I had researched impressed me as deeply as this quarter-cup of liquid, which represents, quite literally, the water cycle. Amniotic fluid is a distillation of rainwater and groundwater. It contains the sap of apples, the juice of oranges, the tea I drank a few hours earlier, and the milk I poured over my cereal that morning. In short, this beautiful fluid came into my body from the surrounding environment and, in turn, creates the first environment that my unborn child—essentially a marine mammal—lives within and drinks from. (Later ultrasound images show her gulping and swallowing it.)

Body: earth—we exist in exquisite communion. This intimate connection also means that whatever chemical contaminants are found in the rivers and underground aquifers of the surrounding landscape are also found within the miraculous aquarium of a pregnant woman's body. We now know more about the relationships between our bodies and our environments than we ever have before. What we know points to the conclusion that toxicity is present in the mundane landscapes of everyday life, as much as it is present in waste dumps and near incinerator stacks. We know that because our air, food, and water are frequently contaminated with trace amounts of toxic chemicals, all of us suffer from toxic exposures to some degree, regardless of our attempts to limit these exposures by choosing where and how we live. This knowledge should be sufficient to bring us into a new era of chemical regulation, and of urban design and landscape planning. It isn't the same old story we heard in the 1970s; it's a new story, and one that requires our attention in new ways. The news is that we are *all* connected to our environments in physical and biological ways—it isn't just the unlucky residents of Love Canal or Times Beach who are affected by carcinogens. Yet despite all that we know, in many ways we are just beginning to understand the ramifications of this communion between our bodies and the earth, air, and water.

As it turned out, I had sufficient reason to worry about waterborne contaminants in my own environment. At the time I became pregnant, I was teaching at a small college in rural central Illinois in a town that draws its drinking water from a reservoir that is filled by various creeks and streams that traverse some of the most pesticide-intensive agricultural fields in North America. Pesticides and fertilizers are detectable in the finished tap water of this community, and pregnant women are routinely advised by their doctors not to drink the water there.

Then there's the lawn. On a warm Saturday in spring, my husband and I took a drive in the countryside. When we stopped in a small town for gas, I asked Jeff to drop me at the corner upwind from the gas station while he refueled. Hoping to avoid inhaling benzene and other gasoline additives known to be both carcinogenic and teratogenic (birth-defect causing), I found myself standing on a little sidewalk surrounded by tiny plastic flags, each warning the reader to beware of the grass—it had just been sprayed with pesticides. This corner property belonged to the village funeral home; its sweeping lawn was a perfect greensward with not a dandelion in sight. As it had recently rained, rivulets of rainwater ran from it and streamed toward the storm sewer, which eventually drains into one of the tributaries of the nearby Mackinaw River, which itself drains into the Illinois River and then the Mississippi, providing drinking water for tens of thousands of men, children, and women (some of them pregnant) who live downstream.

Ecological Roots

In *Living Downstream*, I advocate that everyone at some point in their lives should go in search of their ecological roots. Just as an awareness of our genealogical roots offers us a sense of heritage and cultural identity, our ecological roots provide a particular appreciation of who we are biologically. Searching for one's ecological roots means asking questions about the physical environment we have grown up within and whose molecules are woven together with the strands of DNA inherited from our genetic ancestors. This search is made easier by a suite of laws codified as the Emergency Planning and Community Right-to-Know Act (EPCRA), which passed the U.S. Congress in 1986 over intense industry opposition.

The linchpin of EPCRA is the Toxics Release Inventory, which requires that certain manufacturers report to the government the total amount of each of some 654 toxic chemicals released each year into air, water, and land. The Environmental Protection Agency (EPA) then makes these data public information and available on the Internet. When I requested paper copies of this information from the EPA for my own home county, I received hundreds of pages of computer printout itemizing the toxic emissions for area industries during the years since 1987, the year this information first became available. Reading through this list was an amazing education, and my return to my hometown to document these toxic sites—some of them within a mile of the house I grew up in—is a journey I recount in *Living Downstream*. For example, in 1991, area industries released 11.1 million pounds of toxic chemicals into air, water, and land. Among the known and suspected carcinogens released were benzene, chromium, formaldehyde, nickel, ethylene, acrylonitrile, butyraldehyde, lindane, and captan.

Using right-to-know laws, I learned that traces of benzene as well as the dry-cleaning fluid perchloroethylene are detectable from time to time in the finished tap water drawn from my hometown drinking water wells. Dry-cleaning fluids have been linked in some studies to bladder cancer, the kind of cancer I was diagnosed with at age twenty. I learned that a pesticide formulator located near my high school has, over the years, released hundreds of pounds of pesticides considered carcinogenic into both the air and the sewer system. The local landfill, I discovered, accepts hazardous waste from as far away as New Jersey. This is the landscape where I grew up. Its constituent molecules are now the molecules of my flesh, bones, fat tissue, blood, and skin. They are currently being woven into the body of my daughter.

Cancer and the Environment: The Evidence

In my research on cancer etiology, I have taken a close look at the connections between cancer and the environment. How much evidence do we have for such a link, and how should we react in the face of it? Essentially, I found no one study that constitutes what we in the scientific community would call absolute proof. I did, however, uncover many well-designed, carefully constructed studies that, all together, tell a consistent story. Each is like a puzzle piece; when placed together, they form a startling picture. It is one, I argue, that we ignore at our peril.

One line of evidence comes from cancer registries, which measure the incidence of cancer in our population. These reveal that non-tobacco-related cancers are rising in incidence—among all age groups, from infants to the elderly, among all ethnicities, and among both sexes. The increase in cancer dates back to World War II, which marks the beginning of an exponential rise in the production, use, and disposal of synthetic chemicals, many of which are known or suspected human carcinogens. Changes in hereditary patterns cannot account for these increases in cancer; we are not developing more tumors because we are suddenly sprouting new cancer genes. These data are all age-adjusted, meaning that they are corrected for the impact of aging. Neither can improved detection account for these increases. We see the most swiftly accelerating rates among those cancers for which there are no effective screening tools. These include childhood cancers—up 10 percent in the last decade alone; testicular cancer among young men, which has tripled in incidence rate since World War II; non-Hodgkin's lymphoma, tripling since 1950; brain cancers among the elderly—up 54 percent in the last two decades. Some of the apparent rise in breast cancer is undoubtedly attributable to the introduction of mammography in the early to mid 1980s. However, statisticians who have audited the breast cancer data believe only about 20–40 percent of the apparent rise in breast cancer over that past three decades can be explained by the increased availability of this screening technology. Breast cancer rates began rising long before mammograms were widely available. Moreover, the two groups of American women suffering from the fastest increases in breast cancer incidence are precisely the two groups *least* served by mammography: African Americans and the elderly.

Another line of evidence comes from computer mapping, which clearly shows that cancer is not a random tragedy. Bladder and breast cancers are highest along the eastern seaboard—from Maine down to Washington, D.C., and around the Great Lakes. These are the two most heavily industrialized regions of the United States. By contrast, non-Hodgkin's lymphoma is highest in the Midwest and Great Plains, corresponding to the region of highest agricultural use of pesticides. Overall, cancer rates are higher in U.S. counties with contaminated groundwater and toxic waste sites.

Interesting patterns exist in both New Jersey and Pennsylvania, for example. In 1984 in Pennsylvania, a cluster of bladder cancers was documented among the inhabitants of Clinton County, where a 46-acre toxic waste site is contaminated with known bladder carcinogens. Similarly, in New Jersey, communities located near toxic waste sites have significantly elevated mortality from stomach and colon cancers. In twenty-one New Jersey counties, risk of breast cancer rises the closer a woman lives to a dump site. Whether the route of exposure is air pollution or water contamination or both, we don't yet know. Interestingly, the clustering of environmental health problems is not limited to cancer. A study published in the journal *Environmental Health*

Perspectives in August 1997 shows unusually low birth weights among new-born infants born between 1971 and 1975 to mothers who lived near the notorious Lipari landfill in southwestern New Jersey. These years represent the landfill's peak period of leaching chemicals into drinking water and air. Birth weights rebounded after the landfill was closed. Elsewhere, in neighboring areas farther from Lipari, birth weights held steady through this same period.

A third line of evidence comes from our own bodies. We know with certainty that a whole kaleidoscope of chemicals linked to cancer exists inside all of us. Pesticide residues, industrial solvents, electrical fluids such as PCBs, and the unintentional by-products of garbage incineration—namely, dioxins and furans—are now detectable in breast milk, body fat, blood serum, semen, umbilical cords, placentas, and even in the fluid surrounding human eggs extracted from women who are undergoing in vitro fertilization. Residues of household pesticides and wood preservatives are commonly found in the urine of American schoolchildren. We do not know with certainty what the cumulative effect of these multiple routes of exposures is, even though we are honing in on the various biological mechanisms by which these chemicals damage genes, disrupt hormones, impair immunity, and otherwise place a healthy cell on the pathway to tumor formation.

A fourth line of evidence comes from animals. Whenever wildlife gets cancer, there is almost always evidence of environmental contamination. When the same species are found in cleaner habitats, they almost never have cancers. The classic example here is freshwater fish with liver cancer, which is now epidemic in many inland rivers and lakes. Beluga whales in the St. Lawrence Seaway, a very polluted stretch of water, are another example. These whales are so full of toxic chemicals that when they are found stranded and dead, they must be disposed of as toxic waste. This population of whales suffers from intestinal cancers as well as cancers of the bladder, ovary, and salivary gland.

Experiments conducted on certain clam populations in Maine provide further clues. Off the coast of northern Maine, populations of soft-shell clams in certain clam beds suffer from high rates of gonadal cancers—cancers of the ovaries and testicles. They are known to be exposed to 2,4-D, an herbicide sprayed in nearby blueberry bogs, which apparently runs from these fields into the estuaries where the clams live. Zoologist Rebecca van Beneden at the University of Maine in Orono brought these clams into the laboratory. She grew some in clean aquariums and others in aquariums contaminated with 2,4-D at levels that approach those in the clam beds. Through these controlled experiments, she was able to actually induce gonadal cancers in these animals. These results, I believe, have an urgent message for the women of Washington County, Maine: Cancer registry data reveal they suffer from higher than expected rates of both breast and ovarian cancer.

From these many lines of evidence, I have concluded that we need a weight of the evidence approach to cancer. Consider non-Hodgkin's lymphoma (NHL). As noted above, cancer registries reveal that its incidence rate has tripled in the last half century. Regional clusters show an association with pesticide use. I also looked at the occupational literature, asking, Are there any occupational groups whose members contract NHL at rates even greater than the general public? Three groups stood out. One was farmers. Another was Vietnam veterans exposed to the pesticide Agent Orange (which contains 2,4-D as well as 2,4,5-T and is known to be contaminated with dioxin). The third was golf course superintendents. When I looked at the animal data on NHL, I found out that dogs whose owners routinely use certain herbicides on their lawn suffer from canine NHL at twice the rate as dogs whose owners do not use lawn chemicals. Finally, I looked at the genetic data, asking, What kind of cellular damage is associated with pesticide exposure, and do we ever see this kind of damage with NHL? Dr. Vincent Garry at the University of Minnesota has published an elegant series of studies showing that pesticide applicators have an unusual mutation in high frequency. Called a DNA inversion, this mutation refers to a piece of a chromosome that has broken off, flipped upside down, and reattached itself. The only other population in which he has found high frequencies of this particular mutation is NHL patients.

None of these studies alone—of mutations, of dogs, of occupations, of clusterings, of time trends—provides absolute proof of a connection between NHL and pesticide exposure, but all together they tell a consistent story about the possible relationship between one kind of cancer and one group of chemicals.

The Precautionary Principle

I believe that we have enough information to act now, that uncertainty about the details and the need for more research should not be used as an excuse to do nothing. More data is never a substitute for good judgment. Consider that the U.S. Surgeon General warned us in 1964—on the basis of animal studies and statistical associations alone—that smoking causes lung cancer. The definite proof for that link came only last year when the exact gene-mutating chemical in tobacco smoke (benzo-a-pyrene) was finally isolated and identified along with the precise gene (p53) it damaged. How many more people would have started smoking or failed to quit, how many more would have died, had we waited thirty-two more years for the final proof before taking action to warn the public and ban smoking in public places? We must likewise find the courage to exercise judgment in the face of good but partial evidence and to embrace the precautionary principle, which dictates that people should not remain in harm's way while the wheels of scientific proof-making grind slowly on.

The precautionary principle is currently enjoying increased prestige among policy makers and scientists. In January 1998, I was invited to join an international group of scientists, lawyers, government officials, physicians, labor leaders, authors, community activists, and at least one urban planner for a conference on the precautionary principle and its implementation. Snowbound in the elegant confines of Frank Lloyd Wright's Wingspread House in Racine, Wisconsin, we spent a weekend hammering out the following consensus statement:

Wingspread Statement on the Precautionary Principle

The release and use of toxic substances, the exploitation of resources, and physical alterations of the environment have had substantial unintended consequences affecting human health and the environment. Some of these concerns are high rates of learning deficiencies, asthma, cancer, birth defects and species extinctions, along with global climate change, stratospheric ozone depletion and global worldwide contamination with toxic substances and nuclear material.

We believe existing environmental regulations and other decisions, particularly those based on risk assessment, have failed to protect adequately human health and the environment—the larger system of which humans are but a part.

We believe there is compelling evidence that damage to humans and the worldwide environment is of such magnitude and seriousness that new principles for conducting human activities are necessary.

While we realize that human activities may involve hazards, people must proceed more carefully than has been the case in recent history. Corporations, government entities, organizations, communities, scientists and other individuals must adopt a precautionary approach to all human endeavors.

Therefore, it is necessary to implement the Precautionary Principle: When an activity raises threats of harm to human health or the environment, precautionary measures should be taken even if some cause and effect relationships are not fully established scientifically. In this context, the proponent of an activity, rather than the public, should bear the burden of proof.

The process of applying the Precautionary Principle must be

open, informed and democratic and must include potentially affected parties. It must also involve an examination of the full range of alternatives, including no action.

Issues for Designers and Planners

Urban designers, engineers, landscape architects, and building architects are all involved in the construction of the landscapes in which we live. By setting policies, selecting technologies, and designing space, these professions affect the range and amount of toxic chemicals that are woven into those landscapes. The green lawn and the incinerator smokestack are ubiquitous features of even the most mundane urban and rural settings, so much so that they are often invisible to us in the background of our lives. The lawn and the smokestack are also good examples of the everyday kinds of toxicity that have made cancer a common element of life in industrialized countries. While there are many ways in which planners and designers can help to promote the health of our landscapes, I focus here on the lawn and the smokestack as examples because their commonness makes them both prosaic and profound.

The Lawn

About 80 percent of U.S. households use pesticides of some kind. About 50 percent of all households use yard and garden weed killers. These kinds of uses place us in intimate contact with pesticide residues, which can easily find their way into bedding, clothing, carpets, and food. Yard chemicals tracked in on the bottoms of shoes can remain impregnated in carpet fibers for years because pesticide residues persist much longer indoors in the absence of sunlight, wind, running water, and soil microbes that break these chemicals down when they are outdoors. Recent findings show that toddlers experience significant exposures to pesticides by crawling on carpets and ingesting house dust—even more so than by ingesting pesticide residues on food. Several studies link childhood cancer to home and garden pesticide exposure. In Denver, children whose yards were treated with pesticides were four times more likely to have soft tissue cancers than children living in households that did not use yard chemicals. Yard and garden weed killers have also been associated with brain tumors in children. Lawn chemicals also run into surface water and sink into groundwater. In addition to benzene and dry-cleaning fluids, my hometown drinking water wells contain traces of one such weed killer—possibly leaching into the aquifer from an overlying artificial lake where it is used to control pond weeds (which themselves have become a problem due to fertilizer runoff from surrounding yards).

Golf course superintendents die more often from cancer than do the general population. Like farmers, they also exhibit excess rates of lymphoma (as noted above) and cancers of the brain and prostate. The reason for these excesses is not known. Nor is the threat of pesticide drift from golf courses into surrounding communities well understood. However, we do know with certainty that a typical golf course uses four times more pesticides per acre than an agricultural field. Nevertheless, organic golf courses are possible— and, in fact, are already a reality in certain communities in California and Colorado. Pesticide-intensive turf management is not the only way to create a fairway or a green. The handful of organic golf courses that do exist employ grass species that are suited to the climate, as much native flora as possible, and mowing practices known to minimize weeds and pests.

It is past time to rethink the chemical-green lawn.

The Smokestack

Dioxin is classified as a known human carcinogen—one of the most potent ever identified. Our main route of exposure is through food, and the main source of food contamination with dioxin is caused by garbage incinerators. Never manufactured on purpose, dioxin is created when certain kinds of organic matter are placed together with chlorine in a reactive environment. In the inferno of a trash incinerator, many common synthetic products may serve as chlorine donors for the spontaneous generation of dioxin: paint thinners, pesticides, household cleaners. The major source, however, is a kind of heavily chlorinated plastic called PVC—polyvinyl chloride. Toys, appliances, shoes, car parts, and many types of construction debris are all commonly made of PVC.

Even the newest, most state-of-the-art incinerators send traces of dioxin into the air. These molecules cling to bits of dust and are eventually carried back to earth where they coat soil and vegetation. These chemical contaminants are then consumed by us directly or are first concentrated in the flesh, milk, and eggs of farm animals. Dioxins are stubbornly persistent in the body and resist metabolic breakdown. Hence, they magnify in concentration as they move up the food chain. Because breast-feeding infants occupy one rung higher on the food chain than adults, nursing newborns receive the highest dietary loads of dioxin of anyone.

It is time to envision communities where recycling of trash rather than incineration or landfilling becomes the easiest, most economical, and most convenient option. Since discarded food stuffs can also serve as a source of organic matter in the creation of dioxins during trash incineration, there is also an important role for garden composting to play.

In the case of hospitals, which are the biggest incinerators of PVC plastic, new designs and new community relationships must be forged. In

December 1997 I had the great opportunity to tour the waste management operation at the Medical Center Hospital in Burlington, Vermont, which is a recycling hospital and a model for other health care facilities. Less than one truckload of pathological waste is incinerated each year, twenty percent is sterilized and landfilled, and the remaining 80 percent of the hospital's garbage is recycled. No plastic ever meets the flame. Food waste from the cafeteria and coffee shop is composted and used as fertilizer on nearby organic gardens, and the produce is sold back to the hospital.

Final Thoughts

As I write this, awaiting the birth of my daughter, I am mindful that cancers among infants and toddlers are very much on the rise. For example, between 1973 and 1995, children from birth to four years showed an 18 percent rise in leukemia, a 32 percent rise in kidney cancers, a 37 percent increase in soft tissue cancers, and a 53 percent increase in brain and nervous system cancers.

A new study by the esteemed researcher E. G. Knox provides the most detailed picture yet of the close association between childhood leukemias and environmental toxics. Knox and his colleagues mapped the home residences of all 22,000 children who had died of leukemias and other cancers in England, Wales, and Scotland between 1953 and 1980. Using atlases and business directories, the Knox team also charted the locations of every potential hazardous site ranging from power plants to neighborhood auto body shops. They then combined the two maps. Their findings reveal that children face an increasing risk of cancer if they live within a few kilometers of certain kinds of industries—especially those involving large-scale use of petroleum or chemical solvents at high temperatures. These include oil refineries, airfields, paint makers, and foundries. The danger is greatest within a few hundred meters and tapers off with distance. Among children who had moved within their short lifetimes, the relationship was stronger for their birth address than it was for their address at the time of their death. This result strongly suggests that very early—probably prenatal—exposures to environmental carcinogens create the threat of cancer in children.

It is time to design communities with the health of embryos in mind. The designers and planners who take this challenge and provide us with new prototypes will be practicing their professions in a profoundly different way. Likewise, the educators who begin their courses by talking about the impact of the built environment on the health of embryos will be teaching in a way that is both profoundly ethical and eminently practical. Their students will learn about the relationships between our bodies and our landscapes that cannot be severed. The visions those students someday provide as professionals may recognize our embodied relationships with landscape as one of the fundamental qualities of being human.

Note

All cited studies are discussed and referenced in S. Steingraber, *Living Downstream: An Ecologist Looks at Cancer and the Environment* (Reading, MA: Addison-Wesley Press, 1997). Updates on the importance of prenatal exposures, golf course hazards, medical waste incinerators, and the emergence of the precautionary principle can be found in the prologue and afterward of the paperback edition of *Living Downstream* (Vintage Books/Random House, 1998). The ecology of amniotic fluid is further described in Steingraber's new book, *Having Faith: An Ecologist's Journey to Motherhood* (Perseus Book Group, 2001).

Two

Design and Planning as Healing Arts:
The Broader Context of Health and Environment

Kristina Hill

Contemporary ideas about health don't support the idea that design and planning are healing arts. And, in turn, most contemporary designers and planners don't claim those skills. Yet earlier in the twentieth century, the terms "health," "safety," and "welfare" provided justification for professional licensing in the urban design professions. They also justified a raft of changes to the form of human settlements, now collectively referred to as zoning. But today's urban residents don't call on designers to prevent disease or assist them in maintaining good health. Likewise, medical doctors don't often make recommendations about the form of cities, or participate in design charrettes. The contemporary models of practitioners in both spheres, those who intervene in the processes of the human body and those who intervene in the landscape, are typically defined too narrowly to overlap. These professional specializations have grown so rapidly, and so effectively, that we routinely miss the obvious truth that health (defined as wholeness of the body and the mind) and environment (a term derived from root words that refer to an encircling world) are intimately related.

What if "health" and "disease" were redefined in ways that included a role for environmental design? How would the design professions respond to a new demand for the "healing arts" of design? In this brief essay, I'd like to suggest a reframing of the concept of health that does two things: (1) It provides a role for design and planning, and (2) it creates links to recent medical and anthropological research. I'm not sure how the design professions will respond to the challenge of becoming healing arts. But it seems clear that the response of educators and students to this opportunity is essential to building a renewed relationship between environmental design and health.

What Definition of "Health" Would Allow Us to See the Role of Environmental Design?

I propose that health can be defined as an experience of mind and body that results from paying positive attention to three sets of relationships:

- relationships among living bodies,
- relationships among processes and organisms within living bodies, and
- relationships between living bodies and the physical earth.

This definition of health is fundamentally ecological, because it is focused on relationships between living organisms and their biophysical environments. It makes room for many different models of healing, including the medical model that is dominant in American culture, traditional indigenous knowledge, and an environmental model that provides a strong role for designers and planners.

Caring for Our Own Health through Our Relationships with Other Living Bodies

David Abram, author of *The Spell of the Sensuous* (1996), stresses the ways in which perception of the "more-than-human world" (p. 14) affects our sense of what it is to be human and to be whole. He relates his observations of the use of indigenous ecological knowledge in traditional healing practices of nonindustrialized peoples in Bali and Nepal. These stories offer insight into the Christian and Western biases of earlier anthropologists' observations, and allow us to imagine afresh an old perspective on healing the spirit and the body. The key is simply not to require that "spirits" must be disembodied.

In one example, Abram (1996) tells the story of living with an indigenous shaman in Bali who cares for the "spirits of the ancestors" (pp. 11–13) by putting out small platters of rice at the edge of a human living space. He sees that this rice is eaten by long lines of ants that make up an abundant population in the nearby tropical forest. At first, Abram feels a sense of superiority in knowing what "really" happens to the rice that seems to disappear after it is offered to the ancestors. Then, as he gets to know his host and his beliefs better, Abram realizes that the key to understanding this practice is to allow the term "spirit" to refer to nonhuman forms of life, and to make a literal connection between dead humans and living ants by realizing that dead organic matter is recycled into new life. The spirits of the ancestors are the forms of life that have reorganized the molecules of dead humans, and the *relationships* (in an ecological sense) among these life-forms and living human beings are critical to the health of that human community. The rice offerings provide a food source that deflects the ants from what could otherwise become an invasion into the human living space and its food supply.

Abram (1996) offers many other examples of ways in which the traditional shaman maintains the relationships between his people and other forms of life as a way of maintaining human health. This may seem to be far afield from the professional activities of design and planning, but in fact there are strong analogies. When the American Society of Landscape Architects met in Boston in 1999, there was an associated exhibit of "eco-revelatory" design projects at the Boston Architectural Center. Many of the designers were on hand to discuss the ways in which their projects revealed natural processes, an attribute required by the exhibit organizers (Brown et al. 1998). Several designers asked questions in this discussion about the relationship between design and science in their colleagues' work. Achva Stein, a landscape architect practicing in Los Angeles, replied that designers must combine an empirical knowledge of the world with an ability to orchestrate magic—and that the role of the designer is therefore like the role of a shaman in a traditional culture. While not all designers may wish to be shamans, Stein's point is that this is the potential of our work.

Unfortunately, when we design and plan for landscapes, we are not talking only about managing our relationships with ants. That would be comparatively easy. What we face in the global environment is a complex and difficult task of managing our relationships with microscopic forms of life, bacteria and viruses, that have life spans much shorter than our own and are therefore able to evolve rapidly to adapt to new conditions. Two well-known landscape architects, Frederick Law Olmsted and Ian McHarg, have addressed this task in their approaches to urban design. Olmsted's design for Boston's Muddy River, the Fenway, was in part a response to a public-health problem caused by microparasites. In the nineteenth century, it was standard practice to pipe human wastes into the Muddy River to flush them away from human habitations. But because the river discharged into a tidal estuary, the level of water inside its banks fluctuated, leaving wastes exposed on the banks. The stench that resulted was the primary motivation for redesigning that river landscape, partly because the smell was associated with disease. Children, and the women who cared for them, died in large numbers from diarrhea, a condition that can be caused by exposure to pathogens in human wastes (City of Boston 1885).

The implications of germ theory were not yet widely understood and accepted by nonexperts; thus, Olmsted's design probably responded to the sense of unwholesomeness provoked by the smell of human wastes. Yet the real agent of disease was not the smell, but bacteria such as *E. coli* that live naturally inside human bodies. Regardless of whether Olmsted understood germ theory, his design prevented human exposure to sewage on the riverbanks, thus managing human relationships with bacterial organisms by separating them in space.

Ian McHarg is best known for the high priority he placed on using the knowledge of natural sciences in his landscape planning method. But McHarg, too, sought to understand and manage the geography of human relationships with microparasites. Indeed, the specific type of overlay method for which he is famous may have evolved from McHarg's own encounter with a deadly bacterium. Several years ago, I was rereading *Design with Nature* (McHarg 1969) in preparation for teaching a class and happened also to be reading Susan Sontag's *Illness as Metaphor* (1978). I read McHarg's introduction, in which he talks about his stay in a tuberculosis sanitarium. He wrote that when he moved from an unpleasant sanitarium in Scotland to a different one in Switzerland, where he eventually recovered from the disease, all he brought with him was some clothing and a box of lung X-rays.

When I put the book down and returned to reading Sontag, it was exactly at the point where she was discussing the importance of X-ray images to tuberculosis patients. She recounted how patients would sit together in a sanitarium comparing their X-rays, which they often carried with them in the pocket of a robe or pajamas. The X-ray was a critical diagnostic tool that revealed the progress of the disease, or of the healing process, to the patient and the doctor.

I reopened *Design with Nature* and looked at the gray-scale images of landscapes inside. Was it possible that McHarg's version of the overlay technique was inspired by his experience with medical imaging via X-rays? Did the tool he used to protect the health of landscapes evolve from a tool used to assess human health and disease? The maps showing the density of cases of various human diseases in Philadelphia, many caused by bacterial infections, confirmed McHarg's interest in charting and preventing these unhealthy relationships between human bodies and populations of microparasites.

Years later, I had the occasion to ask Professor McHarg whether he thought that his gray-scale overlay technique had been inspired by his intense familiarity with X-rays as a diagnostic tool. He shrugged it off at the time, but do we always recognize the sources of our own inspirations, years later? It made me wonder what relationships there are or could be today between the imaging tools used to map landscapes and those used to study human bodies. The potential here for designers and planners to participate in epidemiological research or to draw on the inspiration of these technical representations to help the public understand their relationships with the living environment seems enormous.

Emerging infectious diseases add new challenges every year to the current set of epidemics and chronic problems we face, including AIDS, Ebola,

Hanta virus, new strains of tuberculosis, influenza, malaria, cholera, MBE ("mad cow disease"), and Lyme disease. Some of these diseases threaten human populations because of environmental pollution (outbreaks of cholera, for example, are sometimes connected to the disposal of human wastes). Some become problematic in situations of poor human nutrition or of reduced immune response in human hosts. And still others become widespread because of changes in habitat quality for nonhuman species and increased contact between species. This last case is a situation where design and planning are especially relevant.

Human alterations of the environment can change the ecology of pathogens and their host species in ways that increase the population size of pathogens or influence pathogens to switch to new hosts. The origins of Lyme disease provide a useful example. Epidemiologists and ecologists have discovered that the bacterium that causes this disease, *Borrelia burgdorferi*, is transmitted by the deer tick, *Ixodes scapularis*. Historical changes in human land-use patterns have influenced the spatial distribution of this tick and its most common host species, the white-tailed deer. In the eighteenth and nineteenth centuries, much of the northeastern United States was deforested by humans. The white-tailed deer population is thought to have declined drastically during this deforestation period (Dobson et al. 1997). The deer population remained relatively large on Long Island (New York) however, and consequently, so did the population of the deer tick. As reforestation occurred in the Northeast, the deer population spread back into the region, bringing the deer tick and the bacterium that causes Lyme disease with it. The habitat alterations that accompanied suburban development in the Northeast may have triggered a number of effects in this process, including an increase in the deer population size, a related increase in the deer tick's population size, and greater contact between the tick and human hosts (Barbour and Fish 1993; Dobson et al. 1997). Lyme disease was reported in 1992 to be the most common arthropod-borne disease in the United States (MMWR 1992).

Similarly, some researchers believe that the virus that causes AIDS in humans originated in the green monkeys of West Africa, whose habitat suffered degradation from an encroaching human population (Anderson 1991; Dobson et al. 1997). Whether this habitat degradation affected the infection rate among monkeys is unknown, but it did lead to increased contact between humans and green monkeys. Similarly, Hanta virus outbreaks that occurred during the past decade in the American Southwest are likely to have resulted in part from a loss of rodent predators, caused by changes in human land use, and a corresponding increase in rodent populations near humans (Levins et al. 1993; Epstein et al. 1997). Fish kills and severe human illnesses have resulted from contact with toxins produced by a tiny organism known as

Pfiesteria piscicida. The increasing nutrification of shallow ocean waters on the east coast of the United States, caused primarily by runoff from agriculture, has spurred the population growth of this organism (Burkholder 1999). Water-based tourism and the fishing industries of North Carolina and Maryland, in particular, create situations where humans come in contact with this organism and its toxins.

Garrett (1994) has chronicled the pattern of emerging infectious diseases over the last few decades. Paul Epstein and Andrew Dobson have written at length about the relationship between these diseases and human alteration of the environment, and I refer readers to their chapters in the excellent book *Biodiversity and Human Health* (Epstein et al. 1997). In general, the message I take from this research is that we must safeguard the habitat quality of animal species that live around us, and the integrity of their ecological communities, if we wish to maintain our own biological health.

The idea that habitat quality for other species can have an indirect impact on human health is directly related to David Abram's stories of the shaman preserving his community's health by maintaining a desirable relationship with ants. In my experience, planners and designers are largely unfamiliar with these interrelationships among biological populations that have drastically different life spans. Likewise, landscape ecologists do not typically address these complex host-parasite relationships in their studies of optimal spatial patterns for biodiversity (Forman 1996). Yet this constitutes a clear opportunity to justify habitat conservation in the course of regional development by arguing for human health and self-interest. No one can know what public health surprises may await future generations if we do not limit human degradation of remaining natural areas.

Relationships among Processes and Organisms within Living Bodies

The examples above were intended to explain how health depends on relationships *among* living bodies, yet there is a strong overlap with the idea of health as a state supported by relationships between organisms and processes *within* living bodies. Microparasites such as bacteria and viruses live within the human body as well as outside it, and can be both beneficial and detrimental to human health. The internal ecology of the human body is quite heavily influenced by flows of organisms across the boundary of our skins, belying the myth that we exist in a world apart from "nature." The Czech playwright and statesman Václav Havel once wrote that the human body is just a particularly busy intersection of molecules (Vadislav 1987). We exist at all only because of the busy interactions among much smaller forms of life in our intestines and elsewhere, and we can cease to exist if those interactions change in detrimental ways.

Systemic functions such as the ability of our bodies to respond to stress, whether this stress is caused by changes in the size of microparasitic populations within our bodies or by psychobiological effects, are also subject to environmental influences. Research has shown that increased UV-B radiation, a result of a thinning global ozone layer, can lead to suppression of the human immune system's response to infectious diseases (Kripke 1990; Jeevan and Kripke 1993; Kerr and McElroy 1993).

On the positive side, Robert Ulrich's pioneering research (1984) identified the supporting influence of views of parklike settings on human abilities to recover from illnesses in hospitals. Others have demonstrated that there is a relationship between the human endocrine system, the hormonal system that responds to psychological and physical stress, and the ability of the immune system to respond to infection and disease (Weijant et al. 1990; Gebhardt and Blalock 1992). The use of environmental design to reduce stress levels seems very likely to support the human healing process; not doing so, conversely, seems likely to deter healing. Nancy Gerlach-Spriggs and her co-authors (1998) have documented what is known about the relationship between physical settings and the healing process. Multiple lines of evidence reported in their work suggest that this relationship can be potent. Environmental design can be a factor in human recovery from stress and disease, as well as in preventing disease.

Clare Cooper Marcus and Marni Barnes (1995) have also offered insights into the restorative potential of gardens, along with insights about the design of restorative spaces. Daniel Winterbottom and his landscape architecture students at the University of Washington are building such spaces at cancer recovery facilities and children's AIDS care facilities. It will take careful work to establish scientifically whether these facilities add to healing, even in settings where that seems intuitively obvious. Tragically, if the immune responses of human beings do become impaired by global environmental changes, it will be increasingly necessary for physical planners and designers to create restorative environments to support biological healing.

Fortunately, we don't need a health disaster to argue for the use of environmental design in support of natural healing. What we do need is to pay attention to epidemiological trends, and to remember what is already known about the relationships between perception, the endocrine system, stress, and the immune system. We also need to speak a professional language that allows us to address human "wholeness" in its multiple dimensions, including the social, spiritual, physical, and psychological. The planning and design of restorative spaces is part of a broadened definition of the healing arts that directly involves landscape architects and building architects. This is a goal that can also engage planners and engineers, who could provide space for this "restorative infrastructure" while laying out the other forms of infrastructure that support human settlements.

This infrastructure of restorative spaces should accompany other forms of infrastructure that support the growth of urban areas, as human populations continue to increase. The goal would be to provide an antidote to stress, and in so doing, to support the internal systemic functions of the human body that maintain the health of processes and living organisms within the body itself. If we can come up with the dimensions for such an infrastructure, proposing its size, shape, and desirable qualities, members of the design professions could participate in the healing arts at a landscape scale as well as at individual sites of healing. The emerging field of psychoneuroimmunology is likely to offer more specific evidence of the potential importance of this "healing infrastructure" over the next several decades (Rabin 1999; see also the journal of psychoneuroimmunology, *Brain, Behavior and Immunity*).

Relationships among Living Bodies and the Physical Earth

Finally, I propose that our health is supported by a third component: relationships among human bodies and the physical earth. This idea has been put forth eloquently by poet and ecologist Sandra Steingraber in her book *Living Downstream: An Ecologist Looks at Cancer and the Environment* (1997) and in her essay in this volume. She includes a brief parable in her book about people who live alongside a river. The people living along the river see others trying frantically to swim out of it, trapped and drowning in the swift waters. They jump in over and over again to rescue these drowning people, even as the number of people going by in the water grows larger and larger. Oddly, the people living alongside the river never think to go upstream to find out who is pushing these poor people in! Steingraber uses this parable to point out the irony of our society's response to cancer. Even as we expend ever-increasing resources to rescue people from cancer and congratulate ourselves on improvements in our medical technologies, we seem loathe to ask ourselves why cancer incidence rates are increasing and to confront its causes. Both her book and her essay focus on charting the environmental contamination and exposure pathways that play a role in the incidence of cancer.

Steingraber has gone upstream in search of answers and come back to tell the rest of us that the spatial pattern of contamination that results from human use of toxic substances, along with their movement in the flows of water, air, and soil that surround us, can make us ill. The most striking aspect of her work, however, is the way in which her presentation of existing research persuades readers that there is no absolute boundary between their bodies and the earth around them. By what we eat, drink, and breathe, we take in the molecules that once were part of the soil of American prairies, along with the molecules of pesticides and herbicides that can give us cancer. Steingraber points out that our bodies literally are made of these molecules, and that therefore it can be no surprise that our health depends in part on our relationship with the physical materials of that earth, both as nutrients and as carcinogens.

Designers and planners have many opportunities to participate in charting and preventing human exposures to toxic chemicals. In my own experience, park systems can be designed to help identify, draw attention to, and treat widespread contamination of groundwater by carcinogens (Hill 1998). Designers can also work to reveal the processes associated with contamination and the remediation of contaminated landscapes (Bargmann and Levy 1998). More generally, planners and designers who practice in industrialized areas frequently deal with sites that have low levels of contamination, known as brownfields. Our work with these sites is a good example of the opportunity for design and planning to be healing arts, influencing the health of urban residents.

Reading the typical urban newspaper might give the impression that dangerous environmental contamination is a thing of the past, something that happened in Love Canal or Times Beach but not in our own communities. Yet incidents of contamination occur frequently in every suburban landscape as a result of accidental fuel oil spills, the use of toxic compounds by homeowners, and other releases of toxic chemicals. As we do more testing of urban and suburban groundwater, we are likely to discover exposure pathways in places that we did not suspect were contaminated. While the concentrations of contaminants may be low, their cumulative impacts can be substantial. Frequently, these levels of contamination are labeled "background levels" in investigative reports and therefore no cleanup is required. This issue is likely to come up more frequently on our professional radar screens over the next several decades, as we begin thinking more about the cumulative exposures that lead to increasing rates of cancer.

How Can the Design and Planning Fields Reemerge as Healing Arts?

These ideas about the relationships between human health and the environment are similar in some ways to ideas used to promote open-air parks in the nineteenth century, when slum living conditions were seen as part of the source of epidemic diseases (Clarke 1973; Crantz 1989). But today they come with a new language and rhetoric as a result of changing perspectives in social, medical, and ecological research. If we as designers and planners accept the challenge of placing our professions among the healing arts, we must seek to understand the critical dynamics that contemporary researchers believe lead to disease in human populations.

To do so, we will have to learn some of the language of medical and epidemiological research. We're fortunate that gifted writers like Sandra Steingraber are able to translate that language. Steingraber's essay in this volume uses two symbolic features of our contemporary landscape, the lawn and the incinerator smokestack, to draw connections between the physical earth and human bodies. Her essay is important because it argues that the widespread, "low levels" of environmental contamination that result from activities such as maintaining lawns and burning solid wastes are ethically unacceptable—even though many of us have come to accept them as a by-product of urbanization. I recommend readers to her book, *Living Downstream* (1997), to find an in-depth and beautiful exploration of a more complete set of these connections.

Citations

Abram, D. 1996. The spell of the sensuous: perception and language in a more-than-human world. Pantheon Books, New York, New York, USA.

Anderson, R. M. 1991. Populations and infectious diseases: ecology or epidemiology? Journal of Animal Ecology **60**: 1–50.

Barbour, A. J., and Fish, D. 1993. The biological and social phenomenon of Lyme disease. Science **260**: 1610–1616.

Bargmann, J., and S. Levy. 1998. Testing the waters. Pages 38–41 in B. Brown, T. Harkness, and D. Johnston, editors. Eco-revelatory design: nature constructed/nature revealed. Landscape Journal, Special Issue: Exhibit Catalog.

Brown, B., T. Harkness, and D. Johnston. 1998. Eco-revelatory design: nature constructed/nature revealed. Landscape Journal, Special Issue: Exhibit Catalog.

Burkholder, J., 1999. The lurking perils of *Pfiesteria*. Scientific American, August, **281**: 42–50.

Chivian, E., M. McCally, H. Hu, and A. Haines, editors. 1993. Critical condition: human health and the environment. A report by Physicians for Social Responsibility. MIT Press, Cambridge, Massachusetts, USA.

City of Boston Health Department. 1885. Annual Report of the Health Department. Boston, Massachusetts, USA.

Clarke, R. 1973. Ellen Swallow: the woman who founded ecology. Follett Publishing Company, Chicago, Illinois, USA.

Crantz, G. 1989. The politics of park design: a history of urban parks in America. MIT Press, Cambridge, Massachusetts, USA.

Dobson, A., M. Campbell, and J. Bell. 1997. Fatal synergisms: interactions among infectious diseases, human population growth, and loss of biodiversity. Pages 87–100 in F. Grifo and J. Rosenthal, editors. Biodiversity and human health, Island Press, Washington, D.C., USA.

Epstein, P., A. Dobson, and J. Vandermeer 1997. Biodiversity and emerging infectious diseases. In F. Grifo and J. Rosenthal, editors. Biodiversity and human health, Island Press, Washington, D.C., USA.

Fiennes, R.N. 1978. Zoonoses and the origins and ecology of human disease. Academic Press, London, UK.

Forman, R.T.T. 1996. Land mosaics. Cambridge University Press, Cambridge, UK.

Garrett, L. 1994. The coming plague: newly-emerging diseases in a world out of balance. Farrar, Straus & Giroux, New York, New York, USA.

Gebhardt, B., and J. Blalock. 1992. Neuroendocrine regulation of immunity. Pages 1145–1149 in I. Roitt and P. Delves, editors. Encyclopedia of immunology, Academic Press, London, UK.

Gerlach-Spriggs, N., R. Kaufman, and S. Warner. 1998. Restorative gardens: the healing landscape. Yale University Press, New Haven, Connecticut, USA.

Grifo, F., and J. Rosenthal. 1997. Biodiversity and human health. Island Press, Washington, D.C., USA.

Hill, K. 1998. Ring parks as inverted dikes. Pages 35–37 in B. Brown, T. Harkness, and D. Johnston, editors. Eco-revelatory design: nature constructed/nature revealed. Landscape Journal, Special Issue: Exhibit Catalog.

Jeevan, A., and M. Kripke, 1993. Ozone depletion and the immune system. Lancet 342: 1159–1160.

Kerr, J., and C. McElroy, 1993. Evidence for large upward trends in ultraviolet-B radiation linked to ozone depletion. Science 232: 1062–1064.

Kripke, M., 1990. Effects of UV radiation on tumor immunity. Journal of the National Cancer Institute 82: 1392–1396.

Levins, R., P. Epstein, M. Wilson, S. Morse, R. Slooff, and I. Eckhardt. 1993. Hantavirus disease emerging. Lancet 342: 1292.

Marcus, C.C., and M. Barnes. 1995. Gardens in healthcare facilities: uses, therapeutic benefits, and design recommendations. Center for Health Design, Martinez, California, USA.

McHarg, I. 1969. Design with nature. Doubleday/Natural History Press, Garden City, New York, USA.

Morbidity and Mortality Weekly Report (MMWR). 1992. 40: 505.

Rabin, B. 1999. Stress, immune function and health: the connection. Wiley-Liss, New York, New York, USA.

Schettler, T., G. Solomon, M. Valenti, and A. Huddle. 1999. Generations at risk: reproductive health and the environment. MIT Press, Cambridge, Massachusetts, USA.

Sontag, S. 1978. Illness as metaphor. Farrar, Straus & Giroux, New York, New York, USA.

Steingraber, S. 1997. Living downstream: an ecologist looks at cancer and the environment. Addison-Wesley, New York, New York, USA.

Ulrich, R. 1984. View from a window may influence recovery from surgery. Science 224: 420–421.

Vladislav, J. 1987. Vaclav Havel, or, living in truth: twenty-two essays published on the occasion of the award of the Erasmus Prize to Vaclav Havel. Faber Publishing, London, UK.

Weijant, D., D. Carr, and J. Blalock. 1990. Bidirectional communication between the neuroendocrine and immune systems: common hormones and hormone receptors. Annals of the New York Academy of Sciences 579: 17–27.

PART III

Education for Practice

How can design educators guide students toward knowledge and skills that most effectively relate design and ecology? What may be the keys that hold long-term value for design education and practice? This section contains three chapters that explore the conceptual frameworks needed to realize an ecologically based approach to design and planning in education and how that foundation can lead to more meaningful practice. In Chapter 9, "Ecological Science and Landscape Design: A Necessary Relationship in Changing Landscapes," designer Joan Iverson Nassauer considers how ecologists and designers have collaborated in teaching and practice, and how this may point toward a deeper integration of ecology within landscape architecture. Chapter 10, "On Teaching Ecologic Principles to Designers," by planner Carl Steinitz, proposes a pedagogic framework to bring ecological models into the design process at different educational levels and examines how this would affect the teaching of design and planning studios. Designer Michael Hough asks "What does a designer need to know about a place?" and considers how the answers can provide underpinnings for teaching ecology to design students in Chapter 11, "Looking Beneath the Surface: Teaching a Landscape Ethic."

CHAPTER 9

Ecological Science and Landscape Design: A Necessary Relationship in Changing Landscapes

Joan Iverson Nassauer

Landscape design is cultural action about nature, and landscape design constructs ecosystems. Because design always affects ecological processes—even when designers are not attentive to these effects—design has a necessary relationship with ecological science. How design curricula and professional cultures embody this relationship is conceptually incomplete and subject to surprising contention. While no one would dispute that landscape architects need to know about nature, what they need to know and how they should learn it is much less clear. Typical approaches that design curricula have taken toward ecological science could be characterized as ambivalence, inspiration, and integration. Each of these is problematic. None adequately fulfills the expectations of students who come to landscape architecture inspired by the promise of learning to design with nature. None recognizes the depth and variety of ecological knowledge that can inform design. None realizes the potential for design to bring ecological health to changing landscapes. Design curricula may realize this potential if they clearly embrace a core definition of design as cultural action, while they also educate students to practice design in a continuous exchange with ecology.

Ecology and design are two very different ways of looking at and prescribing action in the landscape. Ecology is scientific study, and design is creative cultural action. While they are different, they share something fundamental to both: the landscape. This creates obvious grounds for mutual interest and collaboration, and, at the same time, it creates a basis for misunderstanding and competition between different views of the same subject. Imagine Henry Cowles and Jens Jensen exploring the Indiana dunes together early in the past century and you have a picture of the collaboration that ecology and design can have (Grese 1992). Now, imagine a landscape architecture student entering professional school infused with the desire to support nature with design, but soon absorbing the tacit knowledge that some students excel in design while others pay attention to science and planning, and you have a picture of a different relationship. The picture of Cowles and Jensen shows a great ecologist and a great landscape architect, each accomplished in his own discipline, each seeking its potential, and each undeterred by its conventional limits. The second picture suggests that the student's education taught her that ecology was at best an alternative niche in landscape architecture; it had no necessary relationship with the cultural action of design. Both pictures represent real aspects of the relationship between ecological science and design in landscape architecture today.

To explore how students could use the landscape foundation shared by ecology and design, we might look again at Cowles and Jensen. They were accomplished in their own disciplines. Their experience in their own disciplines led them to viewpoints and questions for which they valued the insights of the other. How might design pedagogy lead students to formulate viewpoints and questions that need ecological critique, development, or answers? How might curricula engage them in practice of this exchange? Cowles and Jensen expanded the intellectual potential of their own disciplines and were undeterred by conventional disciplinary limits. Students begin their professional education with a deep desire to search for their own potential, and when they begin, they are free from nearly all conventions. Often, professional design culture and curricula change that. Design education often acculturates students as much as it gives them new knowledge. To consider what students learn in professional programs, landscape architects and planners should examine not only curricula but also the culture of design as it characterizes ecological science.

Past relationships between ecology and design in landscape architecture curricula can be broadly characterized as (1) ambivalence about the necessity of ecology for design, (2) ecology as a source of inspiration for design, and (3) integration of the substance of ecology into design. These relationships exist in curricula around the country, sometimes in the same place at the same time. While the characterizations simplify reality, they are intended to point to areas of contention and conceptual ambiguity in design curricula. Considering them may suggest more valid, complex, and intentional relationships between ecology and design for future curricula.

Ambivalence

Today it would be difficult to find a prominent designer who does not pay homage to the importance of ecology for design. Yet designers' behavior, in the content of professional courses and curricula, in the priorities and alliances of professional organizations, and in their own designs, sometimes belies the words. Students' experience of curricula may lead them to conclude that ecological science competes with the knowledge and methods particular to design, potentially limiting the intuitive, place-based products of design with analytical methods and universal generalizations valued by science, possibly replacing time for studio or graphics with science coursework. Scientific knowledge may be seen as dully formulaic or exhaustively factual in comparison with the holistic, artistic revelations of design. When design and ecology are even implicitly set in such a dichotomy, students perceive that they must make a choice between landscape architecture based on ecology and landscape architecture as design.

Design programs need to explore this dichotomy so that they can dispense with it. Certainly, H. W. S. Cleveland, Jens Jensen, or Frederick Law Olmsted Sr. would not have seen design and ecology as mutually exclusive. Their education was not formally limited to design topics; it was broadly based on discussions with others in the humanities, arts, and sciences. More recently, landscape architecture has again claimed ecology as a fulcrum for the meaning of design (e.g., Olin 1988; Spirn 1988a, b; Riley 1988). Pushed by public interest, landscape architects have returned "wild" nature to even the most urban venues. Yet making ecology part of design curricula has encountered surprising resistance. Why?

In part, design and planning have been blinded by their own stereotyping of ecology, which has tended to limit ecological applications to the analysis of regions. Thirty years ago, Ian McHarg set an indelible imprint on the professions with his stunning statement, *Design with Nature* (1969). While his philosophy pointed to the centrality of human perception of nature, young landscape architects of the time gravitated toward the book's concrete methods and images, maps of suitability and vulnerability analyses. Analysis was the part of McHarg's message that entered the professional mainstream. For designers, ecology emerged firmly attached to regional analysis. With influential exceptions (Ferguson 1987, 1998; Hough 1984, 1995; Morrison 1979) that led some landscape architects to a dawning awareness of the value of native plants and surface stormwater management, the ecology of the site typically was limited to rather superficial analysis. At the site scale, ecological factors were described as constraints to development rather than systems or processes with spatial characteristics.

This separation of site from region and design from analysis expressed landscape architecture's ambivalence toward science. Geographic information systems (GIS), the international growth industry incubated by regional analysis, is the most obvious object lesson of this separation. When Howard Fisher collaborated with Carl Steinitz to begin applying computer mapping to landscape analysis at the Harvard University Graduate School of Design in 1965 (Steinitz 1993), landscape architecture graduates of Harvard and the University of Pennsylvania began to export its concepts and technology to many landscape architecture programs. But these new computer mapping courses played to mixed reviews by faculty and students. Some saw this powerful new technology opening doors to a broad future for the profession. Others wondered what entering code into a computer terminal could possibly have to do with design.

The GIS paradigm tsunami that eventually swept up geography, and much of planning, forestry, wildlife biology, and ecology had its contemporary beginnings in landscape architecture. McHarg's book popularized basic methodological possibilities for applying GIS and inspired many but not most designers of an era. For some, making maps describing the implications of ecology remained an alternative to design, not a part of it. Computer-aided mapping lived but languished in landscape architecture programs that were uncertain what it had to do with design. For those who thought about design as what a person can touch, the space a person can enter, and for those who approached design primarily with intuition, computer-aided mapping sometimes seemed a foreign language, and its ecological roots seemed equally alien. Regional scale analysis maintained a presence in design curricula, but it was not at their heart, and ecology was marginalized with it.

Meanwhile, computer-aided mapping mushroomed into GIS course series and graduate programs in geography. GIS has fundamentally changed the way agriculture, forestry, industry, governments, planning, and landscape architecture work. Numerous extremely successful types of GIS enterprises were initiated by landscape architects Jack Dangermond, Lawrie Jordan, and Bruce Rado (e.g., ESRI, ERDAS) (Steinitz 1993), but these enterprises are not viewed as landscape architecture—only related to it. In the 1990s many design programs had to rush to catch up with the GIS technology that had been initiated but not nurtured in landscape architecture. In the design professions today, the relationship between ecological analysis and ecological design, and the relationship between region and site is more apparent than a decade ago; but the essential ambivalence that kept landscape architecture from seizing the moment of intellectual, entrepreneurial, and design opportunity with GIS may continue to predispose design to a related kind of blindness to its relationship with ecology.

Inspiration

In another characterization, ecology may be welcomed into the professional curriculum and into the studio as an inspiration for design. This relationship recognizes that ecology affects sites as well as regions and that ecology can inspire form as well as delimit analysis. It replaces ambivalence with an interested conversation between acquaintances. Drawing on ecology for inspiration recognizes that ecology is not design, and design is not ecology. It assumes that there is something to be gained by bringing ecology to design while holding it at a distance: near enough to make selective impressions. Such a relationship may evoke wide enthusiasm; it offers much and risks little, but it leaves design teaching and curricula with the ambiguity of acquaintance. How much do designers need to know about what they find interesting in ecology before they can use it? In what ways should they be attentive and true to the source of their inspiration, ecological science, and in what ways are they free to interpret the source? Finally, do design curricula implicitly trivialize the content of ecology when they approach it as mere inspiration for design? Each of these questions confronts designers with the potential that what they say, their rhetoric, will not be true to its source, ecology.

At a time when ecological rhetoric sells everything from herbicide to vacation tours, designers will be sorely tempted to allow their rhetoric to exceed their knowledge. The dangers of this temptation are both ethical and ecological. The popular authority that landscape architects enjoy as designers of nature colors their pronouncements and their designs with ecological beneficence, almost regardless of their deeper knowledge or intentions for ecological health. How can professional curricula prepare designers and planners to use their power of expertise with integrity: describing what they know and do not know about the ecological function of their designs, designing for both ecological function and cultural perception while distinguishing between them?

Even if designers use ecological rhetoric with the greatest integrity, they must consider ecological risks. As proponents of sustainable development have concluded, anyone who changes the landscape must be extremely cautious when they intervene in any ecosystem because, even when experts think they understand, they may be wrong (Holdgate 1996). Cautious intervention does not mean doing nothing or making only small gestures. It means asking questions of people who have the greatest knowledge, weighing and judging multiple answers, and monitoring the effects of design. It also means being prepared to respond to unintended consequences (Gunderson, Holling, and Light 1995). Using the rhetoric of ecology comes with considerable responsibility. It implies a definition of design that extends far beyond the time of construction. Is the responsibility so great that professional curricula should integrate ecology with design? Is it enough to be pleasantly inspired by ecology, or does design need more than an easy acquaintance?

Integration

Perhaps landscape architects need to become sufficiently expert in ecology to draw from it directly, but what is sufficient, and at what point do designers and planners become almost ecologists and not quite designers anymore? Central to landscape architects' identity is their view of themselves as integrators and generalists. How much they need to know about what they are integrating is less clear. If design curricula set their students off on a Sisyphean challenge to learn all of ecology, they exhaust students' creative energy on an impossible task. Even ecologists do not attempt to know all about the natural world. Landscape architects who attempt such a feat may distract themselves from the important work of making the landscape while they ceaselessly tread the slope of preparation to act. If ecology is integrated into design curricula in a way that leaves students feeling that they need to know it all, it may prevent them from achieving any sense of mastery, as designers or as ecologists. Or worse, it may cause designers to suffer from a kind of generalist's hubris, believing that they can do all things well—including ecology.

Alternatively, the breadth of ecological knowledge that some design curricula seek can be achieved successfully in interdisciplinary processes. Work that involves many different areas of natural science and many different people can focus students' energy on how to draw upon the knowledge of others, as well as how to teach others to draw on their knowledge. Recognizing the value of interdisciplinary work would lead curricula to teach students how to work iteratively between ecology and design to recommend action (Lyle 1985). This approach can reinforce the traditions of design. It asks design to continue to deepen its traditional strengths in knowing human experience, imagining new landscapes, and communicating their possibilities. It also stimulates those traditions by engaging designers in an exchange that takes ecological knowledge so seriously that it depends on ecologists to be part of the design process.

Lessons from Ecology

Both the history and the vanguard of ecology can teach designers something about how to bring ecologists into design processes. When ecology began as a discipline in the nineteenth century, about the same time as landscape architecture, it grew from the work of eighteenth-century naturalists who knew nature well in the place where they were (Slaughter 1996; Worster

1977). Rather than mastering an abstract science, the naturalists looked long and closely at particular landscapes and sought to understand what they observed. Similarly, Cowles and Jensen made forays into the landscapes near their home, Chicago, to observe and discuss what they saw. Ecology may be brought into an appropriate relationship with design place by place, as designers and ecologists carefully look at landscapes together. This is not integration but rather a tightly woven interdisciplinary relationship (Lyle 1985, 1994; Nassauer 1995a; Poole 1994).

In such a relationship, design balances ecology by seeking the qualities of individual places. Ecology balances design by comparing the particulars of place with more abstract models of ecological processes, seen or unseen. In the course of looking at places together, both ecologists and designers begin to notice some of the same landscape characteristics, to share a language, and to respect that some knowledge of the other discipline will remain foreign to them.

If students learn both how to see intuitively and how to notice ecologically, they are likely to see the necessity for discussion between designers and scientists. Such a discussion is not a simple dialogue; ecology and design each embody complex interpretations and perspectives. There are many ways for designers to see and for scientists to analyze, and ecological expertise does not predictably lead to a particular view of a place. In the end, science, like design, requires judgment. Discussions that enlist ecological judgments are happening now—in interdisciplinary practices, in public agencies, and in landscape architecture research. Frequently, landscape architects initiate and lead such discussions. Yet, this very practical form of discussion generally remains peripheral to design curricula, a good idea but not taught as a skill or critically examined. Certainly, examples and methods for interdisciplinary work have not been a centerpiece of design education.

The object lesson of GIS should loom large as landscape architecture programs assess how they will go forward from this moment of opportunity. Within the past decade, while landscape architects have been working with ecologists in practice, and thinking and writing about the relationship of design and nature, the vanguard of ecology has extended to include several burgeoning approaches that advance ecology from analysis to action. While landscape architects have participated in developing these interdisciplinary approaches, the entire discipline of ecology is learning and changing from them. Conservation biology (Noss and Cooperrider 1994), ecosystem management (Grumbine 1994), landscape ecology (Forman and Godron 1986; Risser 1987), and restoration ecology (Jordan, Gilpin, and Aber 1987) each point beyond biophysical research and analysis to understanding and motivating human perceptions and behavior that change the landscape. Ecology has been rocked by these approaches, and, while none has been accepted without skepticism and debate, the ideas and their discussion have been brought into the normal business of ecology. Federal, state, and local land management and conservation policies have been reoriented around these ideas. Increasingly, the popular press and citizens use their vocabulary. Yet these influential approaches to changing landscapes remain optional for the design disciplines.

Concepts and ideas in conservation biology, ecosystem management, landscape ecology, and restoration ecology are or could be as much about cultural action as they are about ecological function. Ecologists, knowing the limitations of their knowledge, have frequently called for knowledge of culture that will support action. Ecologist Jane Lubchenco (1998), president of the American Association for the Advancement of Science, dubbed humans "a new force of nature" in her presidential address, "Entering the Century of the Environment: A New Social Contract for Science." Design programs should respond in kind. Operating from the strength of their traditional core, design as cultural action, they should recognize the necessity of ecological knowledge with a new ecological contract for design. Landscape architecture, like ecology, should behave differently as a discipline five years from now. Conservation biology, ecosystem management, landscape ecology, and restoration ecology should grow and be challenged within the design professions as well as within ecology.

Interdisciplinary Discussion

For landscape architecture to evolve and lead in new interdisciplinary approaches, consultative interdisciplinary discussion should become a regular, practiced part of design programs. A strength of studio pedagogy is that it includes repetition for students to develop practiced, discerning abilities. Consultative interdisciplinary discussion should become one of those practiced abilities. Just as students draw differently in their first year of studio, they might consult with ecologists differently before they have begun to hone their understanding of cultural perceptions or their ability to make form. Early on, they will hear and inquire of ecology differently, but they should know that the discussion is of the essence and practice it. Later, they will have better questions for ecology because they will know more about it, and they will have better answers for ecologists who are themselves interested in cultural action. Just as students are taught the essential and somewhat amorphous knowledge of making landscape form, they should be taught the essential and somewhat amorphous knowledge of how to consult with others who have rigorous knowledge of ecology.

To teach designers to be leaders and expert participants in interdisciplinary work with ecologists, design curricula could include the following experiences for students.

Discuss places.

Students can learn how to do this in interdisciplinary discussions about the places they are designing. By discussing both real places and the imaginative speculations of students' design ideas, ecologists can see new implications of their science and new questions for research. Thinking together about changing landscapes challenges both ecologists and designers by requiring them to take a normative stance, considering what should happen to a real place. By observing a real landscape, ecology and design rather naturally find questions for each other. For example, landscape architects should know how to look at an ecosystem, scanning for plants and animals that belong there. Knowing what should be there allows them to ask questions about anything that looks unfamiliar. For landscape architects to notice ecological patterns with sufficient acuity to formulate good questions for ecologists, they need to learn how to describe the natural world, drawing on morphological taxonomies much as the eighteenth-century naturalists did. Landscape architecture students will not become taxonomists, but engaging taxonomists in discussions of places can help students learn how to form the right questions about what they are observing.

Discuss pattern and form.

Ecological processes form landscapes, and design affects ecological function. Design students should use discussions with ecologists to match design intentions with their ecological effects, and to mesh cultural traditions and preferences with their ecological implications. Landscape ecology established and continues to open new ground for discussions of pattern and form because its makes pattern, or landscape structure, a fundamental ecological concept (e.g., Forman and Godron 1985; Dramstad, Olson, and Forman 1996). It enriches designers' ways of pattern-finding and pattern-making in landscapes at all scales, and it demonstrates the ecological wisdom of the old landscape architecture tradition of working across scales.

Discuss from the position of culture.

If curricula teach students to know that their primary responsibility is to be experts about human experience and the cultural construct of nature, and to propose imaginative possible landscapes, designers will have better questions for ecologists, and they will have more to offer in the discussion. Being cultural experts does not mean that designers should not learn about ecology. It means that they should learn the vocabulary of ecological concepts, how to recognize what belongs in a landscape, how to have a discussion with ecologists, and what they should expect to learn and produce from such a discussion. It also means that design and planning programs should draw deeply from the humanities and social sciences as well as from the design disciplines to give students truly expert understanding of the human experience of landscapes. Design and cultural perceptions of landscape have been integral to landscape ecology for the past twenty years (Nassauer 1995a). Conservation biology and ecosystem management have explicitly incorporated the perceptions of indigenous people (Grumbine 1992) and organizational behavior (Westley 1995) as key aspects of their interdisciplinary knowledge base. Ecological restoration is coming to grips with the necessity of linking the plant ecology of restoration with larger cultural meaning (Jordan 1994). Designers must be prepared to speak with broad knowledge and authority about human experience of landscape.

Respect knowledge and learn skepticism.

Students will be more powerful participants in interdisciplinary discussion if they learn how to respect and borrow from rigor as defined in science. Landscape architecture's drive to invent frequently overwhelms its need to know. Design students should bring more rigorous cultural knowledge to design. Design pedagogy also should clearly articulate an iterative design process that includes intervals for consulting experts in the biophysical sciences (Lyle 1985). Building an interdisciplinary discussion with ecology will require designers to learn to pause long enough to probe, skeptically judge, and learn from ecologists. This pause should allow designers to be equally skeptical of their own assertions about design and culture. To enable designers to compare the answers they receive from ecologists as well as to consider their own knowledge of culture, the working vocabulary of design must be expanded to include such fundamental scientific standards as validity and reliability. Learning from science and from the humanities, landscape architects and planners can adopt a more critical stance from which to judge the ecological and cultural effects of their work.

Exercise caution and humility.

Recently, many ecologists (Holdgate 1995; Holling 1995; Gorham 1997) and even some economists (Hannon 1992) have called for humility and caution not only when people change the landscape, but also when scientists describe it or make predictions about it. There is a growing consensus that a key reason to protect indigenous species and ecosystems is that humans do not know enough about them to calculate the implications of their change or loss. On this point as well as others, designers can learn directly from ecology to be cautious in disturbing ecosystems that are working. Design might also emulate ecology and learn to be cautious in disturbing settled places and local cultures that are working. Perhaps the burden should be on those who conceive landscape change to know what is not working or may not work in the future before they disturb a system that has sustained itself. Designers and planners can learn from restoration ecology's strategy of emulating indigenous ecosystems; borrowing from the landscape language of local people, using their familiar landscape symbols and aesthetic types, can help designers innovate to achieve larger ecological values in design (Nassauer 1997). Designers should be cautious with culture. Rather than perpetuating professional design culture's congenital infatuation with novelty, design curricula also might teach students the value of respectful observation and a more subtle ingenuity.

Conclusion: Interdisciplinary Practice in Design Education

Design education can articulate a necessary relationship between ecology and design by making consultative interdisciplinary discussions a regular experience for students. Practicing interdisciplinary consultation builds on the core traditions of design and planning without marginalizing ecological knowledge. It makes ecology more than an inspiring acquaintance for design. While it acknowledges the necessity for ecological knowledge of landscapes, consultative discussion recognizes that expert knowledge in design and ecology can rarely be integrated in a single person.

Landscape architecture should change as students regularly practice discussion with ecologists, just as conservation biology, ecosystem management, landscape ecology, and restoration ecology are changing ecology. A curriculum that articulates a necessary relationship with ecology might place ecology students and landscape architecture students in consultation in studio together. It would place landscape architecture students in science courses from which they could learn a morphological vocabulary of landscapes that would enable them to recognize what belongs in urban and indigenous ecosystems and to formulate questions for ecologists about what they notice. It would give landscape architects the knowledge to use rhetoric about nature thoughtfully and ethically, and to respect what they do not know as much as the power of their imaginations.

Finally, a landscape architecture curriculum that builds a clearer relationship with ecology should sharpen not blur students' understanding of design as cultural action. It would require studio pedagogy as well as curricula to underscore designers' obligation to understand the culture that gives landscape meaning and enriches human experience. Designers who can offer more well-founded insights into human experience of the landscape will be more credible partners in discussion with ecologists, as well as collaborators with other disciplines. A curriculum that advances a necessary relationship between design and ecology might also critically examine the culture of the design professions. If designers distinguish the habitual from the essential in their professional cultures, they may be more likely to recognize and create opportunities for growth and change in the professions. Building consultation with ecologists into the design process is such an opportunity.

At a time when people have begun to care that the countryside is shrinking, species are rapidly vanishing, and global climate is changing, the design professions are increasingly recognized for their ability to protect nature in the course of landscape change. Landscape architecture nearly always has claimed to do this and intended to do this. Yet designers have achieved widely varying degrees of success. Often, cultural necessity, described as economic pressures or client desires or the dominance of cars or the limits of codes and ordinances, has been offered as a rationale for ecological compromise. If design curricula recognize the necessary relationship between design and ecology, such compromise often may be replaced by invention of landscapes that meld ecological processes into valued places for human experience. By knowing how to learn about the ecological processes that function in a place, designers will have a stronger basis for invention. By being more firmly rooted in ecological knowledge, designed landscapes can deepen public understanding of nature rather than confuse it with unfounded rhetoric. By knowing their own responsibility for cultural understanding and landscape innovation, designers will more convincingly prove their own necessity in determining the future of changing landscapes. Landscape change could be the product of discussion that enables both design and ecology to influence the future with more powerful and meaningful effect.

Citations

Dramstad, W. E., J. D. Olson, and R. T. T. Forman. 1996. Landscape ecology principles in landscape architecture and land use planning. Island Press, Washington, D.C., USA.

Ferguson, B. K. 1987. Water conservation methods in urban landscape irrigation: an exploratory overview. Water Resources Bulletin 23: 147–152.

———. 1998. Introduction to stormwater. John Wiley and Sons, New York, New York, USA.

Forman, R. T. T., and M. Godron. 1986. Landscape ecology. John Wiley and Sons, New York, New York, USA.

Gorham, E. 1997. Human impacts on ecosystems and landscapes. Pages 15–31 in J. I. Nassauer, editor. Placing nature: culture and landscape ecology. Island Press, Washington, D.C., USA.

Grese, R. E. 1992. Jens Jensen: maker of natural parks and gardens. Johns Hopkins University Press, Baltimore, Maryland, USA.

Grumbine, R. E. 1992. Ghost bears. Island Press, Washington, D.C., USA.

———. 1994. What is ecosystem management? Conservation Biology **8**: 27–38.

Gunderson, L. H., C. S. Holling, and S. S. Light. 1995. Barriers and bridges to the renewal of ecosystems and institutions. Columbia University Press, New York, New York, USA.

Hannon, B. 1992. Measures of economic and ecological health. Pages 207–221 in R. Costanza, B. G. Norton, and B. D Haskell, editors. Ecosystem health. Island Press, Washington, D.C., USA.

Holdgate, M. 1996. From care to action. Taylor and Francis, Washington, D.C., USA.

Holling, C. S. 1995. What barriers? what bridges? Pages 3–35 in L. H. Gunderson, C. S. Holling, and S. S. Light, editors. Barriers and bridges to the renewal of ecosystems and institutions. Columbia University Press, New York, New York, USA.

Hough, M. 1984. City form and natural processes. Van Nostrand Reinhold Company, New York, New York, USA.

———. 1995. Cities and natural processes. Routledge, New York, New York, USA.

Jordan, W. R., III. 1994. "Sunflower forest": ecological restoration as the basis for a new environmental paradigm. Pages 17–34 in A. D. Baldwin, Jr., J. DeLuce, and C. Pletsch, editors. Beyond preservation: restoring and inventing landscapes. University of Minnesota Press, Minneapolis, Minnesota, USA.

Jordan, W. R., III, M. E. Gilpin, and J. D. Aber. 1987. Restoration ecology: a synthetic approach to ecological research. Cambridge University Press, Cambridge, UK.

Lubchenco, J. 1998. Entering the century of the environment: a new social contract for science. Science **279**: 491–497.

Lyle, J. T. 1985. Design for human ecosystems. Van Nostrand Reinhold, New York, New York, USA.

———. 1994. Regenerative design for sustainable development. John Wiley and Sons, New York, New York, USA.

McHarg, I. L. 1969. Design with nature. Natural History Press, Philadelphia, Pennsylvania, USA.

Morrison, D. G. 1979. Prairie grasses, Monarch butterflies, rosehips: the "wild" moves in on the backyard. Landscape Architecture **69**: 141–145.

Nassauer, J. I. 1995a. Cultural principles for landscape ecology. Landscape Ecology **10**: 229–237.

———. 1995b. Messy ecosystems, orderly frames. Landscape Journal. **14**: 161–170.

———. 1997. Cultural sustainability: aligning aesthetics and ecology. Pages 65–83 in J. I. Nassauer, editor. Placing nature: culture in landscape ecology. Island Press, Washington, D.C., USA.

Noss, R. F., and A. Y. Cooperrider. 1994. Saving nature's legacy. Island Press, Washington, D.C., USA.

Olin, L. 1988. Form, meaning, and expression in landscape architecture. Landscape Journal **7**: 149–168.

Poole, K. 1994. Ecology as content: a subversive (alternative) approach to ecological design. In Ecology, aesthetics, and design. Scholarly papers presented at the 1994 ASLA Annual Meeting and Expo. ASLA, Washington, D.C., USA.

Riley, R. B. 1988. From sacred grove to Disney World. Landscape Journal 7: 136–147.

Risser, P. G. 1987. Landscape ecology: the state of the art. Pages 3–14 in M. G. Turner, editor. Landscape heterogeneity and disturbance. Springer-Verlag, New York, New York, USA.

Slaughter, T. P. 1996. The natures of John and William Bartram. Vintage Books, New York, New York, USA.

Spirn, A. W. 1988a. Nature, form, and meaning: guest editors' introduction. Landscape Journal 7: ii.

———. 1988b. The poetics of city and nature: towards a new aesthetic for urban design. Landscape Journal 7: 108–126.

Steinitz, C. 1993. GIS: a personal historical perspective. GIS Europe, June: 9–22.

Westley, F. 1995. Governing design: the management of social systems and ecosystems management. Pages 391–427 in L. H Gunderson, C. S. Holling, and S. S. Light, editors. Barriers and bridges to the renewal of ecosystems and institutions. Columbia University Press, New York, New York, USA.

Worster, D. 1977. Nature's economy: a history of ecological ideas. Cambridge University Press, Cambridge, UK.

CHAPTER 10

On Teaching Ecological Principles to Designers

Carl Steinitz

This chapter presents the framework within which I have organized the teaching of ecological principles to designers, and which I have used for many years to integrate lectures, studios, and research. For more than thirty years, I have been teaching in the Department of Landscape Architecture of the Harvard Graduate School of Design. During this time, I have visited almost all of the landscape programs in North America and Europe, and many in the rest of the world. I certainly comprehend the great variety of institutional settings from which the subject of this book is derived and in which its findings will have influence. I am sure that there is no single and appropriate set of conclusions, and I am absolutely certain that my experiences at Harvard have limits to their transferability. Nonetheless, I hope that my contribution will be of interest, use, and perhaps of influence.

This contribution must be seen in the context of my personal experience. I entered this field from its edge, bringing some ideas, but without substantial prior education or experience in either landscape architecture or ecology. In retrospect, I was fortunate, curious, energetic, somewhat iconoclastic, seriously interested in teaching and a broad range of major environmental issues, and at an institution that valued and supported my personal and academic "research and development." I always had very good students and collaborating faculty. And I learned much of what I now think I know in large part from these other people. In short, I am a consumer of ecology, not a producer.

I appreciate Herbert Simon's (1969) definition of design, and especially when the word "design" is seen as an active verb:

Everyone designs who devises courses of action aimed at changing existing conditions into preferred ones.

Surely all of us can relate to this definition.

In "Design Is a Verb; Design Is a Noun" (Steinitz 1995), I argued that both ecology and art arc defined in human terms. They are different, they can be in conflict and frequently are, but they can also be symbiotic. The sad fact is that all too often our field can be seen as dividing between the two conflicting cultures—art and ecology. However, in my view, "design" as a noun should be an idea made tangible, but also more than that; it should be a social communication that is experienced and understood. As George Santayana reminds us,

> When creative genius neglects to ally itself to some public interest, it hardly gives birth to wide or perennial influence. Imagination needs a soil in history, tradition or human institutions, else its random growths are not significant enough and, like trivial melodies, go immediately out of fashion.

A related point is made in a commentary on *Exodus,* chapter XXV,

> The true artist possesses the power to inspire others. A light that cannot kindle other lights is but a feeble flame. The core of art is its teaching and ennobling influence not only on other artists, but on humanity.

Some might argue that the primary objectives of landscape architecture are aesthetic, others that they are ecological, and others that they are relationships between ecology and perception. My view is that because landscape architecture is the result of design as a verb, of an anthropocentric process of intentional change, its primary aims and decision criteria are social relationships. The primary means of design, the materiality and the organization of experience, are the appropriate roles of ecology and perception (Steinitz 1995). Thus, regardless of whether design is directed toward intentional change or intentional conservation, it has the primary social objective of changing people's lives by changing their environment and its processes, including its ecological processes.

The teaching of design "as a verb and as a noun" is a very difficult task, and in recent years I have become a critic of the ways in which we teach (speaking broadly and not ad hominem). I have written and lectured on the subject (Steinitz 1990, 1993, 1995), and I have formulated a framework both for design and for education that tries to sharpen the questions that we pose to our students. I have found the framework to be both robust and useful in organizing my academic activities, and I think that it is germane to the issues posed by the organizers of this volume. I know that it is familiar to some readers, but it may be worthwhile to repeat the short description for others.

My proposed framework (Figure 10-1) organizes six different questions, each of which is related to a type of theory-driven answer or model. The framework is "passed through" at least three times in any project: first, downward in defining the context and scope of a project—defining the questions; second, upward in specifying the project's methods—how to answer the questions; and third, downward in carrying the project forward to its conclusion—getting the answers. The six questions with their associated modeling types are listed downward, in the order in which they are usually considered when initially defining a landscape project.

I. How should the state of the landscape be described; in content, boundaries, space, and time? This level of inquiry leads to Representation models.

II. How does the landscape operate? What are the functional and structural relationships among its elements? This level of inquiry leads to Process models.

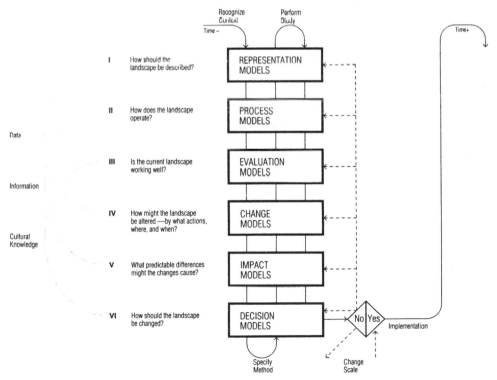

FIGURE 10-1.
A Framework for Design.

III. Is the current landscape functioning well? The metrics of judgment, whether health, beauty, cost, nutrient flow, or user satisfaction, lead to Evaluation models.

IV. How might the landscape be altered; by what actions, where, and when? This is directly related to level I, in that both are data; vocabulary and syntax. This fourth level of inquiry leads to Change models. At least two important types of change should be considered: change by current projected trends, and change by implementable design, such as plans, investments, regulations, and construction.

V. What predictable differences might the changes cause? This is directly related to level II, in that both are based on information; on predictive theory. This fifth level of inquiry shapes Impact models, in which the Process models (II) are used to simulate change.

VI. Should the landscape be changed? How is a comparative evaluation among the impacts of alternative changes to be made? This is directly related to level III, in that both are based on knowledge; on cultural values. This sixth level of inquiry leads to Decision models. (Implementation could be considered another level, but this framework considers it as a forward-in-time feedback to level I, the creation of a changed representation model.)

Note that the six levels have been presented in the order in which they are normally recognized. However, I believe that it is more important to consider them in reverse order, both as a more effective way of organizing a landscape planning study and specifying its method (which I consider the key strategic phase) and as a more effective educational approach. A design method for a project should be organized and specified upward through the levels of inquiry, with each level defining its necessary contributing products from the models next above in the framework.

VI. Decision—To be able to decide to propose or to make a change (or not), one needs to know how to compare alternatives.

V. Impact—To be able to compare alternatives, one needs to predict their impacts from having simulated changes.

IV. Change—To be able to simulate change, one needs to specify (or design) the changes to be simulated.

III. Evaluation—To be able to specify potential changes (if any), one needs to evaluate the current conditions.

II. Process—To be able to evaluate the landscape, one needs to understand how it works.

I. Representation—To understand how it works, one needs representational schema to describe it.

Then, in order to be effective and efficient, a landscape-planning project should progress downward at least once through each level of inquiry, applying the appropriate modeling types:

I. Representation
II. Process
III. Evaluation
IV. Change
V. Impact
VI. Decision

At the extreme, two decisions present themselves: no and yes. A "no" implies a backward feedback loop and the need to alter a prior level. All six levels can be the focus of feedback; (IV), "redesign," is a frequently applied feedback strategy.

A "contingent yes" decision (still a "no") may require a shift in the scale or size or time of the study. (An example is a highway corridor location decision made on the basis of a more detailed alignment analysis.) In a scale shift, the study will again proceed through the six levels of the framework, as previously described.

A project should normally continue until it achieves a positive "yes" decision. (In my area of application, a "do not build" conclusion can be a positive decision). A "yes" decision implies implementation and (one assumes) a forward-in-time change to new representation models.

While the framework and its set of questions looks orderly and sequential, it frequently is not so in application. The line through any project is not a smooth path: It has false starts, dead ends, serendipitous discoveries—but our activities do pass through the questions and models of the framework as I have described it, before a "yes" can be achieved. The same questions are posed again and again. However, the models, which are the answers, vary according to the context (Steinitz et al. 1996).

The framework in Figure 10-1 can be recognized as both a scientific research process and as a simulation model, and these are how I present it to my students. The framework has been useful in organizing studios, advising doctoral students, and structuring case study papers by masters students. I have also used it to organize large interdisciplinary research programs. For three landscape planning examples that combine research with teaching, and with major ecological components, see the following on my Web site <http: www.gsd.harvard.edu/info/directory/faculty/steinitz/steinitz.html>:

Alternative Futures for Monroe County, Pennsylvania
Alternative Futures for the Region of Camp Pendleton, California
Alternative Futures in the Western Galilee, Israel

Figure 10-2 shows how the six questions and their related model types can be used to describe or structure a multilevel academic program in our and related fields. It recognizes three levels of professional education: entry level (typically a bachelor's or first-professional master's-level degree), postprofessional (typically a master's degree), and research-professional (typically a doctoral degree, although this level is intended to recognize creative and nonacademic professional practice).

A professional-entry educational perspective typically teaches (wrongly, in my opinion) what is purported to be "The Design Method," a conservative path from level I through level VI (albeit frequently called "data-analysis-synthesis-evaluation"). A postprofessional education approach is more likely to be speculative, recognizing diversity of method and the need to fit the approach to the problem, a "thinking path" from VI through I followed by action from I through VI. The academic level of critical scholarship and that of creative practice may take an iconoclastic attitude toward the state of theory itself. From this viewpoint, any level is an appropriate starting point or focus in the reframing of questions and in the search for answers, and thus for new methods, models, and theory. The framework for professional education also describes both the sources and prime characteristics of each type of model at each educational level, as well as the manner in which design problems are normally given to students and design methods decided upon. The education framework in Figure 10-2 purposely omits "content." When applied to a curriculum for teaching landscape planning, the content might include aspects listed in Figure 10-3.

From the perspective of this book on the teaching of ecology to designers (in Simon's definition), a focus on the horizontal line that describes Process Models may be most useful. I think that our entering students require an understanding of ecological principles as robust "rules of thumb." These can be achieved through site visits, case studies, survey courses, and readings such as Carroll (1972), Hendler (1977), and Dramstad, Olson and Forman (1996), and they can certainly be tested and used in introductory-level studios.

Postprofessional master's-level students should be held to a higher standard. Here, more specialized courses are an appropriate vehicle for learning, such as courses in hydrology and wetlands, soils, wildlife management, and human ecology. These can also be tested and applied in studio, and preferably in interdisciplinary activities on large, complicated, and multiscale projects. While holding postprofessional students to a higher standard, and offering them a "deeper" experience, we should also hold their studio work to a higher standard. Their designs should be comparatively evaluated by the very same models that we seek to incorporate into their design thinking. This is not something normally or easily done by our predominantly informal jury-based evaluation system. I do it in each of my studios.

LEVEL OF EDUCATION

LEVEL OF INQUIRY	Professional Entry ↓ given problem	Postprofessional ↓ select problem	Research Professional ↓ seek problem	
I. Representation Models	introduced basic	specialized in-depth	invented experimental	source char.
II. Process Models	common knowledge "rules of thumb"	researched diagrammatic	empirical replicable	source char.
III. Evaluation Models	as told simple	as experienced prof. judgment	as sought informed	source char.
IV. Change Models	precedent archetypes	experience adaptations	hypothesis innovations	source char.
V. Impact Models	case studies reasonable guess	formal models rationale	experiments evidence	source char.
VI. Decision Models	profession + faculty conservative	faculty + mentor speculative	mentor + self theoretical	source char.
	given method ↑	select method ↑	create method ↑	

FIGURE 10-2.

A Framework for Professional Education.

LEVEL OF INQUIRY

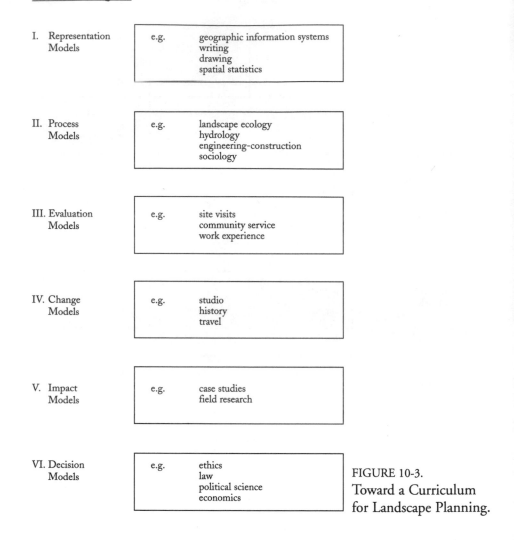

| I. Representation Models | e.g. | geographic information systems
writing
drawing
spatial statistics |

| II. Process Models | e.g. | landscape ecology
hydrology
engineering-construction
sociology |

| III. Evaluation Models | e.g. | site visits
community service
work experience |

| IV. Change Models | e.g. | studio
history
travel |

| V. Impact Models | e.g. | case studies
field research |

| VI. Decision Models | e.g. | ethics
law
political science
economics |

FIGURE 10-3.
Toward a Curriculum
for Landscape Planning.

For doctoral-level students, we should require an even higher standard. At this level, testing via empirical models is expected but not sufficient. Doctoral students should be adding to our ecological knowledge and its applicability. Some good examples are the dissertations of Sharon Collinge, Kongjian Yu, and Kristina Hill.

Sharon Collinge (1995), in her thesis "Spatial Arrangement of Patches and Corridors in the Landscape: Consequences for Biological Diversity and Implications for Landscape Architecture," completed a series of field experiments with insects in a native grassland near Boulder, Colorado. These tested hypotheses regarding the influences of fragment spatial characteristics and patterns of land conversion on species loss, recolonization, and individual movement patterns. While field experiments with native grassland insects provide specific information on terrestrial insect response to fine-scale changes in habitat spatial structure, this research may serve as a model system for increasing our understanding of the ecological implications of particular spatial patterns of landscape configuration at broader scales (Collinge 1995, 1996).

In his dissertation entitled "Security Patterns in Landscape Planning: With a Case in South China," Kongijian Yu (1995) investigated whether there were certain landscape patterns that can most or more effectively safeguard landscape-related processes while maximally providing possibilities for changes. The thesis proposes the concept of security patterns (SPs)—portions and positions that have, or potentially have, strategically critical influences on landscape processes. Ecological, visual, and agricultural SPs are explored. For example, the dynamics of the ecological processes are represented by accessibility surfaces based on calculations of minimum cumulative resistance. Four structural components can be identified on the accessibility surfaces: buffer zones, intersource linkages, radiating routes, and strategic points. These four components, specified by quantitative and qualitative parameters, together with the identified sources, compose the ecological SPs at various security levels. SPs such as these can be used to develop various change alternatives within the same security level. They can constrain change models and they can be a basis for spatial bartering (Yu 1995, 1996).

Kristina Hill (1997), in her thesis "The Representation of Categorical Ambiguity: A Comparison of Fuzzy, Probabilistic, Boolean, and Index Approaches in Suitability Analysis," compares four different ways of representing categories in a habitat suitability analysis: (1) index categories, which use a single function to define graded membership in categories; (2) Boolean categories, which use a binary membership function to define membership in discretely bounded categories; (3) probabilistic categories, which also use discretely bounded categories but allow for stochastic uncertainty in the assignment of events or measurements to these categories; and (4) fuzzy sets, which define graded membership categories and allow them to overlap along the same measured variable. The theoretical similarities and differences between these four methods are presented, with tests of their ability to represent the uncertainties associated with the spatial distribution of a bird species, the black-capped chickadee (*Parus atricapillus*). If a land manager must predict the abundance of a species from only vegetation data, the most successful approach in the forested landscape of the Massachusetts case study was to use three overlapping fuzzy sets that cover the range of variation present in the vegetation variables. In general, the use and construction of models that focus attention on the ambiguity of category definitions may promote more rigorous application of these models, and a more informed dialogue between managers, planners, and ecologists on the conflicts inherent in conservation and development, than would a single-function model derived from multivariate methods (Hill 1997).

An Example from the Core Curriculum

Most of the issues associated with teaching ecological principles to designers undoubtedly center on first- and post-professional students. I am responsible for the coordination and production of a considerable portion of the fourth semester of our core curriculum for first-professional degree, master's-level students. Prior to this semester, all students have had a course in landscape ecology (taught by Richard Forman), and most have had additional relevant academic experience, either in their prior bachelor's-level education or at Harvard. However, their studio experiences have generally not emphasized ecological concerns and, if anything, have favored "artistic" design expression at a project scale. I might add that very frequently, their design proposals are more ecologically impoverished than the original site conditions were. Little serious attempt is made to consider the possibility of design that elegantly addresses both ecological and expressive concerns, although in recent years that prospect has increasingly been considered.

During this fourth semester, I teach a course entitled "Theories and Methods of Landscape Planning," which is structured by my framework and consists of lectures and case study reports. See Steinitz 1998a and 1998b.

The lectures are scheduled so that I discuss the theories and methods in the same period that the students in the studio, which I coordinate, are applying them. The studio is organized into nine stages, which range from a regional to a project scale and back. See Steinitz 1998b.

Each of the nine subproblems of the studio schedule is focused around one of the questions in my framework, and the sequence of answers, albeit subject to constant revision, cumulate toward a relatively complete multiscale project.

The second studio "problem" requires the students to model and evaluate about a dozen site processes. Pairs of students, using a geographical information system on central Massachusetts, assess the vulnerability—the risk of potential harm from the impacts of land-use change—of the entire study area. The students must define, research, and model the process, and their "product" is a concise paper, a lecture to the entire class, and a GIS model and its resultant risk assessment, the final five-level map, which is the Evaluation Model. These are evaluated by knowledgeable faculty (for example, Bob France, a coteacher of the studio) and the students, some of whom are quite advanced in the study and application of ecological principles. The resulting work is then published and shared via Web site by all participants in the lecture course and the studio. The models and their resulting maps are a major source of criteria for the third studio problem, which is the site selection for a new community, the expansion of an existing town, and the design of a spatial conservation strategy for the town and its region. The maps are brought to the site region and "tested" via the direct experience of the students, who use their evaluations to guide and "validate" their siting decisions. They are also used as the basis for impact models that the students also define, and that are used later in the studio to comparatively evaluate all the design proposals.

The (current) list of Process Models leading to Evaluation Models is the result of discussions among the teaching team and also reflects a departmental strategy to emphasize ecological issues and concerns during this semester. Thus, the list in Table 10-1 represents a collective judgment rather than a rigorously theoretical perspective.

It is important to note that the list of processes is not limited to the eco-

TABLE 10 1. Process models/evaluation models

Wetlands: A variety of federal, state, and local legislation prohibits disturbance of wetlands, including bogs, marshes, lakes, streams, floodplains, and estuaries. Policies provide for peripheral buffer zones within which construction cannot occur. Once damaged, can wetlands be restored or re-created?

Surface-Water Quality and Erosion: Landscape construction and maintenance increases storm runoff and erosion. Particular soils are more erosion-prone than others, and soil erosion may result in sedimentation and degradation of nearby lakes and streams. Residential and agricultural areas also influence nutrient input to adjacent water bodies. How can development be sited to minimize detrimental changes in surface-water quality?

Groundwater Quality and Septic Systems: Landscape construction, maintenance, and human occupancy alter groundwater recharge, flow, and quality. Groundwater may be the drinking water source of your development. How can the proposed development avoid lowering groundwater quality?

Soils and Agricultural Productivity: Legislation requires conservation of soils well suited for agricultural production. These soils may also be appropriate for building sites.

Energy/Microclimate: New England can be very cold and windy in winter, and this affects heating energy costs, outdoor comfort, agricultural productivity, road icing, etc. Where are these exposed and/or colder areas? Should they be avoided? Where is it very hot in summer?

Biodiversity: Where are the most "biologically rich" areas of Petersham?

(continued on next page)

TABLE 10-1 *continued*

Connectivity: In the face of continued habitat loss and isolation, many landscape ecologists have stressed the need for providing landscape connectivity, in particular wildlife corridors. Despite continuing debate over the effectiveness of corridors in enhancing biodiversity, there is a growing body of research that cites the positive net benefits accruing from incorporating linkages between habitat patches.

Fragmentation: The loss and isolation of habitat is often defined in terms of fragmentation. Fragmentation is not only a human-induced phenomenon but can be due to natural disturbances as well. The spatial scale at which fragmentation occurs is a key concept for identification of alternative strategies to preserve and maintain biodiversity (see pp. 16–20 in Principles Section of L.E. Guidelines Document).

Rare and Endangered Species: The Endangered Species Act protects threatened and endangered species of plants and animals and the habitats in which they are found. What rare or endangered species are found in Petersham, and where do they live?

Historic and Cultural Resources: Petersham is rich in European colonial history. What about pre-European history and recent "history"? What are the historic and cultural resources of these towns, and should they be integrated with your design or avoided?

Visual Quality/Genius Loci: The urban landscape has developed over time into a "New England character" that is highly valued by current residents and tourists. What is it and where is it? Can it be conserved?

Cost of Big Buildings: What is the cost of constructing a large building on your site? Are there ways to maximize economic efficiency?

Sewage Treatment: A critical aspect of the future development of Petersham is the provision of effective sewage treatment. What are some options, and what are key locational factors?

logical realm. As can be seen, several are social or economic or physical, and in addition to this list, each of the design teams has several other important and socially derived criteria to meet in their site evaluation and design activities. In other words, while there is a distinct and necessary role for ecological processes, it is neither unique nor is it necessarily a dominant one in all design decisions. In fact, there are different priorities that must be assessed by the students as a function of the issue at hand. A conservation strategy may be predominantly ecological in orientation, but the siting of a religious community might be dominated by its tenets of belief or the needs of its rituals. And traffic and taxes do count.

It is also important to note that the level of modeling that we consider appropriate to this studio activity focuses on the risk to the ecological and other processes of the physical aspects of change that generate impacts (which can be positive or negative). This utilization of rules of thumb aims to prevent problems via siting and design, and also guide proposed changes. This is in contrast to the more spatially and quantitatively complex modeling on which our more specialized courses focus and that are the basis of our research activities, most advanced studios, and doctoral-level research.

The success of this approach is usually obvious. While these process-modeling activities are undertaken, they are the total focus of the entire academic enterprise. Students are attentive and energetic in the highly structured activities. They know that their products will be directly necessary and useful to their following studio activities and to the comparative evaluations of their studio designs, and they have high motivation to listen and learn from their colleagues. In large part, the influence of ecological (and nonecological) processes on the products of design (as a verb) is because of the students' direct role in defining and implementing the models. For many of the students, this activity provides impetus to enter the more specialized courses offered by my faculty colleagues. These efforts are certainly not "stand alone"; they are totally integrated into the broader enterprise of landscape planning and design.

As for the design "product," I will suggest that the reader look at the several examples that are on the Web and especially those from the studios on the Camp Pendleton Region, which were conducted between Harvard and Utah State University. Several designs at four different and linked scales are represented: the large region, a third-order watershed, a large development project, and any typical house lot. All the designs are substantially influenced by the several ecological models that described and evaluated the conditions within which the studios were set.

I end with a quote from Lowry (1965) that still summarizes why I think that students at all levels learn best when modeling processes that they need to understand in studio, in research, and in practice:

> Above all, the process of model building is educational. The participants invariably find their perceptions sharpened, their horizons expanded, their professional skills expanded. The mere necessity of framing questions carefully does much to dispel the fog of sloppy thinking that surrounds our efforts at civic betterment. (p. 166)

Citations

Carroll, A. 1972. The developer's handbook. Connecticut Dept. of Transportation, Hartford, Connecticut, USA.

Collinge, S. K. 1995. Spatial arrangement of patches and corridors in the landscape: consequences for biological diversity and implications for landscape architecture. Dissertation, Harvard University, Cambridge, Massachusetts, USA.

———. 1996. Ecological consequences of habitat fragmentation: implications for landscape architecture and planning. Landscape and Urban Planning 36: 56–77.

Dramstad, W. E., J. D. Olson, and R. T. T. Forman. 1996. Landscape ecology principles. Pages 1–40 in Landscape architecture and land-use planning. Harvard Graduate School of Design, American Society of Landscape Architects, and Island Press, Washington, D.C., USA.

Hendler, B. 1977. Caring for the land: environmental principles for site design and review. American Society of Planning Officials, Chicago, Illinois, USA.

Hill, K. 1997. The representation of categorical ambiguity: comparison of fuzzy, probabilistic, Boolean, and index approaches in suitability analysis. Dissertation, Harvard University, Cambridge, Massachusetts, USA.

Lowry, I. 1965. A short course in model design. Journal of the American Institute of Planners 31: 158–166.

Simon, H. 1969. The sciences of the artificial. MIT Press, Cambridge, Massachusetts, USA.

Steinitz, C. 1990. A framework for theory applicable to the education of landscape architects (and other design professionals). Landscape Journal 9: 136–143.

———. 1993. A framework for theory and practice in landscape planning. GIS Europe: 42–45.

———. 1998a. Planning and design of landscapes. GSD 1212, <http://www.gsd.harvard.edu./info/directory/faculty/steinitz/steinitz.html>.

———. 1998b. Theories and methods of landscape planning. GSD 3307, <http://www.gsd.harvard.edu./info/directory/faculty/steinitz/steinitz.html>.

———. 1995. Design is a verb; design is a noun. Landscape Journal 14: 188–200.

Steinitz, C., et. al. 1994. Alternative futures for Monroe County, PA. Harvard University, Graduate School of Design, Cambridge, Massachusetts, USA

———. <http://www.gsd.harvard.edu./info/directory/faculty/steinitz/steinitz.html> Alternative futures for Monroe County, Pennsylvania
Alternative futures for the region of Camp Pendleton, California
Alternative futures in the Western Galilee, Israel

Steinitz, C., M. Binford, P. Cote, T. Edwards Jr., S. Ervin, R. T. T. Forman, C. Johnson, R. Kiester, D. Mouat, D. Olson, A. Shearer, R. Toth, and R. Wills. 1996. Biodiversity and landscape planning: alternative futures for the region of Camp Pendleton, CA. Graduate School of Design, Harvard University, Cambridge, Massachusetts, USA.

Yu, K. 1995. Security patterns in landscape planning: with a case in South China. Dissertation, Harvard University, Cambridge, Massachusetts, USA.

———. 1996. Security patterns and surface model in landscape ecological planning. Landscape and Urban Planning 36: 1–17.

CHAPTER 11

Looking Beneath the Surface:
Teaching a Landscape Ethic

Michael Hough

The Shire conference on teaching ecology in design and planning programs held in July 1998 was a significant event for environmental education. It raised some key questions about what students should know about the institutional and societal context within which education takes place: How does one engender a thoughtfulness in students about the nature and purpose of design? How important is the development of environmental values in teaching? How will students be prepared to contribute their skills to the environmental crises facing society in the decades to come? Is an education in ecology just one specialization in a wide array of approaches to design, or is it basic to all design? A glance at some of the professional glossy magazines and coffee table books that cross my desk might suggest the former is currently truer than the latter.

My perspective on the subject at hand has been shaped by three experiences: in the early years of the undergraduate program in landscape architecture at the University of Toronto, where the familiar ground of spatial design tradition frequently lacked critical analysis and broad environmental perspectives; at York University's Faculty of Environmental Studies, where the integration of diverse approaches to environmental issues and self-motivated learning are hallmarks of the program, but where awareness of the spatial experienced landscape is frequently lacking; and in 38 years of professional practice, which has brought together both academic and practical experiences and shaped my thinking.

When one considers the twentieth-century values and priorities that shaped the urban environment most of us live in, it becomes a matter of urgency to determine what the design professions should be contributing in the decades of the twenty-first century. The list is long: the need to make them civilizing places to live in, regenerating ecologically degraded urban landscapes, coming to terms with their social and multicultural diversity, protecting the ecological integrity of larger landscapes—the national parks and protected natural areas of the North American continent and the conflicting visions between use and beautiful landscape. The development of a design philosophy that recognizes diversity and the differences between one place and another is, as Eugene Odum (1969) has suggested, central to the maintenance and enhancement of environmental and social health. When we recognize that the overwhelming trends in information- and knowledge-based technologies are increasingly focused on computer screens and simulated reality, these larger environmental issues speak to the urgency of grounding students in an experiential understanding of place and environmental values. The nurturing of environmental ethics is also centered on exposure to the literature on environmental thought—the exploration of the relationships between the human and the nonhuman in nature (Evernden 1992; Livingston 1994; Seddon 1997). Anne Spirn admirably addresses the topic of constructions of nature in Chapter 2.

The Role of the Natural Sciences in Design and Planning

My focus on the practice and teaching of ecology is based on a number of premises. First, the discussion of *what* should be taught in design and planning programs is less important than *why and how*, since literacy in the natural sciences does not, by itself, make thoughtful designers. The argument clearly is not against science, whose application is urgently needed at a time when the physical world we live in is everywhere under threat. It's about a number of things. First, there's a need to inculcate a critical perspective on how places can be understood—initially by direct observation and subsequently by scientific investigation. Second, the conventions of formal design too frequently ignore a scientific perspective that provides the basis for integrative thinking. Third, the teaching of ecological principles is a necessary component of all design curricula, irrespective of focus. It is, therefore, as much about an *informed* environmental ethic—doing the right things for the right reasons—as it is about technical scientific knowledge. It will affect every aspect of student work in later life, and the values and priorities they bring to the issues that will confront them in the future. Fourth, the term "ecology" involves both the physical and the biological processes of the land that include the concept of causality—the processes of climate, geology, geomorphology, hydrology, soils, plants, and animals reflect the historical reasons for a place's identity and its sense of wholeness (McHarg 1969). Fifth, one of challenges of the collaborative process lies in overcoming the arcane language that has traditionally been invoked to ensure that no one outside the various professional communities understands what we mean. This puts an onus on effective dialogue between the disciplines as an basis for transdisciplinary work.

The premise suggested here, and echoed by James Karr (Chapter 6), is that being well versed in knowledge or methods is less important than having familiarity with the interdisciplinary process. Particularly compelling from the perspective of wholeness is Sandra Steingraber's description of the intimate connection between the chemical contaminants found in rivers and aquifers and those found "within the miraculous aquarium of a pregnant woman's body" (Chapter 8, p. 192), and the myriad of other interconnections between natural systems and human activities.

The Influence of Unpredictability

The natural sciences clearly have technical and practical application to the designer's search for an ecologically sustainable basis for practice—of fitting human needs and places into an ecologically functioning framework, where the latter is left unimpaired by the former. But the sciences can also teach us about other more profound ideas that have to do with unpredictability in natural systems and how this applies to the design and planning disciplines. "There is a growing realization among resource scientists and managers that nature is capricious and that a great deal of uncertainty underpins theory about the dynamics of populations and communities" (Nudds 1999, p. 179). The recognition in the scientific community of unpredictability of natural systems has close parallels with the evolutionary processes of the city whose form and growth change and adapt in response to unforeseen economic and social conditions over time, and over which designers and planners have little control. An answer to this apparent conundrum may be found in the notion of adaptive resource management, a process developed by biologists to deal with the uncertainty that accompanies management models based on ecological theory. Because knowledge about nature is imperfect, managers need to learn while doing (Nudds 1999); the principle being that small incremental steps are the best way to determine whether a particular course of action will achieve its initial goal. Learning facilitated by feedback obtained from monitoring allows the manager to modify and adapt the goals of the management plan accordingly.

There's a need to rethink conventional wisdom and embrace the principle of unpredictability in the design of human habitats and places. Things never stay the same. Forty years ago, Jane Jacobs (1961) observed that cities are laboratories for experimentation and change and are far too complex to be understood by planners, designers, or anyone else. Robert Fulford, author of *The Accidental City* (1995), echoes my own conclusions: that while those in charge may imagine that they're responsible for creating cities, complex and stable neighborhoods are *not* the product of grand designs or fast-track developer master plans. They are created by the people who live there, and through a multitude of small decisions and choices made by many thousands of individuals.

The processes of change, whether social or natural, seem to respond to similar forces. Similarly, people often use designed projects in ways that are quite different from the designer's original plans. In fact, it may be said that the ability to predict human behavior is one of the myths perpetuated by the design professions. William Whyte (1980), who spent many years finding out how people use the civic spaces in New York City, clearly showed that the most dynamic and successful places are those where people themselves take charge, in often surprising ways (Figure 11-1). Design intervention may be no more than establishing a spatial structure, providing a focal point in a

FIGURE 11-1.
Toronto City Hall square. Built in the early 1960s as an open plaza and skating rink, and defined by a colonnade at its edges, the square has evolved to include a weekly market, a portable band shell, a restaurant, and a monument to peace. Its simple framework has allowed many adaptations to take place, enriching what was once a space into a social place for numerous civic functions. (Photo by M. Hough.)

square, or seats that can be moved around at will. How local neighborhood streets are used is also revealing. Officially, streets are the responsibility of roads departments, and it's illegal for children to be playing there. But they do anyway, because streets are often the best places for all kinds of activities (Hough 1995). How people use streets, in fact, is exactly what they were originally intended for—to *reinforce* social life in local places.

To do this with any confidence means designing one step at a time, basing decisions first on general visions for the future, and second, on what is known at any one time. This is the basis for community design where realizing the goals of citizen organizations is an ongoing, long-term process whose outcomes cannot be predicted. (I examine this issue later in the chapter.) The idea of adaptability also focuses attention on the need to keep up with new ideas. As Carolyn Adams observes (Chapter 5), this is central to the evolution of ideas in one's own disciplines as well as keeping a finger on the progress of others to remain relevant. It is equally necessary for practitioners if they are to develop local solutions for local landscapes. For the practice of design and planning, therefore, a key principle is about doing as little as possible. The pressure to do as *much* as possible often appears to be endemic to the land design disciplines (Hough 1990.) Doing as *little* as possible involves the idea that the greatest diversity and identity in a place often comes from minimum interference. This does not mean that planning and design are irrelevant or unnecessary to a world that if left alone would take care of itself. It implies, rather, that change can be brought about by giving direction or setting a framework within which people can create their own social and physical environments (Hough 1990).

Green Networks

Appropriate frameworks for evolving human communities, or landscapes, can mean bylaws or ordinances that permit change. The restoration of meadow or woodland communities, for instance, can best develop under adaptive management regimes, unlike the universal green carpets that dominate most urban parks and that are maintained to stay exactly the same from year to year. Frameworks can also mean infrastructure—the services that settlement requires in order to function. Sewers, water, hydro, city streets and sidewalks, and public monuments are typical examples. An environmental perspective suggests that this definition should be extended to include *green infrastructure*—the greenways, corridors, and natural areas that can become an organizing framework for urban form and future growth—a very different paradigm from conventional land-use planning (Figure 11-2). Anne Spirn provides an interesting example in her discussion of Olmsted's approach to the Boston Fens and Riverway. Here he established a "skeleton of woods and wetland, road, sewer, and public transit [that] structured the growing city and

FIGURE 11-2.
Air view of Toronto's Don River. It is one of a system of natural corridors and continuous habitats that link its headwaters in the Oak Ridges Moraine to Lake Ontario. Discontinuities and fragmentation of natural habitats are the most crucial environmental issues facing urban regions. (Photo by M. Hough)

its suburbs" (Chapter 2, p. 38). The more recent theory and practice of land-scape ecology provides the scientific basis for reconnecting and restoring fragmented habitats in urban areas (Forman and Godron 1986). It has become one of the most useful tools for environmental planning and is central to teaching in design and planning programs.

Several other examples in practice will illustrate. A few years ago a Toronto study was commissioned to develop a greenway study for an outlying suburban municipality whose lands were largely rural. It proposed a network of protected natural and recreational corridors and restored habitats along existing river systems and man-made linear service ways (Figure 11-3) that would be in place *before* development occurs (Gore & Storrie, Hough Stansbury Woodland et al. 1993). The study had some significance for the municipality, because the single most destructive consequence of urban growth is the loss of natural habitats that occurs when their *continuity* is fragmented into isolated parcels. As an ecological planning concept, green infrastructure is relevant to abandoned brownfield sites in the inner city from multi functional perspectives, bringing together a range of environmental, social, and economic incentives (Hough, Benson, and Evenson 1997).

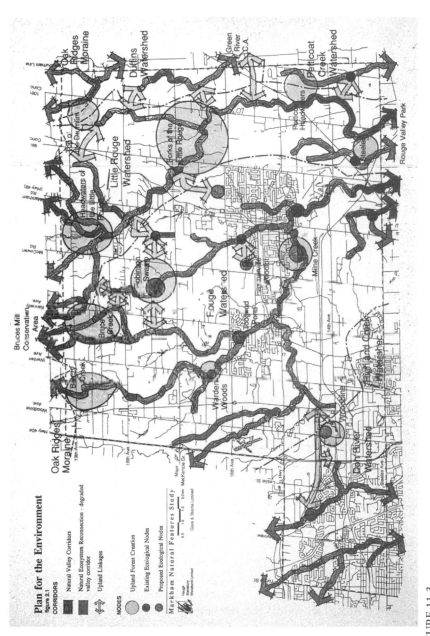

FIGURE 11-3.

A strategic "green" plan for an outlying municipality in the Greater Toronto Area. Protecting and restoring net-works of corridors and parks in urbanizing rural areas is a fundamental long-term measure to maintain habitat continuity *before* development occurs. (Source: Gore & Storie, Hough Stansbury Woodland et al., 1993)

In the industrial Ruhr Valley in Germany, the need for economic renewal has become most obvious in communities where once prosperous heavy industries have become obsolete. They have left behind large tracts of derelict land, polluted and channelized rivers, massive unemployment, and a depressed economy. In the Emscher region, an area of some 800 square kilometers and a population of 2 million people, major restructuring has begun *not* by stimulating the economy or creating jobs, but by *ecological* renewal. In the German view, acquiring much needed green space and restoring landscapes and water quality to environmental health is the practical and necessary basis for restoring jobs and economic health (Emscher Park International Building Exhibition 1991). The city of Minneapolis has also found conclusive evidence that parks and green areas protect the long-term tax base of neighborhoods. A study conducted by the city's parks department found that the market value of properties increased by 30 percent over five years. (Minneapolis Parks and Public Works Commission 1985).

Ecological Design as an Integrative Process

Design is inherently an integrative activity. As Joan Nassauer suggests (Chapter 9), it reinforces the traditions of design by setting up interdisciplinary discussion. It is also reflected in the much used "Ecosystem Approach," a process that brings together many traditionally competing agendas to establish a condition of health (Royal Commission on the Future of the Toronto Waterfront 1990). While this may sound dangerously like another professional buzzword, it is an essentially pragmatic view that recognizes the interrelationships between environment, economy, and community (Barrett and Kidd 1991). It establishes a powerful argument that economic well-being and environmental health are fundamentally interdependent. This approach to solving design problems is collaborative, concurrent, and integrated rather than consecutive and disconnected—an all too common way of thinking that inhibits the development of creative solutions.

The ecosystem approach also implies a nature-first view that must function as a basic structuring element defining human development and modifications to the land. This is the concept of landscape as green infrastructure, discussed earlier in this chapter, which has major implications for maintaining human and environmental health and for reestablishing a sense of place, at local level and regional scales. As the necessary basis for decision making, therefore, green infrastructure has essential didactic connections to the land-based disciplines.

Community Design

The increasing involvement of citizens as *partners* in decision making is probably one of the most significant manifestations of recent social change

(Hough 1995). It also represents a shift from the reactive citizen protests against government initiatives (such as the urban expressways that threatened North American cities in the 1960s and 1970s) to one that is proactive. Community and neighborhood groups are themselves initiating and designing projects at local and regional scales and establishing partnerships with government and the private sector. This manifestation of citizen empowerment will increasingly influence the designer's role in community design and planning. It implies a change from the usual public participation process required in various planning acts, to assisting community and interest groups to realize their own local needs. The consultative process is very different from the "we know best" approach that has long dominated the consulting field and government bureaucracies, where encouraging community involvement may be little more than harnessing citizen energies to fulfill predetermined agendas. Community design also includes an understanding of the peculiar recreational and spatial needs of different social and cultural groups; the diverse ways in which they shape their own environments; their ways of seeing nature; their aesthetic priorities. There's an implicit recognition that the designer makes only the first step in shaping the public landscape. What follows after that must be community driven for it to succeed.

There are major implications for planning and design here. As a process of community facilitation, the designer requires very different skills. They lie in an ability to facilitate the efforts of many interest groups and find commonalties between conflicting community goals. Community design is also about working collaboratively, with a range of professional expertise, with government agencies, as well as with the community, a key point explored by Joan Nassauer (Chapter 9). It's about having enough technical knowledge to ask pertinent questions, and the ability to navigate through complex relationships between the biophysical sciences, soil and water remediation, heritage protection, infrastructure and city services, land ownership, approvals agencies, and political agendas (Hester 1990).

Among the most interesting citizen organizations I have worked with is the Task Force to Bring Back the Don, a diverse group of Toronto citizens who in 1989 decided that Toronto's urban river (Figure 11-4) was a disgrace to the city and required community action to return it to health (Task Force to Bring Back the Don 1991). The issues were numerous and daunting. For over two hundred years, the lower river valley has been subject to deforestation, industrial development, and urbanization. Railway lines were routed through the valley early in Toronto's growth. The river's lower reach was channelized. The port industrial area, begun in 1912, replaced a large and productive lacustrine marsh through which the Don flowed to Lake Ontario. And in the mid-1950s the valley became a transportation corridor with the introduction of expressways, transmission lines, billboards, and municipal services. Aquatic life, birds, and animals disappeared as natural habitats were cleared from the valley. The entire watershed of

FIGURE 11-4.
The lower Don River as it was prior to the 1990s—an environmentally degraded channelized waterway. Bringing the river back to life has been the task of citizen action over ten years. (Photo by Task Force to Bring Back the Don)

36,000 hectares is over 80 percent urbanized, and municipalities upstream contribute most of the stormwater pollution to the lower part of the river.

Storm drainage is also the root cause of flash floods in urban rivers, as direct runoff replaces the water storage and infiltration that occurs in floodplains and wetlands under natural conditions. At its mouth, the Don in its straitjacket looks more like a sewer outfall than a river. Developments surrounding its lower reaches turned their back on a once natural setting. Consequently, the river has become a gap between places, not a place in itself. And since it was next to impossible to get into, no one knew or valued it.

The task force had four goals—reconnect the river to its mouth, restore natural habitats, improve water quality, and regain public access to restore its sense of place. Over the last ten years the task force and tens of thousands of volunteers have made astonishing progress with reforestation, wetland creation and protection of industrial heritage, and establishment of access points. In addition, other initiatives came about as a result of this work. For instance, the Don and Humber River Watershed Councils were established, and the city's downpipe disconnection programs were initiated by the Department of Public Works.

The Don Task Force recently celebrated its tenth anniversary, which raises a basic question: What has made this venture successful? Many reasons can usefully shed light on this initiative. First, political savvy. Early in its work, the task force aligned itself with partners, including city council and government agencies, thus ensuring continuity of the work. Second, funding. Multiple sources of funds for projects have been found at all levels of government, private sources, and non-government organizations. In addition, administrative funding and a planning coordinator, supplied by the city for ongoing operations, have allowed the task force to maintain control of its mandate and avoid the burnout that plagues many citizen initiatives. Third, massive support was received from the Toronto community and politicians, which has resulted in the Don's restoration process being enshrined in political decisions that affect the city and its waterfront.

This experience suggests that ecological sustainability is tied to citizen empowerment. Also critical is the fact that the community process for restoring the Don River to a state of health is *not* a master plan. Rather, it's a *strategy* whose purpose is to guide citizen efforts into the future. In this case, the long-term vision for achieving the fundamental goal is to heal the Don (Figure 11-5).

FIGURE 11-5.
The concept of a long-term vision, as opposed to traditional master plans, is the only way of guiding restoration efforts into the future, which is inherently unpredictable. The goals remain constant; getting there involves adaptation and change, in accordance with changing economic and societal conditions at any point in time. Drawing (a) illustrates how the goal of reconnecting the river mouth to the lake could be technically achieved in 1991 (Hough Stansbury Woodland 1991). Drawing (b) shows an alternative approach that was more realistic and implementable ten years later. The new scheme was a response to further study and knowledge of conditions and changing attitudes and values. (Hough Woodland Naylor Dance Leinster

(a)

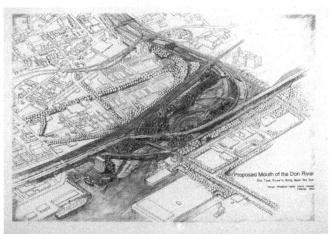

(b)

The process of getting there will change and adapt in response to changing social, economic, and political events over time, and over which no one has control. The vision therefore has no finishing post, because, like the city around it, it's an organic entity in the process of constant change and adaptation.

Linking Urbanism to Nature

Ian McHarg in his seminal book *Design with Nature* (1969) demonstrated the fundamental necessity of linking the natural sciences to regional planning, and his work and methods have had a critical influence on theory and practice. He stopped short, however, of addressing the built environment itself. The emergence of a widespread interest in city nature has brought several factors into sharp focus: the need for an environmental focus that links ecological principles to urban design and city form; and the inherent relationships between people and natural processes. While still in its beginnings, urban ecology will have major implications for how designers contribute to urban health and sustainability. Ecological restoration, designing with climate and energy conservation, the role of vegetation and soils in ameliorating air pollution, storm drainage systems, and the value of urban wilderness to cities are all roles for design that will increasingly preoccupy teaching, practice, and research.

Challenging Design Conventions: The Cult of the Aesthetic

The conventions of aesthetic values, good taste, and civic spirit frequently become anachronistic when one starts thinking ecologically. While no one would argue against the values of an appropriate aesthetic, the conventions of formal design have a way of subverting genuine insights into what lies beneath the surface of the visible landscape. Why is one place perceived to be beautiful and another not, some landscapes special and others commonplace (Figure 11-6)? Such value-laden questions suggest an inherent dilemma between those places we are taught to admire and emulate (the formal civic landscapes of most cities) and the derelict sites that urgently need the designer's attention (those places no longer visited by maintenance personnel). But these questions beg another: Which *are* the derelict sites in the city requiring rehabilitation? The fortuitous and often ecologically diverse landscapes representing evolving natural forces, or the formalized landscapes created by design? In the absence of an ecological view, one answer is certain. All efforts have traditionally focused on suppressing the former and nurturing the latter (Hough 1995).

The notion that an emerging plant community, or the apparent visual

FIGURE 11-6.

The contradiction of environmental values in cities—between the formal landscapes we have been taught to admire and emulate and the places that have evolved on their own. From a biodiversity perspective, compare Figure 11-6, a suburban neighborhood, with Figure 11-7, a wetland that has emerged on its own. Which is the site requiring rehabilitation? (Photo by M. Hough)

chaos of a thriving ethnic neighborhood, needs improving or tidying up is too easily imposed on urban places (Figure 11-7). An environmental focus provides us with the underlying framework for understanding and acting on the inherent diversity of natural systems from which a design aesthetic can emerge. It is also the basis for understanding the inherent order and structure of natural and human environments—what may be described as vernacular forms created by natural forces or social necessity. It may also reveal the environmental costs that accompany much of the public and private landscape. Awareness of the destructive consequences of keeping the front lawn greener than green with the help of fertilizers and herbicides is rapidly gaining currency among environmentally aware people. The aesthetic order of uniform and single-species street tree planting, so favored by urban designers as an appropriate expression of civic design principles, has long been shown to be biologically untenable. The chestnut blight and Dutch elm disease are persuasive cases in point. In a social context, Jane Jacobs (1961) revealed the inappropriateness of urban planning doctrines in the 1960s that advocated

FIGURE 11-7.
The thriving "chaos" of a fortuitous wetland that has emerged on its own in an abandoned industrial area. In the absence of strict design controls, both this wetland and an ethnic urban neighborhood tend toward increasing natural and social complexity. An environmental view suggests that the diversity of such natural and human communities can form the basis for an alternative approach to design form. (Photo by M. Hough)

the renewal of neighborhoods by demolition. Thus, the challenge for the designer is to find aesthetic expressions for urban design that are based on diversity rather than uniformity—one that comes to terms with the dynamics of natural processes.

Wild Nature: Conflicts between Beauty and Use

Beyond the city, the issues are not dissimilar. The conflict that arises between use and protection in the national parks and protected areas of North America is a constant threat to their purpose and goals—to protect ecological integrity for future generations (Parks Canada 1994; U.S. Park Service 1999). The early national parks were founded at a time when there was little known of wildlife corridors and habitat continuity and size, the consequences of habitat fragmentation from roads and railways, disturbance to wildlife, and natural boundaries large enough to support wide-ranging species. Park boundaries were, in fact, originally drawn to protect wild scenery, not ecosystem functions (Sellars 1997). The problem remains relevant today since the attraction of spectacular, beautiful landscapes (Figure 11-8) continues to draw millions of people to the parks, along with tourism promotion, recreational development, urban settlement, and commercial facilities. Many of the parks in the Canadian Rockies, such as Jasper and Banff, are examples of this inherent conflict. It raises the basic issue that places of outstanding natural beauty attract people and development, and impose an ecological footprint that threatens the integrity of pristine ecosystems (Parks Canada 1994). The limitations imposed on park boundaries by arbitrary scenic criteria and their effect on wildlife was illustrated in a 1987 survey of major mammal populations throughout the western North American parks. It revealed how dramatically park boundaries are failing to protect the diversity of species in the regions in which the parks were established (McNamee 1989). Parks have become isolated islands within surrounding regions of conflicting land uses and are too small to maintain their ecological integrity. This situation was

FIGURE 11-8.
The visual attraction of beautiful protected places, such as Grand Tetons National Park, Wyoming, too frequently becomes a threat to their ecological integrity, presenting conflicts among beauty, human use, and ecological integrity. (Photo by M. Hough)

analyzed in an issue of *Audubon* that described the parks as "islands of extinction" (Quammen 1997).

With the scientific knowledge available today, considerable efforts are being made, locally and internationally, to develop policies and introduce ecologically appropriate protection and management practices—for example, the use of fire (controlled burns) to restore fire-dependent plant communities (exorcising Smokey Bear's mythical message that fire is bad), and the needs to reintroduce predator-prey relationships, limit human activities, and reestablish continuity of wildlands. The Yellowstone-to-Yukon Conservation Strategy, for instance, is an excellent illustration of an initiative by environmental organizations to link all the major mountain parks within a corridor that runs north-south along the Rocky Mountains, forming one very large continuous system (Parks Canada Panel on Ecological Integrity 2000). Efforts to give this corridor protected status involve complex negotiations among a variety of stakeholders, and international cooperation between the United States and Canada (Parks Canada Panel on Ecological Integrity 2000).

The management approaches being taken to protect the wild places of North America are both relevant to urbanizing areas and necessary to natural science education (Steinitz et al. 1996; Parks Canada Panel on Ecological Integrity 2000). They follow broadly similar ecological principles even though the specific issues and conditions are very different. Several points need to be considered. Ecological restoration has become one of the most important undertakings in both the cities and regional landscapes. The key question to be asked, however, is Restoring to what? Does it mean bringing back the natural conditions that existed before *Homo sapiens* intervened? Such a scenario is clearly untenable in the cities, where alterations to the land have been so dramatic that the idea is both fruitless and, arguably, undesirable. In national parks and protected areas, the landscape has been under continuous ecological and human change over millennia—by evolutionary processes, by native peoples, and later by colonization—making it difficult to determine which "original" landscape should be the model for basing restoration objectives. For the science-based disciplines, these matters will undoubtedly continue to dominate discussion as they have in the past. For ecological design and planning in the urban regions, one argument emerges when one takes a process-oriented rather than a static view of ecological restoration. It's about establishing new landscapes that may be different from what was there before. It recognizes the fundamental biophysical characteristics of the local region and allows ecosystem functions to evolve through adaptive management that accepts the uncertainty of outcome. And it recognizes the influence of cultural history that is continually shaping the landscape.

Observing What We See

Over many years of teaching both graduate and undergraduate students, I have concluded that most lack an ability for clear observation and understanding of the spatial and sensory world around them. This is true of non-design students, but is especially so of students of design whose propensity for predetermined solutions to problems has always been a source of personal astonishment and dismay. People's perceptions of what lies before them is greatly influenced by their upbringing and the spectacles through which they see their environment. This was well illustrated in a field course in stream hydrology that was undertaken by students at the University of Manitoba.

Over several decades, their teacher took four groups of undergraduates with various academic backgrounds in biology, engineering, and environmental design to survey the stream dynamics of a small watercourse. As part of the exercise he asked individuals from each group to map their observations of the stream in a way that would be understandable to others in the class. The results were interesting for what they revealed. The biologists saw the stream channel primarily in terms of habitat and organic form. They recognized the curvilinear form of the stream course, the cutting of banks as it changed direction, the deposition of eroded materials, exposed shale, wooded slopes, and floodplain. The behavior of the stream itself, however, was generally ignored. The engineers saw the stream as an almost straight man-made channel, flowing smoothly down its course, uninterrupted by obstructions. Its surrounding environs were noted on their sketches as "active erosion, slump, silt." All were drawn with an engineering certainty and precision that defined problem areas requiring correction. Straighten out the channel, folks, and all will be well—a reflection of attitudes that natural forces are to be controlled or overcome (Chapter 2). The landscape students, with no training in engineering or biology, created fine artistic renderings (often in color) that showed slopes, floodplain, and a thin, wavy stream line, but which revealed almost nothing of the watercourse's behavior or its influence on its surroundings (Newbury 1990).

What these drawings illustrated is how observations of a real place can vary to a marked degree with whoever is doing the observing and how strongly influenced the observers are by their individual backgrounds. Particularly interesting were the designer's sketches, which demonstrated how little of the real place had been observed. They also illustrate the difficulty design students often have in observing and accurately interpreting natural features on the ground, while substituting fieldwork for often meaningless site analysis drawings in the studio.

The Principle of Minimum Effort for Maximum Gain

As I argued earlier in this chapter, the imposition of design solutions and ideologies on places is often at odds with the realities of the place, and reflects a knee-jerk reaction to what the designer's role should be in shaping the land-scape. In the absence of a basic ecological foundation on which design can rest, this may not be surprising. An interesting example of the issue concerns a design problem I gave third- and fourth-year students in landscape archi-tecture for an internationally acclaimed urban wilderness on Toronto's water-front. The Spit, as this site is known, was created from landfill and dredging, and built out into Lake Ontario during the 1960s and 1970s by the Toronto Harbour Commission. Its purpose was to create a new outer harbor, at a time when efforts were being made to revitalize the industrial port. A depressed economy and the demise of the shipping industry, however, made the future of the Spit uncertain. Left in limbo, natural regeneration over forty years has turned raw landfill into a place of extraordinary floral and faunal diversity, becoming one of the most significant wildlife sanctuaries in the Great Lakes region as well as a stopover for spring and fall bird migrations. The citizen group that pressured government to protect the Spit as an urban wilderness has insisted that it should be left to evolve on its own with minimal interfer-ence or management.

The design studio that we undertook to examine this remarkable place was a teaching experiment. To begin with, we spent a day walking and cycling around the site, discussing its biophysical processes, the conflicts between wilderness protection and other uses (boating, marinas, golf courses, vehicu-lar traffic), its history and politics. Following this, the students did a week's design charrette in the studio, graphically illustrating their ideas for its pos-sible future. The resulting sketches were astonishing for their wealth of ideas and visions. They showed proposals for housing, hotels, theme parks, under-water aquaria, and recreation and interpretive facilities. These were discussed and critiqued and put away for future reference. The students were then told that they would spend the rest of the term really getting to know what this place was about; its geomorphology, soils, natural systems, habitats, plant communities and birds, its engineering history, the effects of lake erosion on its shorelines, its visual drama and sense of place, the politics and demands of conflicting uses between proponents of boating, ornithology, hiking, marinas, and accessibility by car.

By the end of term the students had amassed a vast amount of informa-tion, had visited the site countless times, and, most important, had the Spit ingrained in their thinking and experience. They knew the place. I announced a final charrette. The resulting designs were surprising and encouraging. Some designed a trail, a small shed for interpretation, a look-out. Others produced beautiful drawings, photos, and analysis of the place as it was. "It's too special," they said, "there's no reason to change it." All these "solutions" reflected a minimalist approach to the site that recognized its essential character and functions.

This experience taught me several things. First, the teaching of ecology in design has to do with understanding the nature of places as a precursor to making purposeful change, which, I firmly believe, is a far more significant act of creativity than imposing prepackaged solutions on the landscape. Second, minimum effort for maximum gain implies an understanding of the processes that make things work, providing the underlying structure that will encourage the development of diverse natural or social environments—knowing where, how much, and when to intervene. Third, having the humility to let the diversity of natural and social systems evolve on their own where they will is often the most appropriate design solution (Hough 1990).

Ecological Values and Understanding Heritage

The North American view of heritage has traditionally been reflected in a preservationist view that fixes places at specific points in time and that are too frequently divorced from their local and regional contexts. It's a fossilization of history that still remains alive and well today and is a view of the past that has significant implications for designers. It suggests that there are inherent discontinuities between how one thinks about design, in the design conventions of modern city landscapes (horticultural driven) and "natural" landscapes (ecologically driven), in a perception of nature divorced from human affairs. An ecological view must form the essential foundation for revealing history as an ongoing process to be remembered less for its great architecture or beautiful landscapes, but rather for what it teaches about the past, be it good or bad. In the Emscher region of the Ruhr in Germany, for instance, some of the old steel mills dating from World War II have been purposefully protected as reminders of the past. As part of their river restoration program, sections of the once channelized rivers have similarly been protected as demonstrations of the ravaged landscapes that today are being returned to health (Emscher Park International Building Exhibition 1991).

Heritage works when it's part of life processes, when it forms an unselfconscious part of the everyday life of people and neighborhoods. It is, consequently, about the ordinary places, institutions, and environments that are involved in daily life: old buildings that are lived in, art as an ordinary part of human traditions, valued places that have become essential to the rituals of living. Heritage is part of those processes that make life rich, rewarding, and treasured. While museums, archaeological sites, and historic buildings, temples, and gardens are essential components of our culture, it is the ordinary, representative landscapes that are usually left out of the equation, and why it is necessary to recognize the interdependence of human and natural systems as part of everyday life. Thus, understanding ordinary landscapes, based on ecological knowledge, is a fundamental necessity if students are to go beyond the aesthetic of what they see, but may not observe.

In an environmental design course I have taught at the Faculty of Environmental Studies over many years, I take my students to the site of a now defunct brickworks in the Don River Valley that runs through the middle of Toronto. Brick making began at the end of the nineteenth century and was discontinued in the 1960s. The site then became a candidate for heritage status. On our field trips to the brickworks and its associated quarry, we discuss the history of the valley, why the factory was located there, the environmental degradation that occurred as a consequence of this and other industries, and the changing technologies of brick making that can be traced in the various buildings that became sequentially redundant and discarded as new manufacturing processes were introduced. We discuss the geological history of the quarry as the overburden was removed to expose sedimentary deposits revealing the history of the interglacial period, changing climates, flora and fauna that have appeared and vanished over tens of thousands of years, the quarry's impact on the valley, and the natural regeneration that emerged when brick making was abandoned.

We also discuss the relationship between the city and the valley industry that supplied the bricks from which early Toronto was built, the symbolic interdependence of nature and urbanism; strategies for heritage protection that focus on the evolutionary processes that have shaped both early industry and the Don valley; and the counterproductive consequences of turning the brickworks into a museum—a prime agenda of government heritage agencies. Also discussed is the irony that is inherent in the preservation and ecological restoration of sites in cases such as this. The industries that once destroyed the valley ecosystem are now, under the guise of heritage values, contributing to the valley's ecological renewal. The emphasis on a fieldwork approach to ecological design has, in my experience, contributed largely to my students' understanding of natural processes and their links with the urban community. It is a dynamic interaction that reinforces the realities of human and natural evolution. And it reflects a fundamental principle: that landscapes, to have meaning, must be tied to their regional geography, their climate and vegetation, their political and social contexts, and their local urban environment. They are rooted to the notions of continuity and place.

Summary

What are the educational underpinnings for teaching ecology to the design and planning disciplines? The following are a suggested guide to answering this question.

A focus on the underlying processes that shape places.

This is about the exploration of natural history, a term that is significant since it requires that places be understood from a generalist and integrative perspective. It is based as much on the observed, experienced environment and historical evidence of landscape change as it is on science—"peeling back the rubber mask of the present," as a colleague once said, "to gain insights into the past, and the environmental, social, and economic forces that made places what they are today" (Newbury pers. comm. 1991).

Learning how to observe.

In my teaching experience in environmental design, I place the greatest emphasis on field trips to illustrate what books cannot, how natural systems and people behave in real life. We examine naturally regenerating sites and formal parks, gardens, and plazas, and compare the differences in diversity, in human perceptions and impacts on the city. We examine the behavior of streams, how people behave in public spaces. We visit suburban and inner-city neighborhoods, the former planned according to municipal planning codes and bylaws, the latter evolved over generations with little planning intervention. We compare the differences in physical environment (the use of and attitudes toward vegetation, gardens, and personal outdoor space), in human behavior (how adults and children use streets and front yards), the role of the automobile, and so on. The most important element here is the understanding of different constructs of nature and what people of different cultural origins value.

Making visible the processes that sustain life.

Of profound significance to design is the fact that our daily existence is spent in surroundings designed to conceal the processes that sustain life and that contribute, possibly more than any other factor, to the acute sensory impoverishment of our living environment. Examples are legion. The supermarkets are filled with produce grown in warm climates and transported thousands of kilometers to cold ones. The supply of electricity and water appears at the flip of a light switch and the turn of a faucet. The curb and catch basin, which make rainwater disappear without trace below ground, cut the visible links between the hydrological cycle and the sewers that flush stormwater into streams and rivers. We are, consequently, unaware of the degradation that occurs, often great distances from the source of the problem—to plant communities and aquatic life, to beaches that have to be closed after a storm. The problem has been compounded in design schools with traditional approaches to engineering and grading courses. The dictum that "the well-designed parking lot drains water away to the catch basin and underground sewer" is familiar territory for many older practitioners. Ecologically sustainable alternatives are reflecting the other view, that the benefits of "good design"—well-drained streets, parks, and civic places—are paid for by the environmental cost of eroded streams, human-induced flooding, and impairment of water quality downstream (Hough 1995). An environmental view might suggest, therefore, that "bad design" is an attractive alternative to explore.

The concept of process.

Process is common to the functioning of both human and nonhuman nature. The essence of design is one of constant transition—a process that recognizes that the completed project is only the beginning, not the end, of continuous change. This is true, in an ecological sense, of all physical landscapes. It is also true of human habitats and cities that must be seen as a continuum, influenced by people, by economics and politics, and by changing values and objectives. In this sense, every design project is new at any point in time, evolving from one state to the next, for better or sometimes for worse. In a word, the designer and the planner are not, and cannot be, in control.

The role of the designer in public landscapes.

A great deal of my teaching time is spent on getting students in design and related fields to understand the nature of intervention. How far should the designer/planner go in making changes to places in different physical and social settings? Where is she/he needed and where not? These questions suggest that design involves several important paradigm shifts in the way students are taught to perceive their roles in the world of practice. The time has come for an alternative definition of design, from an academic discipline that teaches students to impose their ideologies on the rest of society, to the idea of the interdependence of life processes, which brings with it a level of humility to design intervention—doing only what is absolutely necessary. This may be the essence of an environmental design culture because it's attuned to the very nature of human and natural systems—in the emerging challenges of community design, in an understanding of human behavior, in the shaping of public places and urban landscapes; in the protection of wild landscapes, and the significance of diversity.

Jens Jensen once summarized this attitude to design in a comment to Julius Rosenwald, who had commissioned him to design the Sears Robuck estate. Jensen's mandate called for a minimum disturbance to the wooded ravine on the property. When Rosenwald saw Jensen's bill, he challenged the $1,000 fee for what appeared to be so little work. Jensen (for whom modesty was clearly a fault rather than a virtue) replied that "lesser men would have charged far more just to ruin a beautiful site" (Grese 1992, p. 99).

Citations

Barrett, S., and J. Kidd. 1991. Pathways: towards an ecosystem approach. Royal Commission on the Future of the Waterfront, Toronto, Ontario, Canada.

Canadian Heritage Parks Canada. 1994. Guiding principles and operational policies. Minister of Supply and Services, Canada.

Dramstad, W. E., J. D. Olson, and R. T. T. Forman. 1996. Landscape ecology principles in landscape architecture and land-use planning. Island Press, Washington, D.C., USA.

Emscher Park International building exhibition. 1991. An Institution of the State of North-Rhine Westphalia. Information Brochure.

Evernden, N. 1992. The social creation of nature. John Hopkins University Press, Baltimore, Maryland, USA.

Forman, Richard, T. T., and Michael Godron. 1986. Landscape ecology. John Wiley and Sons, New York, New York, USA.

Fulford, R. 1995. Accidental city. MacFarlane, Walter & Ross, Toronto, Ontario, Canada.

Gore & Storrie Ltd., Hough Stansbury Woodland Ltd. et al. 1993. Town of Markham Ontario Natural Features Study.

Grese, R. E. 1992. Jens Jensen: maker of natural parks and gardens. John Hopkins University Press, Baltimore, Maryland, USA.

Hester, R. T., Jr. 1990. Community design primer. Ridge Times Press.

Hough, M. H. 1990. Out of place: restoring identity to the regional landscape. Yale University Press, New Haven, Connecticut, USA.

————. 1995. Cities and natural process. Routledge Press, London, UK.

Hough, M., B. Benson, and J. Evenson. 1997. Greening the Toronto port lands. Toronto Waterfront Regeneration Trust, Toronto, Ontario, Canada.

Jacobs, J. 1961. The death and life of great American cities. Random House, London, UK.

Livingston, J. A. 1994. Rogue primate. Key Porter Books, Toronto, Ontario, Canada.

McHarg, I. L. 1969. Design with nature. 2nd edition. Natural History Press, Philadelphia, Pennsylvania, USA.

McNamee, K. A. 1989. Fighting for the wild in wilderness. Pages 63–82 in M. Hummel, editor. Endangered Spaces. Key Porter Books, Toronto, Ontario, Canada.

Minneapolis Parks and Public Works Commission. 1985. An evaluation of parks and infrastructure in Hennepin County in relation to property values. Minnesota Parks and Public Works Commission, Minneapolis, Minnesota, USA.

National Parks Services. Draft management policies. No date. National Parks Services, Washington, D.C., USA.

Newbury R. W. 1990. These field teaching experiences were presented by Dr. Newbury at a seminar at the University's Program in Architecture and Landscape Architecture, and recorded in Michael Hough. Out of Place: Restoring Identity to the Regional Landscape. Yale University Press, New Haven, Connecticut, USA.

Nudds, T. D. 1999. Adaptive management and the conservation of biodiversity. Pages

179–291 in Richard K. Baydack et al. Practical approaches to the conservation of biological diversity. Biological diversity conservation. Island Press, Washington D.C., USA.

Odum, E. P. 1969. The strategy of ecosystem development. Science **164.**

Parks Canada. 2000. "Guiding principles and operational policies." Ministry of Supply and Services, Canada.

Parks Canada Panel on Ecological Integrity. 2000. "Unimpaired for future generations?" Conserving ecological integrity with Canada's national parks. Ministry of Supply and Services, Canada.

Quammen, D. 1997. Islands of memory. Audubon (July/August).

Royal Commission on the Future of the Toronto Waterfront "Pathways." 1991. Ministry of Supply and Services, Canada.

Seddon, G. 1997. Landprints: reflections on place and landscape. Cambridge University Press, Cambridge, Massachusetts, USA.

Sellars, R. W. 1997. Preserving nature in the national parks: a history. Yale University Press, New Haven, Connecticut, USA.

Steinitz, C., M. Binford, P. Cote, T. Edwards Jr., S. Ervin, R. T. T. Forman, C. Johnson, R. Kiester, D. Mouat, D. Olson, A. Shearer, R. Toth, and R. Wills. 1996. Biodiversity and landscape planning: alternative futures for the region of Camp Pendleton, CA. Graduate School of Design, Harvard University, Cambridge, Massachusetts, USA.

Task Force to Bring Back the Don. 1991. Bringing back the don report. City of Toronto. (Prime Consultant Hough Stansbury Woodland Ltd.) Toronto, Canada.

USDI, National Parks Service. Draft management policies to guide the management of the national parks system. Washington D.C., USA.

Whyte, W. H. 1980. The social life of small urban spaces. The Conservation Foundation, Washington, D.C., USA.

PART IV

Prescriptions for Change

How can ecology be effectively integrated in design and planning education? The following chapters offer specific recommendations for landscape architecture education through conceptual foundations, interdisciplinary collaboration, courses and teaching, curricular development, and links from education to professional practice. Through their specific focus on the relevance of education to practice, these chapters provide a conceptual model that may be relevant to other design and planning professional programs and to applied ecology programs. Kristina Hill and co-authors establish a focal idea applicable across disciplines and propose ecological understanding as an inclusive intellectual basis for the design and planning disciplines in Chapter 12, "In Expectation of Relationships: Centering Theories around Ecological Understanding." In Chapter 13, "The Nature of Dialogue and the Dialogue of Nature: Designers and Ecologists in Collaboration," Bart Johnson and co-authors identify place as the unifying arena for dialogue among designers and ecologists, and outline a framework for common ground across disciplines that might lead to better education and practice. In Chapter 14, "Interweaving Ecology in Design and Planning Curricula," Ken Tamminga and co-authors look at the ways a program curriculum can stimulate a relationship between design and ecology in a university setting. Jack Ahern and co-authors discuss how the idea of place can set priorities for curriculum and classes in Chapter 15, "Integrating Ecology 'across' the Curriculum of Landscape Architecture." Kathy Poole and co-authors propose a conceptual matrix to facilitate the development of ecological understanding through project- and place-based problem solving and within a framework that bridges education and professional practice in Chapter 16, "Building Ecological Understandings in Design Studio: A Repertoire for a Well-Crafted Learning Experience." René Senos and co-authors investigate the important connections of ecological understanding to professional practice and consider reciprocal links between education and practice in Chapter 17, "From Theory to

Practice: Educational Outcomes in the World of Professional Practice." Kristina Hill and Bart Johnson conclude with a chapter that reflects on the perspectives presented throughout the book. They identify areas of common ground while also surveying the issues that may be points of contention. They end by considering how these shared and differing views might emerge as schools of thought and as a foundation for future dialogue on design theory and practice.

CHAPTER 12

In Expectation of Relationships: Centering Theories around Ecological Understanding

Kristina Hill, Denis White, Miranda Maupin,
Barbara Ryder, James R. Karr, Kathryn Freemark,
Rebecca Taylor, and Sally Schauman

Over the last several decades, many educators and professionals in the design
and planning fields have made a "turn to ecology," where "turn" means a shift
in emphasis and priorities. From the publication of *Design with Nature*
(McHarg 1969) to the publication of contemporary approaches in *Ecological
Design and Planning* (Thompson and Steiner 1997) and *Landscape Journal*'s
"Eco-Revelatory Design" issue (1998), designers and planners have grappled
with bringing ecological knowledge into education and practice. But have
these shifts brought a fundamental change in the way we think and represent
our knowledge in design and physical planning? Have they reorganized our
approach to design and planning theory? Or are they just part of a broad theoretical pluralism, one that encourages students and practitioners to decide
individually whether ecological concerns should be raised with regard to a
given project?

In our discussions at the Shire Conference, we concluded that we're currently missing an intellectual model that would allow us to organize issues of
function, social equity, aesthetics, and ecological relationships, all under one
"umbrella" idea. We'd like to see the design fields move beyond the
dichotomies of nature vs. culture, art vs. science, and ecological design vs. the
rest of design. Our hope is that a new paradigm could help us do that by
starting from a different point of view altogether. Our intention in this chap-

ter is to stimulate readers to bring a paradigm we call "ecological understanding" into design and physical planning. We think that the diverse intellectual models and approaches to professional practice that are introduced in a design or planning curriculum can (and should) be centered around this concept. Ecological understanding is an organizing idea that challenges us to structure and use knowledge from diverse fields, including our own, in a different way.

Ecological Understanding as an Intellectual Paradigm

What Is "Ecological Understanding"?

Ecological understanding is, very simply, the expectation and awareness that human actions have consequences and that an intricate web of relationships connects patterns and processes in the physical, biological, and social environments.

In this sense, ecological understanding is a much more general idea than ecological science. "Understanding" refers to the mental models or cognitive structures that create expectations and awareness within our way of knowing the world; a world that is both inside and around us. Ecological science, on the other hand, is the study of relationships between living organisms and their physical and biological environments. This study follows rules that promote critical thinking and rigor, and allow individual hypotheses to be tested and compared. Ecological scientists try to understand how biological relationships work across different scales of time and space. They make use of a general idea of ecological understanding, as we have defined it, when they try to anticipate the ways in which organisms interact with their environments and with each other. Ecologists expect there to be connections, although they also expect that everything will not be connected in the same way. The consequences of actions that trigger these connections may be immediate and local, or may occur in a different space or time. Consequences may also be singular, when each action has a corresponding effect, or cumulative, when more than one action is required to produce an effect. Frequently, the kinds of human actions that are proposed in design and planning result in all of these types of consequences: singular and cumulative, immediate and delayed, local and distant.

Ecological science seeks to build a body of theory that can help explain natural phenomena. Its methods and standards must be "conservative" in the sense that tests of hypotheses must pass high standards of likelihood before these hypotheses are conditionally accepted as true. In contrast, ecological understanding pursues a different kind of conservatism. It begins with the assumption and expectation that change will bring consequences, some of which we can anticipate and some we cannot. Because of this expectation, ecological understanding would lead a person who adopts this paradigm to try to avoid actions that have irreversible impacts. He or she could be expected to adopt a strategic approach that tries to limit ecological risk in a different way than someone who does not expect there to be consequences. This strategic goal involves a different kind of rigorous thinking than is found in science, where the primary objective is to avoid accepting false hypotheses about cause and effect (Shrader-Frechette and McCoy 1993). Scientific experiments are often designed as an attempt to falsify the hypothesis that an effect exists. This is an appropriate kind of conservatism in an effort to build robust, tested theory. But ecological understanding is useful for doing something quite different. Its primary usefulness is in helping us avoid making irreversible mistakes in the way we alter the environments we live in. Ecological understanding places the burden of proof on the advocate of change, with the expectation that changes have unintended consequences in addition to their intended effects.

Rachel Carson's 1962 book, *Silent Spring,* offers an example of this kind of reasoning. Her book was one of the first to introduce ecological understanding to the general public. It provided memorable accounts of unanticipated connections and consequences in Carson's descriptions of the impact of an agricultural pesticide, DDT. Quite unexpectedly, exposure to DDT affected the reproduction of songbirds. In areas where the pesticide had been used, many songbird species began to lay eggs with thinner shells, making the eggs more likely to break prematurely, killing the birds' offspring. Although the ecological theories of that era might reasonably have anticipated such an impact, the government and industry scientists who tested DDT in laboratories did not frame their investigations to include a potential impact on songbird reproduction. If DDT didn't kill birds directly, the official assumption was that it would not be harmful to them. The evidence of the eggshell effect was tested using scientific methods. But the expectation that there might be such an effect, and the conclusion that it should be avoided, came from a more general cognitive model that influenced the investigator's thinking. This same model subsequently influenced Carson's conclusions, that the use of DDT presented unacceptable risks for ecological and human health.

Thanks to Carson, our knowledge of this particular impact has forever altered our confidence in the harmlessness of chemicals and materials manufactured by humans. But we often ignore the chapters of her book that summarized evidence of the link between pesticides and human cancer, perhaps because it has proven more difficult to demonstrate experimentally. Carson used multiple lines of evidence to reason about effects, and she also used the results of scientific studies. But narrow interpretations of statistical significance would not have led her to the conclusions she presented in *Silent*

Spring. Her work has subsequently convinced many people that the presence of secondary and even tertiary effects in ecological systems should inspire us to be more cautious about the ways in which we attempt to influence those systems. The phenomenon of endocrine disruption (Colborn et al. 1996) provides a 1990s example of yet another emerging understanding that was derived from multiple lines of evidence, and from an expectation and awareness of consequences in a world of complex relationships.

Ecological Understanding in the Practice of Design and Physical Planning

In the physical planning and design fields, our claim to professional expertise depends on our ability to understand and reason about relationships between patterns and processes. This is true whether the objective of our work is to affect the emotions of another human being or to protect the environment of an endangered species. Yet the irreducible complexity of ecological relationships typically requires designers and planners to reason and make decisions with only an approximate knowledge of the nature and behavior of the objects, organisms, systems, and relationships that might be affected by a design, a plan, or a policy. These professionals accomplish their task of reasoning "approximately" using everything from dreams to data (Schon 1987). But as Steinitz (1979) has pointed out, the more people, money, or land that will be affected by our professional judgments and recommendations, the more we are required by our clients and by the courts to reason explicitly, using hypotheses that are likely to be correct. If those effects are likely to be irreversible, we should be even more careful not to risk their occurrence. While ecological relationships are complex, it is not impossible to point out known or likely risks of irreversible damage and to limit the degree to which we expose humans and their life-support systems to new risks.

Many previous authors have noted the need for various kinds of ecological awareness in design and planning that would go beyond "nature preservation," per se (Lyle 1999; Beatley and Manning 1997; Thompson and Steiner 1997; Thofelt and Englund 1996; Smith and Hellmund 1993; Van der Ryn and Calthorpe 1991; Hough 1995; Mitsch and Jorgensen 1989; Spirn 1984; Teymur 1982; Lawrence and Bettman 1981; Veitch 1978; McHarg 1969; Hills 1961; Demerath 1947). Some of this literature has emphasized the intrinsic or aesthetic value of landscapes (Eckbo 1998; Brown et al. 1998; Thayer 1994; Calthorpe 1993). Ecological understanding can allow us to build on this concept of intrinsic value. As a paradigm, it offers the chance for us to avoid some of the negative biological, social, and physical consequences of human decisions that impact human health and prosperity, as well as the health and sustainability of our environment. But the ways in which some of us already use this paradigm must be made more explicit, if we hope to share the wisdom and experience of many of today's designers and planners with new generations of practitioners.

Many options for how to make ecological understanding more explicit are available to educators and theorists in our fields. One might be to reexamine our own literature for ideas that we can join together as formalized theories, theories that would incorporate ecological understanding. This approach would allow us to synthesize and preserve an understanding of the history of related ideas in our fields. In addition, we suggest that there is also a need for a bolder approach to theory. The concept of ecological understanding could be used to re-center a broader world of theory, generated in diverse intellectual disciplines, around an understanding of consequences and relationships. We see that option as desirable because it could open up exciting new debates, and create opportunities for intellectual synthesis.

Building a Cross-Disciplinary Model of Ecological Understanding

The first step in the "recentering" process we've proposed might be to look at the contemporary theories of other disciplines, as well as our own, for signs of a "turn to ecology." By identifying these intellectual trends across diverse disciplines, we may discover specific opportunities to propose syntheses of theory, and to engage our colleagues across the university in stimulating new debates. The multidisciplinary positions formed through this kind of debate could (1) refresh the influence of the design and planning fields on contemporary academic discourse, and (2) stimulate experimental methods that could inform design and planning practice. The opportunity to create new theoretical syntheses using a fundamental recognition of relationships that tie actions to effects (biological, social, perceptual, and physical) might open up a new era in the theory of our fields. These syntheses might create new ways to link design and planning to the work of others in diverse fields—building unexpected bridges between art and engineering, medicine and urban planning, biology and design, law and ecology. Its effect could be to engage theorists from all of these disciplines in the same search for understanding, with a mutual respect for uncertainty.

As an interdisciplinary group of authors, we have brought together six specific examples of how ecological understanding is operating to shift paradigms in diverse academic disciplines. The short discussions that follow are designed to illustrate the potential for a broad, interdisciplinary understanding of consequences and relationships. Our name for that understanding, when it is seen as a paradigm, is "ecological understanding." We also hope that these discussions will provide access to the literature of other fields, as it may be relevant to this paradigm.

THE TURN TO ECOLOGY IN ETHICS AND ECONOMICS (DENIS WHITE). Ecological economics, as described by some of its proponents (Costanza et al. 1997), is conceived broadly as a combination of assumptions about how the world works, of preferences for the future, of methods of analysis, and of institutions using these assumptions and methods to achieve the stated future preferences. The assumptions and preferences that appear to have a widespread consensus in ecological economics are (1) that the earth is a closed and materially limited system, (2) that a sustainable future for humans and other species should have a high quality of life, (3) that irreducible uncertainties in knowledge and the irreversibility of certain processes require a precautionary approach to human-induced change, and (4) that institutions should use adaptive approaches to manage for sustainable futures.

The precautionary principle is one ethical approach advocated by ecological economics. The origins of this idea in the twentieth century appear to have been in German social democracy in the 1930s (O'Riordan and Cameron 1994), followed by legal promulgations, again first in Germany in the 1970s, and later in environmental legislation, and in international agreements (Environment Canada, 2001) <http://www.ec.gc.ca/CEPARegistry/the_act/Introduction.cfm>. The idea of the precautionary principle is simple: Act to prevent possible harm (to the environment) even though complete scientific certainty may not be available.

The precautionary principle is one of the aspects of ecological economics that philosopher Mark Sagoff has questioned in a recent debate with economist Herman Daly (Sagoff 1995; Daly 1995). Sagoff asserts that the "reasons to protect nature are moral, religious, and cultural far more often than they are economic" (1995, p. 618) and that prudence (as in the precautionary principle) has a utilitarian or instrumental justification that too easily can be overturned with changes in technology.

Philosopher Kristin Shrader-Frechette has written extensively on the challenges raised by arguments such as Sagoff's (for example, Shrader-Frechette 1994; Shrader-Frechette and McCoy 1993). Her approach has been to recast the idea of the precautionary principle as a trade-off between minimizing the risk to a developer, for example, by not allowing a harmless development versus minimizing the risk to the public by allowing a development that then causes harm. Posing the policy issue as a weighing of likelihood of harm versus no harm helps to focus the values impinging on designs, plans, and actions. Shrader-Frechette then argues that in an environmental policy situation, scientists as well as policy makers, and, we can add, designers and planners, have a *prima facie* obligation to minimize public risk. This is because, according to many philosophers, it is more important to protect from harm than to enhance welfare, that is, to minimize harm to the public at large rather than to enhance benefits to a smaller number of people. An additional ethical argument is that risks should be born primarily by those to whom the benefits will accrue.

An area of applied economics that is relevant to environmental decision making is attempting to quantify trade-offs using cost-benefit analysis. This set of methods and their associated assumptions are also controversial. Again, Shrader-Frechette (1985) has examined many of the issues. One question she raises is whether cost-benefit analysis should be used at all, given the difficulties in adequately assigning costs and benefits for many environmental goods, services, and values. Another issue is that although many argue that cost-benefit analysis can and should be ethically neutral, it is dependent on underlying philosophical assumptions and values about how to aggregate benefits to society at large. However, according to Shrader-Frechette, the major problem of cost-benefit analysis that ties together many other issues is that of quantification and whether it is appropriate for cost-benefit analysis to represent all costs and benefits by quantitative measures.

The debate about quantification in cost-benefit analysis closes the circle in the discussion of ecological understanding in ethics and economics. Shrader-Frechette (1985) argues that many of the problems with the use of cost-benefit analysis in environmental impact analyses are due to either deliberately ignoring, or simply not recognizing, the ethical assumptions implicit in this type of work. To address this weakness, Shrader-Frechette considers a number of arguments that pertain to the debate between partial and full quantification of cost-benefit analyses. She takes the position that an ethically weighted cost-benefit and risk analysis that considers all issues is the most fair method of judging consequences of human activities affecting the environment. An even more far-reaching conclusion is that of Michael M'Gonigle (1999), who argues that ecological economics cannot be properly conceived without attention to political structures and relationships. Whether one agrees with these positions or not, it is clear that the debates among ecologists, economists, and philosophers on these issues have direct implications for the design and planning professions, since the work of these fields must apply these arguments.

HOPE FOR ECOLOGICAL UNDERSTANDING IN LEGAL STUDIES (JAMES R. KARR). In theory, the law is a thoughtful integration of social, political, and scientific knowledge (Karr 1995, 2001). It is also one of the primary mechanisms available to society in resolving controversies that often inappropriately pit conservation against development. Yet because of the social necessity for political compromise, the language used in drafting legislation is often carefully ambiguous. And, in practice, interpretations of the law rely on past decisions, some wise and some silly, more often than they seek the current state of scientific understanding. The autonomy of individual courts and judges can produce a patchwork of legal reasoning instead of reasoning that is as connected as the natural systems we seek to manage and preserve.

The state of water bodies, for example, reflects the status and quality of their surrounding landscapes. Since water moves across the entire landscape, the by-products of human activities move through this landscape as well— regardless of the political or jurisdictional processes that are created without regard for these spatial and temporal processes. Yet the management of water and other natural resources is divided into endless dichotomies—water quality versus water quantity; point-source pollution apart from non-point-source pollution; drinking water versus wastewater; fish-bearing streams apart from non-fish-bearing streams—that defy science and common sense. Even when legislation reflects real connections, as in a Washington State Water Resources Act of 1971 that expressly recognizes the "natural interrelationships of surface and groundwater," judicial decisions may still counter this intent. The Washington State Supreme Court, for example, has limited the authority of a state agency responsible for water rights to recognize the connection between groundwater and surface water (*Rettkowski v. Department of Ecology* 1993; Dufford 1995) even though a later decision of the U.S. Supreme Court affirmed the intimate connection between water quality and water quantity (*Jefferson County v. Department of Ecology* 1994). Clearly, the connectedness of natural resources ought to be reflected in a connectedness of legal doctrines and public policies.

Legal scholars have begun to call attention to this fundamental flaw (Seltzer et al. 1987; Posner 1995; Rose 1997), but the courts continue to produce ill-informed decisions. A recent decision by a Connecticut Appeals Court, for example, upheld a lower court ruling that trees and wildlife are *not* natural resources (*Paige et al. v. Town Plan and Zone Commission of the Town of Fairfield* 1994). The ruling was based on the definition of "natural resources" in *Black's Law Dictionary*, which defines natural resources as material in its native state that, when extracted, has economic value. The court determined that neither the trees nor the wildlife in the woodland slated for development fit this definition of a natural resource. When legal decisions select from among opposing precedents or use definitions that defy common sense to support a particular ideology, the law becomes "a professional totem signifying all that is pretentious, uninformed, prejudiced and spurious in the legal tradition" (Posner 1995, p. 21).

Lee (1993) called for a view of public policy and the law at their best as a practical "gyroscope" helping to safeguard the interests of both individuals and society. This concept of the law fits well with what we have called "ecological understanding." Courts and policy makers who challenge precedence as the central intellectual model of the law and seek instead to balance it with consideration for future consequences are applying ecological understanding. They are balancing ideologies and the autonomy of the courts against the simple acknowledgment that our choices (including the choice of nonaction) have observable consequences. Most important, they are putting into practice the genuinely "conservative" belief that it is unwise to cause irreversible changes in complex systems.

ECOLOGICAL UNDERSTANDING IN CULTURAL THEORY (KRISTINA HILL). What is a theory of culture? Sociologist Wendy Griswold (1994) noted that in academia "most notions of culture are based on assumptions rooted either in the humanities on the one hand, or in the social sciences, particularly anthropology, on the other" (p. 3). Yet "there is no such thing as 'culture' or 'society' out there in the real world. There are only people who work, joke, raise children, love, think, worship and behave in a wide variety of ways" (ibid.). I use the term "cultural" in this sense of ordinary activities, and use "cultural theory" to refer to theories about human behavior, particularly in a context of relations with other humans, their environments, and other forms of life.

The writings in this multidisciplinary area of theory employ paradigms originating in distinct academic disciplines—history, anthropology, sociology, geography, literature, and others. The tools for analysis that have become especially common in this area of theory include analyses of power in relationships among or within cultural groups (see, for example, Warren 1997; Kelly 1994; Lyotard 1984; Anderson 1990), and analyses of images, language, and rhetoric used to represent the world by members of different cultural groups (see, for example, Spretnak 1999; Heidegger 1993; Harvey 1989; Borgmann 1992; Hacking 1983).

In my view, this attention to power relationships and the consequences of representation (where "representation" means both the representation of reality and the representation of citizens in democratic decision making) could be seen as a "turn to ecology." Both kinds of analysis emphasize interrelationships and the consequences of either human actions or human rhetoric. Indeed, one could argue that these tools of analysis are fundamental to acquiring ecological understanding because they help answer questions such as these:

• How is power distributed in human societies, institutions, and cultural groups?

- How do influential groups and individuals make decisions to act or not to act?
- Where and how do these decision makers perceive the consequences of their behavior?
- To what extent are those social and environmental consequences a result of the distribution of power in the immediate group and among larger groups such as societies?

The paradigms of cultural theory provide insights into the diversity of speakers and voices within cultures, within a single text or narrative, and within the reasoning process of an individual human being. This perception of multiple voices and points of view is often referred to as a dialogic perspective, and contrasted with monologic perspectives that seek to reduce this complexity to identify "true" voices or narratives (Bakhtin 1981; Lyotard 1994). Theories that describe narratives as a form of both communication and representation have made significant distinctions between narratives that claim to be universally true ("metanarratives"), and those that claim believability only within a prescribed frame or context ("small narratives") (Lyotard 1994). Many authors have drawn attention to the presence of multiple voices (or perspectives) within single-authored texts and personal narratives using psychological methods of analysis that identify unresolved differences in points of view that may be held by an individual author or narrator (see, for example, Gilligan 1982; Goldberger 1996). The recognition of greater complexity within narratives, and of the speaker's (or author's) ability to express his or her sense of alignment with different positions during a narrative, is also used in the analysis of the power relationships expressed in narratives (Goldberger 1996).

The explicit grouping of narratives into those that claim universal truth and those that do not also reveals differences in the implied power and influence of those narratives. Claims of universal truth are powerful. The influence of such a claim can be contested if the claim is (1) recognized as a narrative, and (2) the social context of its origins is understood. The methods of analysis developed in cultural theory allow us to discuss bodies of theory as narratives and place them in social contexts. In that sense, cultural theory has allowed us to recognize diversity, irreducible complexities (particularly related to the use of language), and contextual relationships in human culture. These insights parallel the insights of biology and ecology in the world of natural processes.

Cultural theory has also challenged some of the methods and claims of science, particularly on the basis of what can go awry in the use of language and on the social relationships that affect the production of scientific knowledge. These critiques apply to all fields of knowledge, not just to science; but science has been the target of particular criticism because of published claims that the scientific method provides a superior epistemology to that used in other fields. For instance, some cultural theorists have noted that because the acquisition of scientific knowledge takes place in specific social contexts, it is not a universally accessible or satisfying "way of knowing" for all social contexts, or in all human groups (Harding 1991; Rose and Nowotny 1979). Others have critiqued the claims of universal validity that often accompany scientific results by saying, roughly, that scientists cannot truly share and test complex bodies of theory because the language used to describe these theories is ambiguous and imprecise, and is understood differently by different individuals (Hacking 1983).

Some scientists have identified this perspective as nihilistic, saying that it denies the nature of reality and of human perception (Soulé and Lease 1995). But this critique can also be productive, allowing us to question our assumptions about language and, more important, about the relative contributions of different epistemologies. Since language is essential to human perception and communication, the critical question seems to be, How can we acknowledge the ambiguities of our condition and still proceed to reason and communicate with each other?

The opportunity to situate and interpret narratives can help us identify the value of the narratives that comprise ecological understanding. Cultural theory can, for example, help us to perceive the differences between ecological science and ecological understanding. Ecological science builds theory using scientific rules of evidence and a fundamentally conservative attitude toward the acceptance of new hypotheses. As is the case with all scientific narratives, ecological theory would probably be considered an example of a metanarrative by most cultural theorists because it claims that its theories are true everywhere if they are true anywhere. Yet within the restrictive epistemological frame of science, many of the complex interactions of ecology cannot be tested convincingly because they cannot be isolated. In this science, the scientific method often produces theories that cannot be rigorously tested (Peters 1991).

Ecological understanding, on the other hand, relies on what might be thought of as a coarser definition of reality. It is a "small narrative," claiming local or regional believability based on empirical or consistent observations, with no claim to universal validity. In the sense of Lyotard (1994), ecological understanding is something we "perform"—not something we claim as a universal truth. The success of each individual performance depends on the rhetorical skill of the performer, on the authority given to a performer who has personally made many careful observations over time, and on the audience members' agreement with the performer, based on their own observations of the environment. For designers and planners, as well as ecologists working in applied conservation biology, this implies that we cannot hide behind science as "truth." In the application of ecological understanding to conservation and development, we must engage in debate using our observations, persuade using our experience, and acknowledge our social motivations.

AESTHETIC CATEGORIES AND THE TURN TO ECOLOGY (REBECCA TAYLOR). Within the last decade, a study by Kaplan and Kaplan (1989) showed that the public's aesthetic reaction to many ecologically valuable landscapes is negative. Joan Nassauer (1995) argued that this is because ecological quality is culturally defined as well as scientifically defined. She called for designers and planners to explicitly use aesthetic conventions to communicate both what is natural and what is valuable in the landscape. The philosopher Marcia Mulder Eaton (1997) extended Nassauer's general idea, proposing a model of aesthetic experience in which ecological health is considered beautiful. In this model, recognizable ecological structures such as vertical layers of vegetation in a riparian zone would inspire our aesthetic appreciation. For Eaton, aesthetic experience is marked by sustained cultural attention that puts relevant information to use in understanding and appreciating the environment. She draws on the work of colleague Allen Carlson to argue that the categories of information that are aesthetically relevant to the appreciation of the natural world are radically different from those used in art appreciation. Carlson (1981) asserts that we don't really know how to appreciate nature aesthetically because, unlike art, we humans had no hand in creating it. As a result, he argues, our cultural appreciation of the natural world depends on our ability to perceive aesthetic relevance in the organic unity of a landscape and its ecological processes.

The crux of the debate then becomes, To what extent does knowledge and reasoning influence aesthetic experience? Geographer Steven Bourassa (1991) specifically denies the claim that aesthetic judgment is somehow different from other kinds of thinking. He argues that aesthetics deal with what "fits" or what is appropriate, and thus aesthetic judgment is a type of practical knowledge. Bourassa also challenges Kant's perceptual hierarchy (Kant 1790), asserting that perception engages all of the senses. Singling out vision and hearing as the only two senses capable of aesthetic perception is unrealistic and unnecessarily limiting. These arguments challenge our definitions of thought and perception by asking whether thinking isn't an "aesthetic sense," and whether aesthetic perception isn't a form of "thought."

Aristotle, whose writing is commonly cited by Western scholars on the subject of reasoning, portrayed a world in which all things can be definitively classified (Cooke 1938). Yet contemporary cognitive science is challenging this early understanding of categories. Rosch's prototype theory of categorization (as reported in Lakoff 1987) suggests that human capacities for sensorimotor perception, experience, and memory play a unique role in categorization. Her research supports the view that the creation of categories is as dependent on human perceptual and physical engagement with the world as it is on the attributes of the objects themselves. If we consider, as Rosch does, that cognition depends on the experiences that come from having a body, then we must acknowledge that our definition of categories (including aesthetic categories) is continually evolving as a result of our physiological interactions with the world. Rosch's (1996) research calls into question Nassauer's (1997) assumption that "the most powerfully omnipresent form of environmental education is simply viewing the landscape" (p. 78). It suggests that ecological understanding can more fully evolve from a combination of visual, aural, tactile, and kinesthetic experiences.

Rather than rely only on visual frames to bring ecological understanding and values into focus, this scholarship in cognitive science supports a model of aesthetics that considers our physiological interactions with the landscape and recognizes the necessity of place-based experience. This model of experiential aesthetics, described in greater detail by Taylor (1998), suggests that aesthetic theories can be linked to the concept of ecological understanding by focusing on the personal experience of observing consequences, using all of the human body's senses

ENVIRONMENTAL POLICY AND ISSUES OF SCALE (KATHRYN FREEMARK). People are organized as individuals, communities, and states (Hanna and Jentoft 1996). Each political level is embedded in a larger social, economic, and political sphere, and interacts with other dimensions through both cooperation and conflict. In Western societies, the hierarchy of decision making often has five to seven geopolitical levels, with boundaries that bear no direct relationship to the spatial boundaries of ecological units. For example, in ratifying the Convention on Biological Diversity, the government of Canada committed to conservation and sustainable use of biodiversity. To accomplish this, the federal government sets national objectives that are developed and implemented by provincial and territorial governments. Provincial actions are more likely to be effective if planned in the context of ecological or biotic regions. Once adopted, a policy is generally implemented by smaller administrative units such as counties. In general, policy must consider the roles and relative intensity of effort required by smaller units when allocating appropriate resources to get the job done. An "ecological" assessment of environmental policy would look both upward for the context at a larger extent and downward for implications of implementation.

Natural systems also have hierarchical influences on their organization and development over time. In order to be meaningful as a management tool, biodiversity must be measured at different spatial scales and levels of organization (genes, organisms, populations, species, communities, ecosystems, landscapes, and so on—to the biome and biosphere). White et al. (1999) discuss alternative conceptual constructs of this spatiotemporal scaling across levels of biological organization. For example, the relationship among spatial and temporal scales in ecology has traditionally been conceived as moving "up" the hierarchy from genes to biosphere. However, both biodiversity and species interactions at these conceptual levels can be difficult to quantify because populations, communities, and landscapes are open systems that vary widely among species and processes (Turner 1989; Wiens 1989). From a different perspective, levels of organization in this biological complexity could be viewed as conceptual constructs that have no preconceived scale (Allen and Hoekstra 1992). Ecosystems and communities can be compared within a landscape of some given geographic area, as well as at larger and smaller spatial extents. A given landscape can be seen to contain smaller landscapes, while itself being a part of a larger landscape. In practice, the choice of spatial and temporal scales is made so that the biological levels of interest appear most cohesive, explicable, and predictable.

An adequate ecological understanding would require the consideration of at least three conceptual levels and/or scales at once: the scale in question, the scale below it that often provides mechanisms for dynamic processes, and the scale above it that provides context and significance to the observations made at smaller geographical or temporal scales (Pickett et al. 1994). Similarly, the development of adequate ecological understanding will require us to examine the ways that human interests and actors influence each other across multiple scales of human decision making. Theories of political science and ecological hierarchy theory offer opportunities for learning to occur across disciplines, as models are compared.

THE ECOLOGICAL CONSEQUENCES OF POLITICAL RHETORIC (BARBARA RYDER). Design and planning professionals typically enter the working world ill prepared to cope with the nature of politics. Yet it is critical for these professionals to understand the campaigns of political persuasion that affect the attitudes of decision makers. An environmental planning firm may be hired to do comprehensive plans for a city or county planning department, usually in response to a state mandate. The citizen board that reviews the comprehensive plan might include a banker, a housewife, a farmer, an auto mechanic, and a dentist. Seldom do such boards have any background in ecological theory, political science, or land-use planning. Most are involved on these commissions because they believe it is their civic duty to represent the taxpayer. And many take positions on particular decisions out of a sense of history and fairness that is heavily influenced by the political rhetoric to which they are exposed in their communities. The theories and methods of case analysis used in the field of political economy could help designers and planners identify ways in which they can communicate ecological understanding, even in these politically charged settings. One example of such a situation is the influence of the so-called Wise Use movement in community-level decision making.

In the mid-1980s, a backlash against environmentalism centered in the American West congealed into the Wise Use movement, with its focus primarily on public lands (Arnold 1996; Snow 1996). The organizers rallied resource extraction industries and blue-collar laborers around issues such as restrictions placed on logging in the national forests. Financially supported by resource industries, the Wise Use movement has been adept at claiming to represent the workers' interests in communities that have long depended upon resource extraction for their economic survival. The movement has been effective in creating a common vision for these industries and their workers. Despite the robust profits of the companies involved in supporting this movement, Wise Use organizers have convincingly argued that the companies and workers are the beleaguered victims of spotted owl or salmon recovery restrictions. They argue that companies and workers should join together to protect a time honored way of life—be it ranching, farming, lumbering, or mining—in a reactionary strategy to defeat the common enemy, government regulation (Helvarg 1994; Sellen 1997).

While a detailed and complex legal discourse on property rights has been going on since the arrival of European settlers, the "property rights revolution" only began to emerge in the early 1990s (Yandle 1995; Seideman 1997). Since the 1930s, the U.S. Supreme Court has consistently supported the concept of public regulation of private property. Two notable examples can be seen in the 1960s (Freyfogle 1998). First, outdated property rules permitting owners of restaurants and motels to engage in overt racial discrimination gave rise to civil rights legislation. Second, strip mining in eastern Kentucky resulted in badly polluted rivers and degraded communities and thereby stimulated changes in property ownership laws that subjected mining companies to greater regulation. Yet the myth of the absolute right of property lives on (Reiger 1986; Windsor 1996). The significant shift in recent decades is that, partly because of the success of the Wise Use movement, some significant aspects of this debate have moved out of the territory of lawyers and judges and into the voting booths. The popularizers of this grassroots political movement do not need to produce or support legal theories if they can acquire advantage by arguing for their position in popular contexts. In the voting booth, this advantage translates into decision-making power and anti-environmental legislative initiatives (Glendon 1995; Sellen 1997).

The next generation of designers, planners, and ecologists cannot afford to avoid assuming a role in this ongoing political dialogue. If designers and planners are to uphold their professional responsibilities to provide the best possible advice, they will first need to help bring balance back into local political debates. Professionals from the design and planning fields can bring valuable training and insight to these debates, such as presentation abilities, process skills, facilitation skills, and ecological understanding. Their abilities to inventory, analyze, synthesize, and create a negotiated product can help remove the sensationalism of the current rhetoric on both sides, allowing a productive dialogue to begin. But if design professionals are not prepared as facilitators and are not informed by a careful analysis of these regional and national political issues, they will find themselves contributing to projects that work against the ultimate responsibilities of the professions to which they belong.

Bringing Ecological Understanding into a Design or Planning Education

Ecological understanding is a concept that is both simple and complex—simple in its definition, but complex in its development and use. As our six examples briefly illustrate, it is particularly complex when considered in relation to other academic disciplines that support the education of design and planning professionals, such as economics, legal studies, cultural studies, and so on. In each of these cases, ecological understanding is or may soon be transforming intellectual paradigms in other fields, just as those fields hold the potential to influence thinking in ecological science and design. But these mutual influences may not occur unless we "seed" them by making some intentional changes in the ways we teach.

Teaching Design Students to Anticipate Consequences During the Design Process

Our goal is to go beyond the current situation, where the impacts of designs and physical plans are often evaluated by scientists and engineers when the initial stages of the design process are already over. We would like to see designers incorporating an expectation and awareness of impacts in the conceptual phase of their work, so that their initial proposals will synthesize information about effects and relationships, working directly with those ideas.

We had many questions among ourselves about the best way to accomplish that goal. In this section, we try to answer two questions: How does an awareness of consequences give rise to a more strategic kind of thinking about change? How can this awareness itself, and the strategic thinking that is required to take advantage of it, be taught in a professional program?

One increasingly common way of representing ecological understanding in our professions has been to teach students to use environmental impact models (Ortolano 1997; Steinitz 1990). The structure of these models may range from very simple to very complex, depending on the complexity of the phenomena under consideration and the resources available for the modeling effort. But the consistent purpose of these models is to allow a professional to visualize the type and magnitude of impacts that might occur due to some human choice or action. This is the crux of ecological understanding, and it involves everything from heuristics, i.e., rules of thumb, to complex computer models that explicitly represent many variables and many types of uncertainty (see Chapters 10 and 16).

In many if not most cases, the potential for effects to be separated from their causes in space and time makes genuinely understanding and predicting environmental impacts very difficult. As agents of change, professionals in physical planning and design must question their assumptions about what is and is not represented in our models of potential impacts (whether these are mental models or computer models), and simultaneously question our ways of making decisions under uncertain conditions. Ecological understanding, as we propose that it be understood, functions best as a kind of heuristic guide to strategic decision making. Its role is important when ecological knowledge provides relevant facts that might otherwise be ignored as well as in situations where the "facts" about cause and effect are uncertain.

Strategies that put ecological understanding into practice should emphasize several fundamental tactics, including the following:

- A conservative attitude toward change (e.g., use of a precautionary principle to avoid changes that involve high risks, either singly or cumulatively) (Chapter 8; Shrader-Frechette and McCoy 1993)
- Use of heuristics, defined as tools for self-teaching, to approximate the way processes operate in the world around us or within us (Haddon 1970; Detwiler 1981; Patten 1986)
- Creation of checklists of regional priorities for conservation, so that complex knowledge can be transferred to nonspecialists through a prioritized set of concerns (AIA 1993; Steinitz and Binford 1989; Forman 1990)
- Use of impact assessment models in the planning/design process (including both site-specific and cumulative impact models) and in the implementation process (i.e., by monitoring over time to make adaptive management possible) (Holling 1978; Steinitz 1990; Ortolano 1997).

Each of these tactics is described in greater detail below. Following this section, our discussion turns to an example of how these strategies and tactics can be taught in an educational setting common to design and planning programs—the studio.

Use of the precautionary principle.

The precautionary principle places the well-known medical rule of "first, do no harm" at the forefront of our considerations about changing our environments. Steingraber (Chapter 8) describes the need for such a principle in the use of carcinogenic chemicals. By implication, it should also be used in the siting of facilities that use any toxic materials (carcinogenic or otherwise biologically active). Unfortunately, precisely because change is an integral part of most physical and biological systems, it may often be the case that a truly precautionary stance requires more than a "no action" solution. Negative trends such as the contamination of ground and surface waters, the decline of the populations of many animal and plant species, and the consumption of fertile soils for residential and commercial land uses will continue over time if no decisive action is taken. These may require us to propose radical or incremental interventions to slow or reverse these trends. Thus, we are arguing for a precautionary principle that discourages humans from choosing to cause irreversible change—not one that discourages all change. Designers and planners must see the implementation of their proposals as opportunities to test designs as hypotheses in experiments—not as permanent icons of an aesthetic preference or social ideology. Moreover, those experiments should involve as little risk and as much potential benefit as possible to our underlying natural and cultural resources.

Heuristics.

Heuristics, or rules of thumb, are tools for self-teaching. Most models are heuristics in the sense that we use their simplified structure to learn about relationships in the more complex systems they describe. Other types of heuristics may involve lists of rules used to guide decision making, or frameworks that suggest possible solutions to known problems. As an example, Table 12-1 presents a heuristic that relies on ecological understanding. A medical doctor named William Haddon (1970) developed this list while serving as head of the National Highway Safety Board. He referred to his profession as "medical ecology," and developed his list of strategies to reduce the number and severity of highway accidents. While it may be used for that purpose, the exceptional value of this list is that its strategies are generic. They can be applied to any situation where a vulnerable structure is threatened by a release of some kind of energy.

To use this list in design education, consider the example of a hypothetical case of a rare fish and a proposed housing development uphill from its stream habitat (Steinitz, pers. comm.). Haddon's (1970) list of strategies for

TABLE 12-1. On preventing the escape of tigers (Haddon 1970)

1. Prevent the dangerous energy from accumulating in the first place.
2. Reduce the amount of energy that is allowed to accumulate.
3. Prevent the release of the energy.
4. Reduce the rate, spatial distribution, or manner of release of the energy.
5. Separate the energy from the vulnerable structure in space and time.
6. Insert a barrier between the source of dangerous energy and the vulnerable structure.
7. Modify the contact surface of the vulnerable structure.
8. Strengthen the vulnerable structure.
9. Use early detection and quick response to limit the spread of damage.
10. Repair, restore, or replace the damaged structure to achieve pre-contact conditions.

preventing disaster can help generate design options to reduce or prevent an impact on the fish species. This heuristic model can be used by anyone who seeks to prevent dangerous accidents—or, in general, someone who must prevent damage to a vulnerable structure that is caused by some harmful release of energy. The vulnerable structure could be a salmon stream, a building, or a child. Haddon notes in the original article the frequency with which we try to prevent the "escape of tigers" by, in effect, closing the door of the cage after the tiger has already gotten out—that is, by using the strategies at the end of this list, instead of first considering those at the beginning. We call this list a heuristic because it can be used as a tool for self-teaching, if a designer or planner wishes to expand his or her options by asking what might serve as an example of each numbered strategy in an actual problem or situation.

Issue-based checklists.

Because ecological systems and their responses to human actions vary regionally, use of ecological understanding in designs will vary in different circumstances. Different regions may have different specific priorities for conservation or different histories of human influence. Thus no single, overarching checklist can be defined for all parts of the world; generic lists (Table 12-2) may be helpful as heuristics that remind us of general concerns important to environmental sustainability. A regionally specific list would be more useful to designers and planners, and could easily be developed from this generic list. Steinitz (Chapter 10) offers a similar list of what he calls "process-based models," but our version is intended as a general framework for these types of lists. A checklist would be defined in detail for a particular region of the

TABLE 12-2. A generic checklist of concerns for environmental sustainability (modified from a list presented in Forman 1990)

1. **Biodiversity:** Genetic diversity and its manifestation from species to ecosystems diversity, including all organisms from vertebrates and invertebrates to plants and microorganisms. "Umbrella" species whose conservation may benefit many other species may be an important focus. Such special species can include those with a large home range; those that require large, contiguous areas of undisturbed land; and those that occupy habitat types that have been or are being lost to development. Microbial biodiversity may be of increasing importance for bioremediation and human health concerns.

2. **Air Quality:** Includes ambient air quality as well as air quality trends indicated by biological monitoring. Indoor as well as outdoor air quality issues should be considered, such as radon gas or volatile organic chemicals.

3. **Water Quality:** Multimetric biological indexes such as IBI (index of biological integrity; see Chapter 6) can help assess the relationship between changes in physical measures of water quality and their biological impacts. Similarly, an improved understanding of the elements of the urban hydrological cycle will help designers and planners anticipate the impacts of decisions made about each element individually. This should include recent concerns about persistent agricultural chemicals, the by-products of chlorine in drinking water, and the widespread presence of volatile organic chemicals in groundwater.

4. **Water Quantity:** Limits on the amount of fresh water available to urban areas are likely to increase in importance over the next 20–30 years. Designers and planners should be aware of this limitation and be prepared to respond with systems that conserve freshwater resources and reuse wastewater. Watershed-based management of resources and land development should be understood and advocated by designers and planners.

5. **Soil Fertility:** Degraded soils limited the sustainability of many past human settlements. Expanding residential and commercial land uses have put development pressure on many of the most fertile soils in urbanizing areas, while industrialized farming practices have put U.S. soil erosion rates at world-class levels (upward of 15–18 tons per acre per year in many parts of the American Midwest). Opportunities to conserve and cultivate fertile soils are an important element of public education on health and resource use, as are studies of regional "foodsheds" that identify the spatial pattern of food-source areas. The evolving concept of "ecological footprints" can also be useful for identifying issues in a given region (Wackernagel and Rees, 1996).

6. **Human Health:** In the spirit of *Silent Spring*, three recent books have articulated the current public health concerns posed by inappropriate uses of land, water, and chemical substances: *Our Stolen Future*, by Theo Colborn et al. (1996); *Living Downstream*, by Sandra Steingraber (1997); and *Biodiversity and Human Health*, edited by Grifo and Rosenthal (1997). The

last of these three books identifies cases in which microbial diversity and habitat degradation combine to affect human biology. A complex picture emerges from these studies that each region of the globe must consider, and that requires us to set aside many of our assumptions about our current environments. The well-being of human society is threatened by factors as diverse as chemical contamination, emerging and reemerging disease, soil and water depletion, and climate change, as well as the geopolitical unrest that derives from these problems (Chapter 8).

7. **Cultural Cohesion:** This odd "catch-all" category is very broad indeed. It includes thinking about trends or changes in the stories we tell each other about our uses of the land and its resources, as well as socioeconomic trends and political conflicts among groups and ideologies. Designers and planners can contribute to sustaining this dimension of human life by helping to create and maintain a shared understanding of the built and ecological "support systems" we all depend on in the form of infrastructure and by promoting social justice in resource use. In short, we must be cognizant of the extent to which human cultural cohesion is dependent on both the human economy and nature's economy.

world. Our generic version is based on a list generated by Forman (1990) that identified factors relevant to environmental sustainability.

Impact assessment models.

Impact assessment models range from simple qualitative observations to complex quantitative models based on extensive fieldwork, executed by computers, and validated using field measurements. Their primary importance is in the feedback they provide to designers and planners. Clearly, no model could ever adequately represent the actual complexity of the environment; each contains assumptions allowing complexity to be abstracted and simplified. The art of building such models based on ecological understanding is in deciding what potential impacts to consider. Steinitz (Chapter 10) has evolved a framework for including these models in a design or planning process that we find particularly useful.

One of the greatest challenges to the effective development and use of impact models, particularly when those models are intended to embody ecological understanding, is the tendency to define the scope of a particular problem or situation too narrowly. Defining a problem is too often confused with comprehending the complexities of a system. Instead, we would say design has as much or more to do with defining our possible actions. If we apply ecological understanding in design situations, we are more likely to conclude that incremental, impermanent alterations of the environment may be preferable to large, radical changes. A slower pace of change allows us to assess actual impacts and adjust our methods to improve them during a longer process of change. This also requires that we assess the "before and after" conditions of our implemented schemes, by monitoring them over time so that trends and changes can be observed. Monitoring is not explicitly presented as a category in Steinitz's framework for design theory and practice, and may be embedded instead in his category of "process models," but we think it is so critical that it should be thought of as a separate category of theory in addition to the categories Steinitz identifies (Chapter 10).

Cumulative impacts are a special class of impacts. These are impacts that cannot be attributed to any one site or project, but that are evident once many sites have been altered. Some examples include nonpoint surface- and groundwater contamination, changes in the flooding patterns of urban streams, loss of native vegetation, conflicts over safety that arise when human residential areas encroach on the habitat of large predators (e.g., black bear, cougar), the presence of nuisance species (squirrels, opossums, rats), and invertebrate pests (termites, disease-carrying mosquitoes, ticks carrying Lyme disease). Since each site or project is responsible only for the degree to which it alone contributes to these larger problems, no site or project is held accountable—although all contribute to the problem. Our impact models must ensure that we observe, understand, and respond to these cumulative impacts by developing improved spatial prototypes or management strategies using the deepened ecological understanding that we gain from them.

An Example: Strategies for Incorporating These Tactics into a Studio

Currently, design approaches derived from both the Ecole des Beaux Arts and from what has sometimes been called a "modernist" view of design and society often coexist in our curricula. Barry Russell (1995) wrote that there has been a historical "split" between the formal study of precedent as the best guide for design versus the effort to innovate and create "original" design solutions. He identified the former as the long-standing tradition of the Ecole des Beaux Arts in Paris and the latter as "one of the main planks of modernism" (p. 34). This debate continues in many studio-based design curricula, as students and faculty struggle to simultaneously recognize historical examples and reward the expression of personal creativity. Russell suggested that the study of successful precedents leads to an observation of patterns that can be "studied, understood and applied," and that "in this context, an informed acceptance or rejection of ideas can take place and real creativity begin" (ibid.).

We agree with Russell that creativity should more reasonably begin once a designer is sufficiently apprised of contextual issues and priorities to make "an informed acceptance or rejection" of alternatives. But we argue that the source of that information is not only historical precedents but also knowledge gathered in a search for ecological understanding, including heuristics that allow us to come up with new strategies (perhaps using old forms in new ways), checklists of regional conservation priorities, some version of an impact model that addresses both local and cumulative effects, and a precautionary principle that requires caution in the face of irreversible impacts. Regardless of the philosophical attitude toward design and design processes that is promoted in a particular educational program, our professional responsibilities require us to provide the best possible advice to our clients. Giving the best advice, in our view, requires that ecological understanding should be applied at the outset—in defining the problems, generating the solutions, and evaluating their relative "goodness."

A design approach based in ecological understanding is different from both the Beaux Arts' focus on precedents as a model for current designs, and the ahistoricism of rational "problem-solving" approaches to design. But it offers us an opportunity to build a dialectical approach to design from what might otherwise be seen as a dichotomy in our dominant schools of thought. If we learn from ecological precedents as well as cultural ones, and in so doing create a context for our creativity, then we can explore this dialectical relationship as we generate and explicitly evaluate our designs and plans.

To incorporate ecological understanding in our work, we must (1) favor reversible changes and temporary land uses over the irreversible and the permanent, (2) use heuristics to simulate a complex environment and to learn from our experiences, (3) monitor the biological impacts of our projects, and (4) address the role of site-scale changes in the creation of cumulative ecological impacts. When considering strategies for new designs and plans, ecological understanding requires us to listen to the traditional admonishment received by physicians to "first, do no harm" before setting goals for human use and enjoyment of the environment. Moreover, trends that were already set in motion by past development may need to be countered by some action in the present. A "no action" alternative is, in actuality, often a decision to allow negative trends to continue. Professionals who bring ecological understanding to their work will find ways to make every project counter harmful, landscape-wide trends such as increasing water contamination or loss of biodiversity and, if possible, inspire regional attention to them.

How do we teach this question-driven approach to design, with ecological understanding as a critical *and* early source of questions? Next, we discuss some of the ways that a typical studio course can incorporate exercises that promote and deepen students' ecological understanding. We also recommend Chapter 16, on teaching ecology in the studio, for a different presentation of these issues and a more detailed discussion of studio pedagogies.

The studio setting.

We propose a three-pronged approach to introducing ecological understanding in the studio:

1. Adopt a question-driven (not a problem-driven) design process. Studio projects typically begin with a brief from the instructor that states what is to be designed, for what uses, and with what constraints. Students then begin a problem-solving process, either implicitly or explicitly. We suggest that this lesson is too linear, in that it implies that the stages of design are arranged in a definite order, and that if students do them all, they will produce an adequate solution to a problem.

   Steinitz' framework (Chapter 10), as one alternative example, might prompt a studio instructor to begin by asking the students these questions: (a) What are the trends over time in a given region (these should be generated from a checklist of significant cultural and natural issues, such as we described above, that has been adapted to the region of study); (b) who will make the decisions about how to respond to these trends; and (c) how will they make those decisions? The decision makers that the students identify may or may not be the "client" in the traditional sense of that role. In any case, the first step is to begin a process driven by questions. Ecological understanding enters into this process first through the regional issues checklist (as a heuristic device for representing processes and setting priorities), and second through a discussion of the need to make informed decisions based on what is often only partial knowledge.

2. Reexamine the representations we use to describe the world we wish to change. Are vulnerable biological (and cultural) processes or entities represented as well as the political or aesthetic opportunities? Have we taken our cultural biases into account and found ways to represent them explicitly? Have we represented our uncertainties about how these vulnerable processes work on and around the site(s) of interest?

   The next step in the design process would be to ask the students what has been left out of the "representation model" of the world that the class has used so far—for example, using the checklist of significant trends.

And, in another meaning of the word "representation," to ask which individuals or groups of people have been left out of the decision-making processes the students have just discussed. These reflective questions about representation can help students understand a bigger picture than they are likely to perceive if they are simply handed a brief that describes their design or planning problem.

3. Seek opportunities to teach design as a form of approximate reasoning that embraces uncertainties and applies a combination of common sense, ethical reasoning, and strategic awareness to avoid disasters and uncover opportunities for positive impacts on our environments (e.g., using Haddon's [1970] list of strategies, Table 12-1).

Typically, once students in a studio have identified a need that they intend to address, they set about designing a building, a site, a piece of infrastructure, or a larger area. But in order to emphasize ecological understanding, students must anticipate that they will be held to some kind of performance standards in this design exercise that are relevant to the checklist of regional trends that has been introduced. This can be done by creating impact models for estimating site specific as well as cumulative impacts. These models can be very simple or very complex, and can even involve use of the actual models that local decision makers use to test proposed policies or designs. This technique introduces the essential idea that designs should be tested for basic functionality *and* for their "goodness" in relation to larger issues. In addition, these impact models should be updated and improved through postoccupancy or postconstruction monitoring in order to teach the necessity of adaptation and improvement in approximations over time.

At the conclusion of the design exercise, the design proposals could be tested using a review process similar to what actual decision makers might select (or be required to use) as a decision-making process. Critiques based on aesthetic criteria, ecological functionality, sustainability, and so forth should be brought into each review as arguments to which the students must also respond. Criteria for "successful projects" should be relevant to the ever-present checklist of regionally important cultural and environmental concerns. Otherwise, the final and most memorable message taught by the studio will be that the attention students paid to those issues was unimportant.

An additional suggestion for studio that addresses the ability of students to think of design as approximate reasoning is an exercise in which students are asked to confront their narrowness of vision. Midway through a studio problem, a change in conditions can be announced such as a change of client, or a change in the amount of land area available, or a change in regulations that introduces significant new constraints. This new "surprise" project can be a short one, perhaps lasting a week or less, and still have a memorable impact on the students' sense of their own ability to deal with the unexpected. It might also increase the adaptability of their future plans and designs.

Presenting Ecological Understanding as an Interdisciplinary Paradigm

Our experiences and the literature of diverse fields lead us to conclude that a "turn to ecology" is evident in many design- and planning-related disciplines. Many fields are experiencing this "turn," which we have defined here as a more serious consideration of relationships and consequences. Fields as diverse as economics, aesthetics, cultural theories, political science, cognitive psychology, legal studies, and others have begun to consider the potential impact of the broad intellectual implications of ecological ideas on some of their central tenets. Our position is that as material from these diverse fields is introduced into a design or planning curriculum, it can be presented through these new viewpoints in order to center them around ecological understanding. This would reinforce the awareness of physical, social, and biological consequences that students are acquiring in their core courses, and reassure them that they are not the only ones who are thinking about those consequences.

Students should also be encouraged to champion the ideas of ecological understanding in other courses outside their professional education, by making themselves aware of the relevant literature in each field and introducing material from that literature into class papers or discussion. This serves the dual purposes of (1) encouraging design students to take an intellectual leadership role, and (2) creating a common thread of issues that continues across the breadth of supporting and elective courses in a professional curriculum. This common intellectual thread can allow students to more effectively synthesize what they are learning and provide them with a more cohesive, less fragmented understanding of the role of a reflective practitioner (in the sense of Schon 1987).

Design faculty, for their part, could contribute to students' development of a nascent intellectual synthesis by holding panel discussions related to ecological understanding with the faculty who teach those supporting or elective courses. Panel discussions such as these would allow students to see faculty members as role models in intellectual leadership. They might also lead to productive collaborations in teaching or research between the faculty involved. They would also help to update professional school faculty on recent trends in the more traditional academic disciplines—a notoriously difficult challenge for faculty who rarely have time to read outside their field.

We have written here as if design faculty typically teach alone, yet there are many examples of successful interdisciplinary collaborations in studio settings. These can be created between full-time faculty and part-time faculty who have been drawn from a pool of practitioners with an interest in teaching, or by combining faculty across departments and colleges in collaborative teaching. Such strategies may have substantial salary costs since a department would be enlarging its teaching staff for these courses. An alternative is for collaborators to be brought into the studio from the ranks of recent Ph.D. graduates from other fields who would like to gain experience teaching in an applied context such as design and planning. Or courses could be taught jointly by faculty from different departments, while retaining department-specific course numbers and administrative record-keeping. This latter strategy involves more of a coordination problem than a curriculum change, and can lead to very successful interdisciplinary experiences. All of these practices would help to extend the framework of ecological understanding across a broader set of disciplines, and perhaps contribute to the broad paradigm shift of a "turn to ecology" across many fields.

In a hostile institutional environment, administrators could pose significant institutional impediments to the recommendations we have made. But when administrators are open to these approaches, much can be accomplished with only modest levels of resource support. Very little (if any) additional money must be spent to accomplish these pedagogical goals. There may, however, be ideological impediments. Faculty may see the paradigm of ecological understanding as a threat to their preconceived sense of priorities for teaching. We choose to believe that this chapter contains some grains of wisdom that could persuade these colleagues otherwise. We also believe that our colleagues in academia and in the professions will agree with us to a greater extent than they disagree. Our optimism derives from our argument that the development of a profound ecological understanding in the design and planning professions is the most important challenge facing current and future generations of teachers and practitioners. Given the overwhelming evidence of recent human impacts on the biological and physical environment (see Chapters 4 and 6), who could argue anything less?

Citations

Allen, T. F. H., and T. W. Hoekstra. 1992. Toward a unified ecology. Columbia University Press, New York, New York, USA.

American Institute of Architects. 1993. Building connections: livable, sustainable communities. Washington, D.C., USA.

Anderson, E., 1990. Streetwise: race, class and change in an urban community. University of Chicago Press, Chicago, Illinois, USA.

Arnold, R. 1996. Overcoming ideology. Pp. 15–26 in P. D. Brick and R. M. Cawley, eds. A wolf in the garden: the land rights movement and the new environmental debate. Rowman and Littlefield, Lanham, Maryland, USA.

Bahktin, M. 1981. The dialogic imagination. Ed. by M. Holquist, trans. by C. Emerson and M. Holquist. University of Texas Press, Austin, Texas, USA.

Beatley, T. and K. Manning. 1997. The ecology of place: planning for environment, economy and community. Island Press, Washington, D.C., USA.

Borgmann, A. 1992. Crossing the post-modern divide. University of Chicago Press, Chicago, Illinois.

Bourassa, S. 1991. The aesthetics of landscape. Belhaven Press, New York, New York, USA.

Bross, I., 1953. Design for decision. Macmillan, New York, New York, USA.

Brown, B., T. Harkness, and D. Johnston (eds.), 1998. Eco-revelatory design: nature constructed, nature revealed. Landscape Journal, Special Issue: Exhibit Catalog.

Calthorpe, P. 1993. The next American metropolis: ecology, community, and the american dream. Princeton Architectural Press, New York, New York, USA.

Carlson, A. 1981. Nature, aesthetic judgement, and objectivity. Journal of Aesthetics and Art Criticism, Fall: 15–27.

Carson, R. 1962. Silent Spring. Fawcett Crest, New York, New York, USA.

Chernoff, H., and L. Moses. 1959. Elementary decision theory. John Wiley and Sons, New York, New York, USA.

Colborn, T., D. Dumanoski, and J. Meyers. 1996. Our stolen future: are we threatening our fertility, intelligence, and survival? A scientific detective story. Dutton, New York, New York, USA.

Cooke, H. (trans.). 1938. Aristotle's categories: on interpretation. Harvard University Press, Cambridge, Massachusetts, USA.

Corner, J. 1997. Ecology and landscape as agents of creativity. Pp. 81–108 in G. Thompson and F. Steiner, editors. Ecological design and planning. John Wiley and Sons, New York, New York, USA.

Costanza, R., J. Cumberland, H. Daly, R. Goodland, and R. Norgaard, 1997. An introduction to ecological economics. St. Lucie Press, Boca Raton, Florida, USA. 275 pp.

Costanza, R., O. Segura Bonilla, and J. Martinez Alier (eds.). 1996. Getting down to Earth: practical applications of ecological economics. Island Press, Washington, D.C., USA. 274 pp.

Daly, H. E. 1995. Reply to Mark Sagoff's "Carrying capacity and ecological economics." BioScience 45: 621–624.

Demerath, N. 1947. Ecology, framework for city planning. Social Forces, October, 26: 62–67.

Detwiler, P. 1981. Environmental analysis after a decade: If prophecy is impossible, then go for understanding. Public Administration Review 41: 93–97.

Dufford, W. 1995. Washington water law: A primer. Illahee 11: 29–39.

Eaton, M. 1997. Beauty that requires health. In J. Nassauer, editor. Placing nature: culture and landscape ecology. Island Press, Washington, D.C., USA.

Eckbo, G. 1998. People in a landscape. Prentice-Hall, Princeton, New Jersey, USA.

Environment Canada. 2001. <http://www.ec.gc.ca/CEPARegistry/the act/Intro-dution.cfm>

Forman, R. 1990. Ecologically sustainable landscapes: the role of spatial configu-ration. In I. Zonneveld and R. Forman, editors. Changing landscapes: an eco-logical perspective. Springer-Verlag, New York, New York, USA.

Freyfogle, E. T. 1998. Bounded people, boundless land: envisioning a new land ethic. Island Press, Washington, D.C., USA.

Garrett, L. 1994. The coming plague: newly emerging diseases in a world out of balance. Farrar, Straus & Giroux, New York, New York, USA.

Gilligan, C. 1982. In a different voice: psychological theory and women's devel-opment. Harvard University Press, Cambridge, Massachusetts, USA.

Glendon, M. A. 1995. Absolute rights: property and privacy. Pp. 182–188 in John D. Echeverria and Raymond Booth Eby, editors. Let the people judge: wise use and the property rights movement. Island Press, Washington, D.C., USA.

Goldberger, N. 1996. Cultural imperatives and diversity in ways of knowing. In N. Goldberger, J. Tarule, B. Clinchy, and M. Belenky, editors. Knowledge, differ-ence, and power. Basic Books, New York, New York, USA.

Grifo, F., and J. Rosenthal (eds). 1997. Biodiversity and human health. Island Press, Washington, D.C., USA.

Griswold, W. M. 1994. Cultures and societies in a changing world. Pine Forge Press, Boston, Massachusetts, USA.

Hacking, I. 1983. Representing and intervening. Cambridge University Press, Cambridge, UK.

Haddon, W. 1970. On the escape of tigers: an ecologic note. Technology Review 72, May: 45–53.

Hanna, S., and S. Jentoft. 1996. Human use of the natural environment: an overview of social and economic dimensions. Pp. 35–55 in S. Hanna, C. Folke, and K.-G. Maler, editors. Rights to nature: ecological, economic, cultural, and political prin-ciples of institutions for the environment. Island Press, Washington, D.C., USA.

Harding, S. 1991. Whose science? Whose knowledge? Thinking from women's lives. Cornell University Press, Ithaca, New York, USA.

Harvey, D. 1989. Consciousness and the urban experience. Johns Hopkins Uni-versity Press, Baltimore, Maryland, USA.

Heidegger, M. 1993. Selections: basic writings, from being and time to the task of thinking. D. Krell, editor. Harper, San Francisco, California, USA.

Helvarg, D. 1994. The war against the greens: the "wise use" and the property rights movement, the new right, and anti-environmental violence. Sierra Club Books, San Francisco, California, USA.

Hills, A. 1961. The ecological basis for land use planning. Ontario Dept. of Lands and Forests, Research Branch, Toronto, Ontario, Canada.

Holling, C. (ed.). 1978. Adaptive environmental assessment and management. Wiley, New York, New York, USA.

Hough, M. 1995. Cities and natural process. Routledge, New York, New York, USA.

Jefferson County v. Washington Department of Ecology. 1994. 114 S. Ct. 1900.

Kant, I. 1790. Critique of judgement. Trans. by J. Bernard, 1931. Macmillan, London, UK.

Kaplan, R., and S. Kaplan. 1989. The experience of nature: a psychological perspective. Cambridge University Press, Cambridge, UK.

Karr, J. R. 1995. Clean water is not enough. Illahee 11: 51–59.

———. 2001. Protecting life: weaving together environment, people, and law. Chapter 14 in R. G. Stahl Jr., A. Barton, J. R. Clark, P. deFur, S. Ells, C. A. Pittinger, M. W. Slimak, and R. S. Wentsel, editors. Risk management: ecological risk-based decision-making. SETAC Press, Pensacola, Florida, USA.

Kelly, M. (ed.). 1994. Critique and power: recasting the Foucault/Habermas debate. MIT Press, Cambridge, Massachusetts, USA.

Kupferberg, S. 1997. Bullfrog (*Rana catesbeiana*) invasion of a California river: the role of larval competition. Ecology 78: 1736–1751.

Lakoff, G. 1987. Women, fire, and dangerous things: what categories reveal about the mind. University of Chicago Press, Chicago, Illinois, USA.

Lawrence, H. and A. Bettman, 1981. The green guide: Eugene's natural landscape. Northwest Working Press, A. Bettman, Publisher; Eugene, Oregon, USA.

Lee, K. N. 1993. Compass and gyroscope: integrating science and policy for the environment. Island Press, Washington, D.C., USA.

Lyle, J. T. 1994. Regenerative design for sustainable development. John Wiley and Sons, New York, New York, USA.

Lynch, M. and S. Wolgar (eds.), 1990. Representation in scientific practice. MIT Press, Cambridge, Massachusetts, USA.

Lyotard, J. 1984. The postmodern condition: a report on knowledge. University of Minnesota Press, Minneapolis, Minnesota, USA.

McHarg, I. 1969. Design with nature. Natural History Press, Garden City, New York, USA.

Meyer, E. 1997. The expanded field of landscape architecture. Pages 45–79 in G. Thompson and F. Steiner, editors. Ecological design and planning. John Wiley and Sons, New York, New York, USA.

M'Gonigle, R. M. 1999. Ecological economics and political ecology: towards a necessary synthesis. Ecological Economics 28: 11–26.

Mitsch, W., and S. Jorgensen. 1989. Ecological engineering: an introduction to ecotechnology. John Wiley and Sons, New York, New York, USA.

Nassauer, J. 1995. Messy ecosystems as a matter of policy. Landscape Ecology 6: 239–250.

———, (ed.). 1997. Placing nature: culture and landscape ecology. Island Press, Washington, D.C., USA.

O'Riordan, T., and J. Cameron (eds.). 1994. Interpreting the precautionary principle. Earthscan Publications Ltd., London, UK.

Orr, D. 1992. Ecological literacy: education and the transition to a postmodern world. State University of New York Press, Albany, New York, USA.

Ortolano, L. 1997. Environmental regulation and impact assessment. John Wiley and Sons, New York, New York, USA.

Paige et al. v. Town Plan and Zoning Commission of the Town of Fairfield. 1994. 35 Conn. App. 646, 646 A. 2d 277.

Patten, C. 1986. Being roughly right rather than precisely wrong: teaching quick analysis in planning curricula. Journal of Planning Education and Research **6**: 22–28.

Peters, R. 1991. A critique for ecology. Cambridge University Press, Cambridge UK.

Pickett, S. T. A., J. Kolasa, and C. G. Jones. 1994. Ecological understanding. Academic Press, San Diego, California, USA.

Posner, R. A. 1995. Overcoming law. Harvard University Press, Cambridge, Massachusetts, USA.

Rapport, D. J., C. Gaudet, J. R. Karr, J. S. Baron, C. Bohlen, W. Jackson, B. Jones, R. J. Naiman, B. Norton, and M. M. Pollock. 1998. Evaluating landscape health: integrating societal goals and biophysical processes. Journal of Environmental Management **53**: 1–15.

Reiger, J., 1986. American sportsmen and the origins of conservation. University of Oklahoma Press, Norman, Oklahoma, USA.

Rettkowski v. Washington Department of Ecology. 1993. 122 Wn. 2d 219, 858 P. 2d 232.

Rosch, E., E. Thompson, and F. Varela. 1996. The embodied mind. The MIT Press, Cambridge, Massachusetts, USA.

Rose, C. 1997. Property rights and responsibilities. Pp. 49–59 in M. Chertow and D. Esty, editors. Thinking ecologically: the next generation of environmental policy. Yale University Press, New Haven, Connecticut, USA.

Rose, H. and H. Nowotny (eds.). 1979. Counter-movements in the sciences: the sociology of the alternatives to big science. D. Reidel Publishing Co., Boston, Massachusetts, USA.

Rosenberg, A. 1986. An emerging paradigm for landscape architecture. Landscape Journal **5**: 75–82.

Russell, B. 1995. Paradigms lost, paradigms regained. Pp. 34–37 in M. Pearce and M. Toy, eds. Educating architects. Academy Editions, London, UK.

Sagoff, M. 1995. Carrying capacity and ecological economics. BioScience **45**: 610–620.

Schon, D. 1987. Educating the reflective practitioner: toward a new design for teaching and learning in the professions. Jossey-Bass, San Francisco, California, USA.

Seideman, D. 1997. Terrorist in a white collar. Time **25** (June 1990), 60.

Sellen, J. 1997. The discourse of private property rights in contemporary environmental politics." Dissertation, Program in American Studies, Washington State University, Pullman, Washington, USA.

Seltzer, R. Manning, and R. Steinberg. 1987. Wetlands and private development. Columbia Journal of Environmental Law **12**: 159–201.

Shrader-Frechette, K. S. 1985. Science policy, ethics, and economic methodology. D. Reidel Publishing Co., Dordrecht, Holland. 321 pp.

———. 1994. Ethics of scientific research. Rowman & Littlefield, Lanham, Maryland, USA.

Shrader-Frechette, K., and E. McCoy. 1993. Method in ecology: strategies for conservation. Cambridge University Press, New York, New York, USA.

Smith, D., and P. Hellmund. 1993. Ecology of greenways. University of Minnesota Press, Minneapolis, Minnesota, USA.

Snow, D. 1996. The pristine silence of leaving it all alone. Pp. 27–38 in Philip D. Brick and R. McGreggor Cawley, editors. A wolf in the garden: the land rights movement and the new environmental debate. Rowman & Littlefield, Lanham, Maryland, USA.

Soulé, M. and G. Lease. 1995. Reinventing nature? Responses to postmodern deconstruction. Island Press, Washington, D.C., USA.

Spirn, A. 1984. The granite garden: urban nature and human design. Basic Books/HarperCollins, New York, New York, USA.

Spretnak, C. 1999. The resurgence of the real: body, nature and place in a hypermodern world. Routledge, New York, New York, USA.

Steiner, F., G. Young, and E. Zube. 1988. Ecological planning: retrospect and prospect. Landscape Journal 7: 31–39.

Steingraber, S. 1997. Living downstream: an ecologist looks at cancer and the environment. Addison-Wesley, New York, New York, USA.

Steinitz, C. 1979. Defensible processes for regional landscape design. Landscape Architecture Technical Information Series, American Society of Landscape Architects, Washington, D.C., USA.

———. 1990. A framework for theory applicable to the education of landscape architects (and other design professionals). Landscape Journal 9: 136–143.

Steinitz, C., and M. Binford. 1989. Michael's list. Course notes for Landscape Architecture studio. Graduate School of Design, Harvard University, Cambridge, Massachusetts, USA.

Taylor, R. 1998. Aesthetics and cognition: the embodied mind in the ecological landscape. Master's Thesis, Dept. of Landscape Architecture, University of Washington, Seattle, Washington, USA.

Teymur, N. 1982. Environmental discourse: a critical analysis of "environmentalism" in architecture, planning, design, ecology, social sciences, and the media. Question Press, London, UK.

Thayer, R. 1994. Gray world, green heart: technology, nature, and sustainable landscape. John Wiley and Sons, New York, New York, USA.

Thofelt, L. and A. Englund, 1996. Ecotechnics for a sustainable society. Mid-Sweden University, Frösön, Sweden.

Thompson, G., and F. Steiner (eds.). 1997. Ecological design and planning. John Wiley and Sons, New York, New York, USA.

Turner, M. G. 1989. Landscape ecology: The effect of pattern on process. Annual Review of Ecology and Systematics 20: 171–197.

Van der Ryn, S., and P. Calthorpe. 1991. Sustainable cities: a new design synthesis for cities, suburbs and towns. Sierra Club Books, San Francisco, California, USA.

Veitch, I. 1978. Ecological approaches to land use planning. York University, Downsview, Ontario, Canada.

Wackernagel, M., and W. Rees. 1996. Our ecological footprint: reducing human impact on the earth. New Society Publishers, Philadelphia, Pennsylvania, USA.

Warren, K. (ed.). 1997. Ecofeminism: women, culture, nature. Indiana University Press, Bloomington, Indiana, USA.

White, D., E. M. Preston, K. E. Freemark, and A. R. Kiester. 1999. A hierarchical framework for conserving biodiversity. Pp. 127–153 in J. M. Klopatek and R. H.

Gardner, editors. Landscape ecological analysis: issues and applications. Springer-Verlag, New York, New York, USA.

Wiens, J. A. 1989. Spatial scaling in ecology. Functional Ecology 3: 385–397.

Windsor, A. S. Jr. 1996. Another eco alarm. The New American 29 (April): 31–32.

Yandle, B. 1995. The 1990s property rights revolution and U.S. agriculture: the limitations of freedom. Vital Speeches of the Day, 1 April, 365–368.

CHAPTER 13

The Nature of Dialogue and the Dialogue of Nature: Designers and Ecologists in Collaboration

Bart R. Johnson, Janet Silbernagel, Mark Hostetler, April Mills, Forster Ndubisi, Edward Fife, and MaryCarol Rossiter Hunter

Designers and ecologists traditionally have received different training toward different ends. Despite contrasts in process and product, the work of a growing number of individuals from both disciplines has converged on common issues of human inhabitation and use of Earth's finite space and resources, which we share with an estimated 10 million to 100 million other species (May 1992; Wilson 1992). The interweaving of ecological and cultural factors in landscapes that range in context from urban centers to wilderness reserves, and in scale from residential lots to large watersheds, provides strong impetus for joint efforts among designers and applied ecologists. Dialogue and collaboration between the two disciplines holds the potential for shared learning that could reshape how we design and manage landscapes. To achieve this, we must rethink how we educate our students.

In this chapter, we examine the ways in which designers and applied ecologists[1] can mutually enrich their disciplines and practice, and how to support this development in university educational programs. We begin by considering how historically different goals and training have led to divergent, but potentially complementary worldviews.[2] Next, we explore how a question-based framework can facilitate dialogue, and illustrate how key concepts can be incorporated. We focus principally on concepts from ecology useful for designers and planners, but in the process we begin to juxtapose and link concepts from both disciplines. Finally, we consider ways to structure educational programs to better prepare students to fruitfully and frequently engage in such working dialogues.

We propose that the most useful approach to dialogue is through the landscape—shared understandings of real places with rich and singular histories and many potential futures. The key is to emphasize an understanding of *place*, and to harness the combined insights of different fields toward landscape design, planning, and management that recognize the profound interdependence of cultures and ecosystems. Not long ago, almost all ecologists appeared focused on a natural world remote from human influences, and most landscape designers seemed preoccupied with aesthetics and purely human considerations. Far-sighted individuals in both disciplines never lost sight of the interdependence of cultures and ecosystems (Chapter 2), but the emphasis on design as cultural art, and ecology as natural science, may have widened the gulf between disciplines. One way to enhance dialogue among these disciplines is for each to understand how the other looks at the world.

Integrating Divergent Worldviews

Designers seek specific solutions for individual places; scientists seek general principles across multiple cases. To this end, ecologists employ a variety of methods to develop and test hypotheses about how living systems behave that may or may not lead to recommendations for application. Inductive reasoning may be used to generate hypotheses or general theory from examination of observed patterns. Deductive reasoning is used to test and refine hypotheses with methods that include experiments based in the scientific method, analyses of existing places, case studies of "natural experiments," and simulation models. Careful observation and the creative use of a variety of investigative methods are hallmarks of the discipline.

For landscape designers, a design process leads to a physical solution to a set of diverse and often conflicting goals. They build upon design precedents in much the way that scientists build upon tested theory, while modifying and adapting solutions in innovative ways to fit a particular place with all its idiosyncrasies. A design process provides a framework to focus creativity on the goals of the project, inspired and tempered by knowledge of the site and its context. It provides methods to analyze sites, kindle imagination, rethink goals or proposals in the face of new understandings, and open the door for intuitive leaps from descriptive analysis to prescriptive solutions. Design processes include both rational and intuitive stages, and may be best characterized as an alternating current of analytical investigation and intuitive synthesis (Lyle 1985).

As a consequence of their different means for understanding landscapes and their different goals in doing so, ecologists and designers may have very different ways of conceptualizing landscapes. Ecologists may consider the landscape through several different lenses. For instance, they may see it as a community composed of populations of coexisting species or as flows of matter and energy mediated through metabolic couplings and energetic transfers. They may examine the variety of species present and how their different life history traits determine how they persist and coexist over time. They may ponder factors invisible to the naked eye, such as the genetic diversity of populations, or in highly abstract frameworks of hierarchy and scale. Some ecologists may see a landscape less as a set of objects arranged in space than as a bundle of intertwined processes (Norton 1991).

A landscape designer may or may not consider the landscape in these ways. But designers see other things that ecologists may not. While many ecologists have devoted their life's work to understanding one or a few species in detail, most designers have focused on how *Homo sapiens* perceive and interact with landscapes. This understanding relates not only to how humans functionally use landscapes, but also to how people attribute meaning to landscapes and how that influences their use and valuation. In particular, designers view landscapes as places imbued with symbolic meaning and endowed with emotional, cultural, and spiritual significance that can deepen over time.

Thus, neither field is characterized by a single way of looking at landscapes, but rather each embodies multiple and distinctive ways of understanding that are used for different purposes. The core strengths of ecologists include their ways of viewing and evaluating complex living systems, and the rigorous analytical methods used to develop an ecological worldview in terms of both broad principles and specific case histories. Strengths of designers include knowledge of the diverse ways in which humans interact with landscapes as well as procedures for creatively building from descriptive analysis toward spatially explicit prescription while integrating multiple and potentially competing interests.

The benefits of collaboration include the potential to link these different modes of inquiry and types of expertise toward decision making that respects the needs of human cultures and ecological systems. How much each discipline needs to know about the other, and what they need to know, depends on the extent to which we expect each to incorporate the other's thinking independently, or to act collaboratively (Chapters 1 and 9). We would argue that in either case, the foundation is the ability to engage in dialogue. We now examine key concepts that are both reflections of these worldviews as well as their building blocks.

A Question-Based Framework for Dialogue

In recent years several attempts have been made to codify key concepts or principles in ecology (Chapter 6) and, more recently, to do so in relation to planning and management (Carroll and Meffe 1997; Romme 1997; Dale et al. 2000). Yet the question often remains, How do you apply the concepts? Karr (Chapter 6) begins to develop a framework for organizing knowledge about a site and illustrates its use by developing a narrative of place that embodies ecological understanding. In another approach, Dale et al. (2000) offer broad principles and then begin to translate them into guidelines for land managers (Table 13-1). These sets of concepts and guidelines, while useful, are far from comprehensive. Moreover, the dynamic nature of ecolog-

TABLE 13-1. Five principles of ecological science with implications for land use (from Dale et al. 2000)

Principles

1. Ecological processes occur within a temporal setting, and change over time is fundamental in analyzing the effects of land use.
2. Individual species and networks of interacting species have strong and far-reaching effects on ecological processes.
3. Each site or region has a unique suite of organisms and abiotic conditions that influence and constrain ecological processes.
4. Disturbances are important and ubiquitous ecological events whose effects may strongly influence population, community, and ecosystem dynamics in all places.
5. The size, shape, and spatial relationships of habitat patches on the landscape affect the structure and function of ecosystems.

Guidelines

- Examine impacts of local decisions in a regional context.
- Plan for long-term change and unexpected events.
- Preserve rare landscape elements, critical habitats, and associated species.
- Avoid land uses that deplete natural resources.
- Retain large contiguous or connected areas that contain critical habitats.
- Minimize introduction and spread of nonnative species.
- Avoid or compensate for effects of development on ecological processes.
- Implement land-use and management practices that are compatible with the natural potential of the area.

Note: Dale et al. (2000) present five principles dealing with time, species, place, disturbance, and landscape, respectively, that they posit can ensure that fundamental ecosystem processes are sustained. To assist land managers, they also offer a set of guidelines for land managers that arise from those principles.

ical and cultural knowledge means that a robust process for engaging designers and ecologists in dialogue about specific places may remain useful longer than any particular set of concepts.

What do we mean by a framework for dialogue? To engage abstract concepts within a design or planning process, they must not only inform the way designers and applied ecologists think about actual landscapes, but they also need to influence the way these professionals act. We argue that concepts and guidelines are necessary but not sufficient for this purpose. Rather, we need a framework for organizing and integrating key concepts of ecology, design, and other disciplines in relation to the needs of a landscape decision-making process. Why do we propose the framework should be question-based? Concepts are powerful tools for understanding. They are, however, intellectual constructs based on abstractions, generalizations, and simplifications of the real world. By focusing on key questions about actual places, the emphasis is shifted from the concepts themselves to how to harness them for problem-solving. At the same time, it is important to note that a shift to questions means that *how* the questions are framed and *what* concepts are used to guide the inquiry strongly influence whether the dialogue deeply engages ecological and cultural realities.

Steinitz (1990; Chapter 10), Montgomery, Grant, and Sullivan (1995), and the Sierra Nevada Ecosystem Project (1996) each offer question-based frameworks for landscape design and planning (Table 13-2). Note that the frameworks are not explicitly ecological or cultural. With no inherent bias toward either culture or ecology, the approach opens the door for developing a common framework for assessments, a key need in landscape-based planning (Johnson and Campbell 1999). The overall content of the three basic frameworks is similar, and we have used a modified and shortened version of the Steinitz framework (Chapter 10), incorporating some of the more specific aspects of the other frameworks (for instance, history and variability) within the concepts themselves. We address Steinitz's first question of landscape representation later in this chapter. Thus, we have organized the discussion of key concepts into five questions:

How does the landscape work?
Is the current landscape working well?
How might the landscape change or be changed?
What impacts might result from implementing a design or plan?
How should decisions be made?

As described in Steinitz (Chapter 10), the initial answers to each question should be used to refine the answers to the others. For instance, deciding how the landscape works depends in part on the criteria used to ask if it is working well. Critical cross-questioning during the design process should help

TABLE 13-2. Three question-based frameworks for landscape analysis with strong similarities in content but significant differences in how content is organized and how questions are framed

Landscape Design and Planning Framework (Steinitz 1990)

1. How should the landscape be described? (*Representation models*)
2. How does the landscape operate? (*Process models*)
3. Is the current landscape working well? (*Evaluation models*)
4. How might the landscape be altered—by what actions, where, and when? (*Change models: projection, intervention*)
5. What predictable differences might the changes cause? (*Impact models*)
6. How should the landscape be changed? (*Decision models*)

Watershed Analysis Framework (Montgomery, Grant, and Sullivan 1995)

1. How does the landscape work? (*Classify landscape based on structure and process.*)
2. What has been the landscape's history? (*Collect data to assess historic conditions and trends.*)
3. What is the landscape's current condition? (*Collect data to assess current conditions and legacy of past disturbances.*)
4. What are the possible/desirable future states for the landscape? (*Reclassify, generate scenarios, and synthesize across analysis modules.*)
5. How should the landscape be managed to achieve objectives? (*Develop prescriptions for redefined units.*)

Landscape Research Assessment (SNEP 1996)

1. What were the historic ecological, social, or economic conditions, trends, and variability?
2. What are the current ecological, social, or economic conditions?
3. What are the trends and risks under current policies and management?
4. What policy choices will achieve ecological sustainability consistent with social well-being?
5. What are the implications of these choices?

ensure that appropriate concepts and methods are used to answer each question.

In approaching these five questions, we utilize concepts from both ecology and design. While we have devoted the most discussion to concepts originating in ecology, we have begun to draw in complementary or unique concepts from design. Our purpose is to initiate a synthetic conceptual framework that begins to encompass both disciplines (Table 13-3), however

TABLE 13-3. Key concepts in relation to design and planning questions

How does the landscape work?

Umbrella concepts
>   *Individualistic, nonequilibrial systems; Hierarchy and scale; Landscape structure, function, and change; Disturbance and succession*

Priorities in time and space
>   *Landscape history and landscape memory; The ecological theater and the evolutionary play; Key species, key processes, and key events*

Is the current landscape working well?

Landscape health and integrity
>   *Ecological health and biological integrity; Livability and sense of place; Inclusive design; Landscape aesthetics*

Landscape metrics
>   *Indicators of biodiversity and landscape function (species, multimetric, bio-physical, social, and economic); Historic range of variability; Ecosystem autonomy and ecological footprint*

How might the landscape change or be changed?

Imagining and proposing change
>   *Trends and interventions; Imagining change (idea generation, visioning, problems as opportunities, recycling ideas, synthesis, visual representation); Alternative future scenarios*

Managing change
>   *Reserve design and matrix management; Ecosystem management, Urban design theory; Five Rs of restoration; Urban infrastructure and green infrastructure*

What impacts might result from implementing a design or plan?

Types and propagation of effects
>   *Direct and indirect effects; Cascading effects; Cumulative effects; Anthropogenic stressors; Thresholds; Positive and negative feedback*

Key classes of effects and system responses
>   *Simplification; Fragmentation; Stability; Resistance; Resilience; Irreversible impacts*

Sources of uncertainty
>   *Incomplete knowledge; System complexity; Individualistic contexts; Nonlinear responses; Multiple responses; Novel disturbances; Chance events*

How should decisions be made?

Forums and frameworks
>   *Cultural differences; Power and its expression; Collaborative decision making*

Adapting to uncertainty
>   *The precautionary principle; Plans robust to competing models; Adaptable design; Adaptive management; Post-occupancy evaluations*

incompletely. We have included concepts that range from broad frameworks for understanding to more specific concepts that may provide guidelines or protocols for design and planning, even as they continue to be tested. In some cases we will link parallel or related concepts from the two disciplines. In others we find that concepts from one or the other field are sufficiently robust to serve as an overarching framework. Not surprisingly, many concepts do not fit neatly into one question. In all categories there will be concepts left unmentioned that are important to one discipline or the other. Our focus has been to identify the key concepts for integrative dialogue and problem solving. We illustrate how the framework can be used, but make no claim of being exhaustive.

How Does the Landscape Work?

Landscapes are individualistic, nonequilibrial systems with interacting physical, biological, and cultural attributes that are hierarchically organized in space and time. At all levels of hierarchy, they can be understood in terms of the interaction of landscape structure, function, and change.

At the core of understanding how a landscape works is to determine causal mechanisms, or as Karr (Chapter 6) puts it, root causes. One of the main challenges is to understand the frequently tight couplings of physical, biological, and cultural processes in light of ecology's new paradigm of flux and change. At the same time, an emphasis on process should not overshadow our awareness that the living "parts" of the system—salmon, people, prairie grasses, and mycorrhizal fungi—are what give them their unique character and are most often our ultimate concern.

Ecologists have framed the question of how a landscape works in different ways for the purposes of design and planning. Dale et al. (2000) describe landscape dynamics in terms of time, species, place, disturbance, and landscape. Meffe and Carroll (1997a) emphasize evolutionary processes. Karr (Chapter 6) focuses on organisms and their life histories. Pulliam and Johnson (Chapter 3) explore the ramifications of flux, disturbance, and historical contingency. We incorporate these ideas within an organizational "scaffolding" for understanding how a landscape works by beginning with the set of three "umbrella" concepts with which we began this section: individualistic, nonequilibrial systems; hierarchical organization; and landscape structure, function, and change. We follow by considering the importance of history and evolution as links from the past to the future. Finally, we examine concepts of key species, key processes, and key events as a way to prioritize what to pay attention to first in a specific landscape.

Systems, flux, and uniqueness of place.

The idea that landscapes are *individualistic, nonequilibrial systems* is meant to emphasize that each has a unique combination of physical, biological, and cultural attributes that are constantly in flux through myriad interactions among components and with exogenous forces. Arising out of ecosystem ecology (Chapter 3), a systems point of view is a process-function approach that emphasizes the transfer and processing of matter, energy, organisms, and information. It emphasizes relationships. We include cultural attributes because to understand or affect landscape processes, factors that include human aspirations, ethics and values, economic and legal systems, and traditions and taboos play as great a role as human life history traits and survival needs. Thus, for design and planning, knowledge of a landscape's climate, physical character, and the set of organisms that occupies it must be matched with knowledge of legal and policy systems, and of how land-use practices interact with different types of species and environments, and more broadly with ecological processes.

Understanding the ecological niches of species, that is, how they make their livings, provides an entryway into making biological complexity tractable in design and planning. Traits related to fecundity, dispersal, growth rates, production of defensive compounds, and range of environmental tolerances all influence species responses to the environment and each other, including the outcomes of competition, herbivory, and predation. In particular, the recognition of different types of overall life history strategies reveals common patterns of habitat occupancy in space and over time. Among plants, for instance, these types include ruderal (or fugitive), competitive (or equilibrium), and stress-tolerant species (Grime 1979). Although the species that fill those niches, or roles, vary in space and over time, the niches themselves typically are predictable. Using the ideas of niche and life history strategies in the course of becoming familiar with the different types of organisms that occupy a landscape is one way to make abstract ecological concepts meaningful in specific places.

Hierarchy and scale.

Many of the concepts of *hierarchy and scale* are formalized versions of ideas already familiar to designers and planners. These include the importance of site context, and the need to develop design elements at scales ranging from detailed components to overall attributes such as circulation systems. Hierarchy theory goes much further, however. It emphasizes that landscapes are characterized by multiple, ranked levels of organization—orderings of subsystems within systems. By focusing on levels of organization, scales of phenomena (and observation)[3] within and among levels, and relationships between levels, hierarchy theory directly contrasts with a reductionist, or mechanistic, approach to complex systems (Ahl and Allen 1996). Hierarchies can be used to characterize landscapes in multiple ways (Urban, O'Neill, and Shugart 1987), as well as to organize design and planning investigations. The concept of the *triadic structure* of ecosystems (O'Neill 1989) is particularly useful for design and planning (Chapter 3). It says that one must always look one scale above the focal level to understand the constraints on, and the significance of, phenomena at the focal level, while looking one scale below to understand focal-level dynamics. Hierarchy theory offers other important insights for design and planning, as well:

- All ecological processes and types of ecological structure are multiscaled (Allen and Hoekstra 1992). *Thus, to understand an important process, look up and down one level from the focal scale.*
- At different scales of observation, the phenomena of most interest may change, as will the relative importance of different processes. *Ask what processes are important at the scale of interest and how they change across scales.*
- At some scales of perception, phenomena become simpler than at others (Allen and Hoekstra 1992). *When a phenomenon appears intractable, examine it at other scales as well.*
- Landscape phenomena that match human scales of unaided perception are the best known and are the most frequently discussed (Allen and Hoekstra 1992), whether or not they are the most important. *Choose scales of observation based on the phenomena in question.*
- Humans often rescale systems in both space and time. For instance, humans typically create small patches with simple outlines (Krummel et al. 1987). They may also alter the rates and duration of physical and biological processes or the frequency of events, such as changes in the recurrence of floods of certain magnitudes. *Ask how human activities may have altered spatial or temporal scales of phenomena and what the effect may be on the landscape and biota.*

Landscape structure, function, and change.

Landscape ecology is based on the premise that how a landscape works depends in large part on its spatial organization. The idea that landscapes can be understood in terms of structure, function, and change (Forman and Godron 1986; Forman 1995) is a key contribution of the discipline. A fundamental premise is that structure influences function and vice versa, and that their interaction leads to change over time. In the words of D. W. Thompson (1961), form is "a diagram of forces" (p. 11).

Landscape structure is composed of two parts: composition and pattern.[4] *Composition* describes the types of different landscape elements, including their numbers, sizes, and shapes. *Pattern,* on the other hand, is the distribution and arrangement of component parts in space. An overall system often used in analyzing the structure of large *land mosaics* (Forman 1995) is that of *patch, matrix, and corridor* (Forman and Godron 1986). Such a system typically works best when elements have relatively distinct boundaries, less so when one grades continuously into the next. In particular, because humans often create disjunct landscape boundaries, the system works well in human-dominated landscapes.

Understanding how patches and corridors created by humans interact with other types is a key linkage between cultural and ecological understanding. For example, because corridors can serve not only as conduits but also as filters and barriers to movement, many human-created corridors interact with natural corridors in complex ways. Roads serve as barriers and sources of mortality to many types of species (Figure 13-1). Roads can also act as barriers to surface and subsurface water movement. At the same time, road ditches become extensions of natural hydrological systems and may function as de facto first-order streams (Forman and Alexander 1998; Jones et al. 2000).

As in ecosystem ecology, *landscape function* is the transfer and processing of matter, energy, and information. It includes nutrient cycling and other bio-

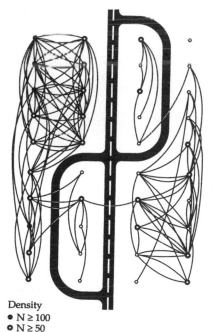

Density
- N ≥ 100
- N ≥ 50
- N ≥ 20

FIGURE 13-1.
Roads that serve as corridors for humans can also serve as major impasses for some terrestrial animals. This diagram illustrates how even a road and parking loops can fragment habitat for forest-dwelling beetles. The lines show movements of beetles that were marked and recaptured (from Noss and Csuti 1997, modifed from Mader 1984).

geochemical cycles, food webs, movement of organisms, flows of water, and so forth. Such flows are, of course, a key focus of human activities: transport of people and goods, electric and gas utilities, disposal of sewage and other wastes. Human-mediated flows often are tightly coupled with other types of flows, such as the use of rivers to dilute pollution and the introduction of exotic pests and diseases through the transport of plant products. While some ecologists frame function narrowly as nutrient cycling and energy flow, it needs to be understood much more broadly by designers and planners working with living systems. For instance, demographic processes affecting reproductive success, genetic processes such as mutation and recombination, metapopulation processes, and competition, predation, and parasitism all are critical ecological processes.

Landscape change is the alteration of structure and function over time through their interaction and mutual influences. Recognizing the ubiquity and importance of change in natural systems is essential to an ecological approach to design and planning (Figure 13-2). How *disturbances* such as floods, fire, and treefall manifest over time as disturbance regimes with attributes of type, spatial extent, intensity, and frequency substantially determines landscape structure, function, and change over time. By causing mortality and changing resource availability, disturbances alter *successional* trajectories, initiating quasi-predictable sequences of plant and animal species and of vegetation physiognomy. Furthermore, because disturbance regimes serve as a selective filter on species composition, changes from historic regimes are likely to lead to the loss of species adapted to the habitats and conditions created by them (Sousa 1984, Denslow 1985, Landres, Morgan, and Swanson 1999). Understanding management activities as the creation or suppression of disturbances—from weeding, mowing, and watering lawns to broad-scale forestry practices—is an important conceptual tool for understanding human activities as ecological processes.

Design theorist Kevin Lynch proposed a nearly parallel system to that of patch, matrix, and corridor for urban systems over forty years ago when he described cities in terms of two element types: a flow system for people and goods, and adapted space (mostly sheltered) for localized activities (Lynch and Rodwin 1958). He discussed these elements in terms of their quantity, density, and grain; the focal organization of key points in the total environment; and their generalized spatial distribution. He then related this spatial organization to functional and aesthetic criteria as they affected people's choices and opportunities for the healthy expression of human life—individual, social, and economic. His description of paths, edges, districts, nodes, and landmarks (Lynch 1960) also is prescient of the patch, matrix, and corridor model and underscores the possibilities for joint theory among designers and ecologists (Figure 13-3).

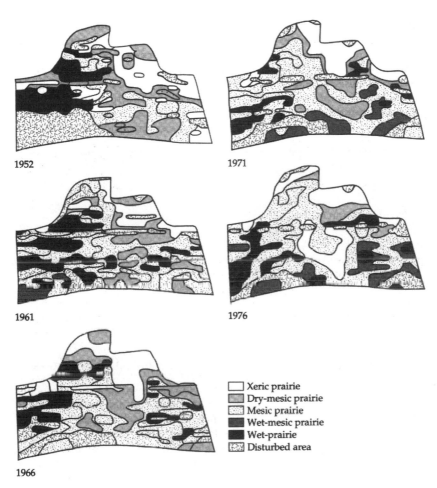

FIGURE 13-2.

Flux and change are the norm in most ecosystems. The observed changes in vegetation on the Green Prairie from 1952 to 1976 were due in part to short-term variability in climate, causing plants to shift to more optimal habitats. At the same time, this landscape would change into a woodland without periodic disturbance from fire or other management (from Cottam 1987).

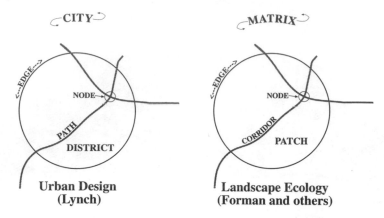

FIGURE 13-3.
Conceptual frameworks for urban design and landscape ecology. Kevin Lynch's system of paths, edges, districts, nodes, and landmarks for understanding cities has much in common with the landscape ecology's system of patch, matrix, and corridor as used by Richard Forman and others and underscores the possibilities for joint theory among designers and ecologists. (Idea from Carl Steinitz.)

Landscape history.

Knowing a landscape's history is critical to understanding how it works, and historical contingency is a key factor in shaping landscapes. How far back in time one needs to look depends on the processes involved, the time spans being planned for, and the types and extent of modification being considered. For example, rivers, streams, and floodplains are not structured by a single type of flood that happens each year, but rather by floods of differing lengths and intensities distributed unevenly over many years. While floods of a magnitude that occur on average every two years may be the primary determinant for stream channel width and depth, fifty- and one-hundred-year floods can reshape the adjacent landscape and stream courses. In many landscapes, humans have substantially altered river flow regimes with dams, channelization, ditching and draining, urbanization, and more. Knowing how humans have modified natural processes and their motivations in doing so is a key step to understanding current conditions as well as potential responses to them.

History is also important for understanding the influences of landscapes on cultures. Because culture is founded on knowledge, traditions, and practices accumulated and reworked over time *in places*, links to the past through the landscape are often critical to maintaining vital cultures as well as to understanding them. Whether based in the long-standing traditions of indigenous peoples or those of several generations of recent immigrants, historic land-use practices and landscape experiences create deep attachments to particular places. The rootedness of a people depends on attachments developed and maintained through continuity of experience and is different from a sense of place, which may be deliberately developed or invoked through reference to that history (Tuan 1977). Because most landscapes have been strongly shaped by human activities, understanding the relationships of people to the land over time is critical to understanding both cultures and ecosystems. As noted in Chapter 2, the origins of the word "landscape" are based in such an understanding.

The concept of *landscape memory* (Chapter 3) encompasses not only how past events (including human activities) help explain current landscape conditions, but also how they may constrain or otherwise influence the future. In similar fashion, memory in the form of cultural imprints on the land, both physical and symbolic, influences people's responses and valuations. Such imprints may be conserved, accentuated, or obliterated by design and planning initiatives. Joint explorations by designers and ecologists of eco-cultural landscape histories could prove fertile ground for collaboration.

The importance of evolution.

For many designers and planners, evolution is an abstract concept far removed from their day-to-day activities. Yet, how evolution plays out, and how humans affect it, are important to people's well-being and also that of other species. Evolution is the cause of increased bacteria resistance to antibiotics, with consequences for human health. The evolution of pesticide resistance can have major impacts on agricultural practices. Through horticulture and gardening, the genes of introduced cultivars can mix with those of native species, altering evolutionary paths.

A basic understanding of evolution is essential for anyone engaged in conservation planning (Allendorf and Leary 1986; Vrijenhoek 1989; Meffe and Carroll 1997a). Successful conservation strategies ultimately must protect the evolutionary processes that have created the complex and diverse biosphere in which we live. If they do not, our efforts, and the biological fabric, may unravel. G. Evelyn Hutchinson (1965) described the natural world as an *ecological theater* that serves as the stage for the *evolutionary play*. Meffe and Carroll (1997b) drew on this idea to propose that the mission of conservation biology is "to retain the actors in that evolutionary play and the ecological stage on which it is performed" (p. 7).

What are the keys to planning for evolutionary processes in conservation? First, sufficient genetic variability within and among populations, which serve as basic evolutionary units, must be maintained for species to have the raw materials to adapt to changing environments (Falk and Holsinger 1991; Meffe and Caroll 1997c). This depends to some extent on population size and the potential for occasional exchange among populations (see metapopulations dynamics in Chapter 3). Rapid loss of population size in particular may result in genetic "bottlenecks," the expression of deleterious recessive alleles, and inbreeding depression, which singly or in combination can reduce adaptive capacity and increase the likelihood of extinction (Falk and Holsinger 1991; Meffe and Carroll 1997c).

Second, species persistence and evolution may depend on the ability to migrate in the course of metapopulation dynamics or in response to climate change. Human domination of landscapes at broad spatial scales has dramatically altered opportunities for species movement and dispersal. Where habitats are fragmented and extensive tracts of agriculture or urbanization impede movement, species may not be able to track suitable environmental conditions under rapid climate change (Peters and Lovejoy 1992; Peters and Darling 1995; Myers 1997). Broad-scale conservation planning must address the need for such movement across latitudinal and elevational gradients.

Third, although most small isolated populations are subject to high rates of extinction (Gilpin and Soulé 1986; Fischee and Stocklin 1997), such conditions may also enhance the likelihood of divergent evolution and speciation (Wright 1931). Thus, a few large protected populations of a species may serve as an "ark" to prevent extinction, but may not be sufficient for continued evolution and speciation over long time periods. The evolutionary play has always used a very large stage. The greater the proportion occupied by simplified human ecosystems or otherwise dominated by human uses, the more imperative it is to consider the long-term implications for evolution, speciation, and extinction. Although preventing extinctions has gained the most attention in the popular literature, the generative process of speciation through evolution is every bit as critical. As Soulé and Wilcox (1980, p. 8) noted, "Death is one thing; an end to birth is something else."

Key species, key processes, key events.

Given the bewildering complexity of how a specific landscape works, how does one distinguish what is most critical to understand? The concepts of *key species, key processes,* and *key events* can focus attention on what may be most important for design and planning decisions, although such conceptual simplifications must be applied carefully. Key species include keystones whose loss may have cascading effects on ecosystem structure or function, as well as other types of indicator species as described later in this chapter. Key processes are those that may be dominant in the system in terms of matter and energy transported or processed, but also those that, like keystone species, may be critical to landscape structure, function, or character despite relatively small flows. Key events include historic disturbances, human-caused or otherwise, that have had a lasting effect through landscape memory but were sufficiently singular and discrete not to be considered as processes.

The idea of key processes, in particular, offers a framework for understanding the interaction of ecological and cultural factors. For instance, road construction may promote development or otherwise increase human presence and impacts. In a broad metaphor, human interventions often serve to change the landscape from a "sponge" to a "pipe," decreasing infiltration and storage and increasing the rate of discharge. The metaphor holds for urbanization with its increase in impervious surfaces, for agricultural practices of ditching and draining fields, and for certain silvacultural practices.

It must always be kept in mind that what is "key" to how a landscape works depends on the standpoint from which it is evaluated. For instance, from the point of view of wildlife habitat, a few trees in a poor urban neighborhood dominated by pavement may make little difference except for a few common urban species. However, from the standpoint of human residents, the effects of having a few trees may be dramatic. Green areas have been shown to support mental and emotional well-being in humans (Ulrich et al. 1991; Hartig, Mang, and Evans 1991; Carr et al. 1992). Recent empirical studies have shown that even small numbers of trees appropriately placed can dramatically increase social activity and strengthen the social fabric of poor urban neighborhoods (Coley, Kuo, and Sullivan 1997; Kuo et al. 1998; Sullivan, Kuo, and Deppoter, in preparation). Thus, to understand how the landscape works, one must think carefully about the criteria that will be used to evaluate it.

Is the Current Landscape Working Well?

The choice of evaluative models and measures, and how they are applied, largely determines how other questions are approached. Selecting a set of concepts and criteria that reflect a deep understanding of the ecological and cultural dimensions of a landscape is essential to a comprehensive evaluation.

Evaluative concepts are means for measuring or describing a system toward normative judgments about desirable outcomes or acceptable conditions. They may then be used for policy, persuasion, and decision making. To evaluate whether a landscape is working well requires an understanding of how its parts and processes comprise a dynamic living system that operates across a range of spatial and temporal frames. To a great extent this requires considering the dominant parts and processes. But it also requires thinking clearly about elements, including species and human social groups, that may have been disadvantaged by past changes or that are vulnerable to the kinds of changes that may occur.

At the heart of an ecological approach to design, one that cares for and attends to whole living systems, is a need to reconcile or at least critically compare ecological and cultural criteria for evaluating landscapes. The same could be said for an approach to applied ecology that is deeply concerned with human needs and aspirations. To this end, we need to consider whether the two disciplines have evaluative standards in common or ones that can be reconciled, or whether they may reflect different criteria applied to different kinds of places.

We begin by examining a set of normative concepts related to landscape health and integrity: ecological health and biological integrity, livability and sense of place, inclusive design, and aesthetics. We then turn to selected evaluative metrics, derived primarily from ecology. Among these we include indicators of biodiversity and landscape function, historic range of variability, and last, ecosystem autonomy and ecological footprint.

Landscape health and integrity.

Both ecology and design use concepts related to the health and integrity of landscapes, albeit sometimes in very different contexts and to different ends. These overarching evaluative concepts are compelling because they offer "big picture" understandings of landscape qualities. Health has implications of vitality and a flourishing condition. Integrity has implications of wholeness and intactness. These strong normative qualities can make the concepts problematic without specific, well-conceived, and agreed-upon meanings. Ideas of ecological health and integrity, in particular, have been treated extensively in the literature (e.g., Costanza, Norton, and Haskell 1992; Woodley, Kay, and Francis 1993; Angermeier and Karr 1994; Callicott 1995; Callicott and Mumford 1997; Rappaport et al. 1998; Karr 2000; among many), and not without controversy.

Karr (2000) offers definitions of health and integrity that link living systems to human well-being in ways highly applicable to design and planning. He describes *biological integrity* as the condition of a place that has its evolutionary legacy intact—with the full complement of its biodiversity components *and* the biogeographic processes that generate and maintain them. Integrity is one end of a spectrum of *ecological health*. A landscape may still be considered healthy by society at levels below that of integrity so long as it is above a threshold of degradation where its condition cannot be sustained.[5] In this respect, health embodies two related attributes: (a) the condition of the specific place being considered and (b) its effect on other places. Thus, human activities should be conceived and executed so that they do not reduce the long-term ability of a place to continue to supply its goods and services for humans or nonhumans. At the same time, activities should not degrade the health or integrity of other locations. These definitions give explicit criteria for evaluating a landscape's current condition, as well as design and planning proposals. Moreover, they define health and integrity in ways that are clearly linked to the dynamic evolutionary and biogeographic contexts of natural systems and, implicitly, to intergenerational equity.

Designers may evaluate landscape health for human inhabitation with concepts of *livability*, and the integrity of people's experience with concepts such as *sense of place*. Livability entails not only whether a landscape meets people's biological needs, but whether it satisfies a host of qualities that together make a landscape a good place to live. Likewise, sense of place refers to the ways in which a landscape resonates for people with its own unique character. It may refer to how a landscape reveals its natural history or cultural heritage. Often the landscapes with the strongest sense of place are those where the relationship of cultural and natural history is deeply evident—sometimes by design. To what extent are current definitions of livability and sense of place compatible with those of ecological health and biological integrity? How can we promulgate understandings of livability and sense of place that do not promote the degradation of ecological health, biological integrity, or human well-being in other places? Since perpetuating ecological health and integrity is essential to the health and well-being of human societies, reconciling definitions of livability and sense of place to ecological constraints and consequences would also seem essential, regardless of whether one considered it an ethical imperative

Livability increasingly has been linked to social and environmental justice. Without attention to effects on minorities, the disadvantaged, and the disabled, livability may simply reflect benefits to a privileged few at the expense of others. Recent concepts of *inclusive* or *universal design* stress the need to create inclusive places that provide equal opportunities and experiences for people of diverse abilities and heritages (Jones and Welch 1999) and support the ways that different cultural groups use and conceptualize space (Harvey 1973; Carr et al. 1992). Universal design poses the question of who is included and who is excluded, providing clear parallels to ecological approaches to design and planning that ask the same question for species.

Environmental issues cannot and should not be treated independently of social, racial, and economic justice. Histories of the environmental movement often have overlooked the contributions of people who have worked on public health and environmental justice in urban and industrial areas (Gottlieb 1993). To fully address how ecological issues manifest in society, designers and planners need to understand how issues of gender, ethnicity, and class have led to disproportionately high exposures to environmental hazards for people of color and others. Recognizing such connections is central to an integrated understanding of what we might provisionally term *landscape health and integrity* with respect to both human and nonhuman inhabitants.

Landscape aesthetics is another area where ecological imperatives and design traditions need to be compared. Aesthetics relate not simply to visual perceptions of beauty, but more broadly to the ways in which people respond to landscapes emotionally and through a variety of senses (Bourassa 1991, Zube 1987, 1990). Even the biblical narrative of the Garden of Eden linked knowledge, sustenance, and aesthetic appreciation (Helphand 1997). Differences in aesthetic appreciation or interpretation of significance are not merely matters of individual taste, but relate deeply to the ways meaning and value are constructed and articulated in different social, political, and cultural frameworks. Designers must contend with the fact that there are always multiple interpretations and meanings in a landscape.

The design disciplines have created a body of aesthetic theory that attempts to explain how people experience and use landscapes (Bourassa 1991). When designers apply these theories to the making of environments, they alter landscape structure and processes in ways that affect ecological health and biological integrity as well as livability and sense of place. What ideas might emerge from discussions of aesthetics between designers and ecologists? Ecologists would be likely to quickly grasp and respond to habitat-derived models such as prospect-refuge theory (Appleton 1975, 1984), even if they disagreed with some of its claims. What might they contribute to information-processing theory (Kaplan and Kaplan 1989), which posits that people prefer landscapes that facilitate and stimulate acquisition of knowledge through qualities of complexity, coherence, legibility, and mystery? And what are the joint social and ecological implications of critical regionalism, which seeks to enhance the "cultural density" (Frampton 1982, p. 76) of urban places and thus intensify people's experience of them (Bourassa 1991)? Cross-disciplinary consideration of these ideas might spawn new or refined theory.

There has been a growing criticism of aesthetic styles that are antithetical to ecological values (Chapter 12). Accompanying this have been thoughtful inquiries into ways in which design can bring aesthetic appreciation more in harmony with ecological health and biological integrity. These include ways to make ecological designs comprehensible by providing recognizable cultural frames (Nassauer 1995), and creating deeply resonant, iconic designs that reveal and celebrate ecological processes (Mozingo 1997). The challenges of building appreciation for different landscape forms and experiences, as well as discovering ways to fulfill ecological functions through accepted aesthetic forms, could occupy a generation of inspired designers and ecologists.

Landscape metrics.

Ecology employs a number of quantitative metrics that themselves have no normative significance, but without which concepts such as health and integrity cannot be evaluated. Different measures of *species diversity*, for instance, are used to assess how species numbers are partitioned within and among habitats. Much of the concern for species diversity in popular interpretations is a response to losses of diversity due to human activities. Norton and Ulanowicz (1992) argue that even the loss of apparently minor species may be important from a long-term point of view. Background species and marginal trophic pathways that seem redundant and even dysfunctional, they say, are in fact spare parts that may come into play when normal ecosystem structures and processes are disrupted. If the disruption continues, these parts and pathways may become prominent components of the ecosystem. For example, a rare herb currently restricted to small crevices on mountain summit rock outcrops in the southern Appalachian Mountains may have been a dominant groundcover during the Pleistocene, and it could again assume this role during future periods of global cooling should it survive its current restriction (Johnson 1995).

To make evaluations of *biodiversity* more usable in environmental policy, Noss (1990) proposed a hierarchical system for monitoring biodiversity based on three primary system attributes, composition, structure (here meaning spatial organization), and function, and across four levels of organization, regional landscape, community-ecosystem, population-species, and genetic (Figure 13-4). He then provided a suite of *indicators* for each of the twelve resulting categories that might be monitored as surrogates for measuring impacts on a larger suite of species or on ecosystem attributes.

One of the most frequently used kinds of indicators is the status of selected species. The categories commonly used include keystone species, indicator species, umbrella species, flagship species, and vulnerable species (Noss 1990; Meffe and Carroll 1997e). On the premise that individual species are rarely good indicators of whole system attributes, *multimetric indices* rely on monitoring many species to directly evaluate biological integrity with respect to relatively intact reference systems (Chapter 6). *Biophysical indicators* of site conditions such as water quality, soil erosion, and site productivity are also used. Moreover, many kinds of *social and economic indicators* are being combined with biological and ecosystem indicators in landscape planning projects (Johnson and Campbell 1999), building awareness of the joint importance of cultural and biological attributes of landscapes for human societies.

A complementary type of metric relates to levels of the variability of conditions. Because variability is an essential attribute of ecological systems, design and management should rarely aim to hold a single composition or pattern constant on a landscape. In fact, doing so can defeat the goal of retaining or restoring a healthy ecosystem. To this end, the *historic range of variability* of a land-

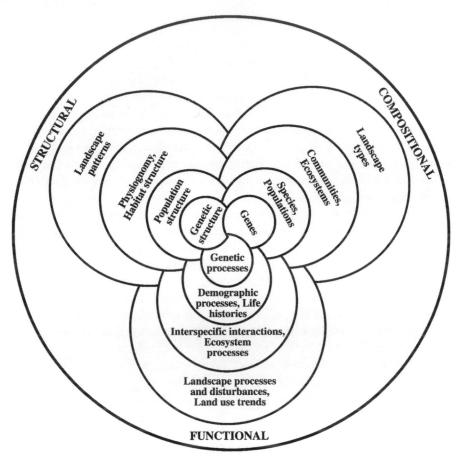

FIGURE 13-4.
A hierarchical system for monitoring biodiversity. To make evaluations of biodiversity more tractable, Noss (1990) proposed a systematic approach to selecting indicators based on three primary system attributes and four levels of organization (redrawn from Noss 1990).

scape has recently been proposed as a useful metric in prescribing land management (Landres, Morgan, and Swanson 1999) and ecological restoration (Hobbs and Norton 1996). Its application has been based on the observation that human activities often push a system outside of its historic bounds, or constrain it to a dramatically more limited range of conditions (Swanson, Jones, and Grant 1997). Because landscapes are relatively resilient to the disturbance regimes under which they have developed, returning a system to conditions within historic values and expanding variability within socially acceptable limits have the potential to at least partially retain or restore species diversity and ecological function. No one may ever be able to accurately measure a landscape's historic range of variability, and no one has yet ascertained the degree to which the metrics of variability are useful surrogates for the effects of the original processes. Yet, the concept still may have utility for design and planning by giving measurable form to the understanding that nature is always changing, but within bounds and limits. It also can help frame other questions: How much and what kind of change may be desirable or acceptable for ecological health? How much fluctuation in ecological conditions can humans accept and still obtain reliable goods and services from ecosystems? In similar fashion, the concept could be conceptually linked to questions of how much social and economic flux is healthy for human communities and within what kind of bounds.

Finally, we suggest that *ecosystem autonomy* and *ecological footprint* are useful concepts for thinking about the spatial expression of health and integrity, both biological and cultural. Recall that one of the precepts of ecological health is not to degrade other places. As described in Chapter 3, the concept of ecosystem autonomy represents a functional approach to evaluating the relative dependence or independence of an area from the inputs of surrounding areas. The ecological footprint of a city, that is, the estimated area of land and water required to provide the resources used by its human population and to assimilate the wastes they produce (Rees and Wackernagel 1994; Wackernagel and Rees 1996), is directly related to ecosystem autonomy. That is to say, the larger the footprint, the less autonomy for that particular site and the greater its impacts on the autonomy of other places. The ecological footprint of a city is commonly 100 to 1,000 times that of its physical footprint (Folke et al. 1997). The larger the footprint, the greater the appropriation of resources toward humans in that area, and away from other species or from people elsewhere.

Note that autonomy considers only the amount of dependence or independence, not whether it leads to the degradation of one place or the other. Similarly, footprint specifies only the equivalent area involved and does not directly account for what happens in the areas actually affected by the city. What is directly relevant to health and integrity is that the massive movement of resources by contemporary human societies has radically changed how ecological footprints and ecosystem autonomy manifest spatially. Historically, both footprint and autonomy were most strongly related to nearby places. Some impacts from development may now be felt most strongly far away, out of the sight and mind of those whose activities are the source of the impacts. In addition to accounting for the materials and energy people employ to support ongoing system functioning, one should also consider the resources used to create infrastructure and the impacts of their production, including extraction or harvest, processing, and manufacturing. Towards this end, autonomy and footprint are useful as tools for thinking about the spatial expression of the environmental impacts of a site, be it a house and yard, a city, or a region.

In combination with understandings of how a landscape works, these types of normative concepts of landscape health and integrity, as well as the specific metrics through which they are evaluated, lay the ground for considering ways in which a landscape might change or be changed.

How Might the Landscape Change or Be Changed?

Considering how a landscape may change or could be changed over time, and whether such changes are desirable, requires relating evaluations of how the landscape is currently working to important ecological and cultural constraints and opportunities. This includes considering (1) the landscape's physical, biological, and cultural conditions and contexts; (2) trends that may drive change in a particular direction; and (3) imaginative ways to respond to these conditions, contexts, and trends to simultaneously foster social vitality and ecological health.

Change in landscapes is not only inevitable, but from an ecological point of view, essential. Ecosystems, however, have functional, historical, and evolutionary limits, and human-induced changes need to be constrained within those limits (Pickett, Parker, and Fiedler 1992; Christenson et al. 1996). Human settlements and social systems also have limits with regard to how much and what kinds of change are healthy (Lynch 1981).

How might designers and ecologists jointly expand their understanding of how landscapes could be changed or conserved? We have suggested an evaluative framework of landscape health and integrity in which understandings of inclusive design, livability, sense of place, and aesthetics are aligned to support ecological health and biological integrity, as well as social justice and cultural diversity. Much of the synthesis required is to join the design profession's strengths for proposing change to meet human needs and aspirations with ecological ways to protect or restore health and integrity in the face of expanding human influences. We will examine several ways in which the strengths of each discipline may be combined to creatively imagine and will critically propose change, followed by ideas of how to manage change over time.

Imagining and proposing change.

As Steinitz (Chapter 10) notes, questions of how a landscape might be altered can be approached through projections of how the landscape may change under current *trends* and by considering *interventions* through plans, investments, regulations, and construction. Two corollaries are important. First, projection of trends may involve multiple scenarios, competing models, or sensitivity analysis as is done in ecological simulation modeling. Second, it is important to note that while some interventions are intended to create change, others may be intended to prevent or minimize it. For example, a management plan may be aimed at controlling the spread of invasive species, or reintroducing fire to prevent the succession of an oak savanna to a closed-canopy conifer forest. Similarly, planners may attempt to conserve historic cultural landscapes or structures. This leads to an expanded set of categories for the *means* of interventions as plans, investments, policy, regulations, construction, and management, and its *goals* to include facilitating, creating, redirecting, or restraining change.

Designers often generate possible design scenarios early in the design process. Rather than constrain themselves to discrete, sequential stages of analysis and proposal development, designers recognize that these processes are highly interdependent. Exploring possible solutions early on can raise key questions about the site and its context, and stimulate deeper, targeted investigations. It also establishes an interplay between rational analysis and imaginative problem-solving early in the project.

To stimulate creativity in imagining change, designers employ an array of techniques. Experienced designers draw upon whatever tool is needed to work through a particular problem or project stage. They may explore multiple design concepts in rapid succession during stages of *idea generation,* or brainstorm ideas in *visioning* sessions with project team members, clients, or other stakeholders. They may look for hidden potential benefits by examining *problems as opportunities* rather than solely as constraints to be "fixed." They may rework or combine selected attributes of different design concepts by *recycling ideas.* They work toward stages of *synthesis* where different analyses are considered in light of desired outcomes and potential conflicts toward an integrative design solution, for instance, by conceiving of ways to give culturally evocative forms to desired ecological functions. They create *visual representations* both as design development tools and to provide stakeholders and clients with images of how different types of proposed changes might look or be experienced.

Some of the most promising approaches to integrating the conceptualization of change with analysis of its consequences are being developed in alternative futures planning. In examples discussed in Chapters 10 and 15, researchers developed a set of *alternative future scenarios* for the landscape with stakeholder participation. The proposals may comprise a combination of trend analysis and intervention scenarios that differ strategically in their approaches to development and biodiversity protection (Hulse et al. 2000). The consequences of each scenario can be analyzed with social, ecological, and economic models, thus integrating proposals for change with impact assessment.

Managing change.

Most approaches for managing change over time address only limited portions of the landscape or selected attributes. Designers and planners may focus on urban or rural areas within policy frameworks such as those of zoning or growth management, while most applied ecologists have focused on areas uninhabited by humans and managed within other regulatory frameworks. An ecological understanding of landscapes requires an integrated view of how to address and sometimes reknit a landscape fabric as a whole system.

Applied ecologists have developed a number of ways to envision change while conserving species, ecosystems, and the processes that sustain them. Landscape ecology, conservation biology, restoration ecology, and ecosystem management each have developed strategic and sometimes spatial approaches for managing change. For example, one of the central features of conservation biology involves concepts for *reserve design*—interweaving areas dedicated to native species and ecological processes with areas of greater human use (Figure 13-5). Franklin (1993), however, emphasized the importance of *matrix management* as a complement to protected reserves, citing the shortcomings of conservation efforts limited to preserves.

Although these ideas were developed for large nature reserves embedded in resource lands, they may work equally well in other settings. For instance, they are readily transferable to urban open-space networks with core natural areas linked by riparian corridors, and surrounded by buffers of passive-use recreational zones. These areas could be further embedded in a matrix of houses and yards made nature-friendly by low-input management, the use of native plantings, and habitat enhancement. What is imperative in both situations is to develop strategies that encompass entire landscapes, and that go beyond treating isolated components of the landscape, whether biotic, physical, or cultural, as if they were independent of one another.

Ecosystem management is an emerging approach to planning that attempts to encompass the needs of humans and other species across large landscapes dedicated to both resource use and biological conservation. Although interpretations of what ecosystem management entails differ, there generally are common themes (Grumbine 1994; Christensen et al. 1996), including the diverse ways that ecosystems provide goods and services for people as well as the importance of retaining the full spectrum of biodiversity regardless of apparent benefits to humans. In particular, ecosystem management represents a shift in policy to emphasize the maintenance of ecological health and biological integrity as the foundation for (and a fundamental constraint on) the goods and services that landscapes can provide.

Theorist Kevin Lynch brought a similar focus to the need for design to address entire cities and regions in a corpus of books and articles on *urban design theory* (for example, Lynch 1960, 1981; Banerjee and Southworth 1995). He stressed not only that well-designed urban environments were essential to the well-being of their residents, but also that design should be based first and foremost on learning from residents themselves how they

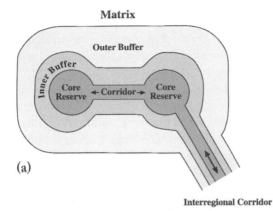

(a)

FIGURE 13-5. Conservation reserve design. One of the challenges of conservation is to move from abstract concepts to real-world proposals. (a) A hypothetical reserve network showing a protected core reserve, surrounding buffer zones with relatively low-impact human uses, and linkages through matrix resource lands that connect reserves (adapted from Noss 1992). (b) The Yellowstone-to-Yukon corridor would connect key national parks and surround them with a large buffer of lands with less stringent land-use restrictions. (Adapted from the Yellowstone to Yukon Conservation Initiative, courtesy Michael Hough.)

experience, use, and shape the landscape in their daily lives. His theories spanned decision-making frameworks, functional theory of the structure and dynamics of cities, and normative theory about urban form and human values. In particular, he emphasized the relationship of urban form to "fundamental human values and rights: justice, freedom, control, learning, access, dignity and creativity" (Banerjee and Southworth 1995, p. 7). If we intelligently bridge established and emerging theory in design and ecology, we may be able to develop robust design and planning theory that more deeply integrates cultural and ecological understanding of whole landscapes, and that includes cities within bioregional frameworks.

Another important framework for conceptualizing change comes from restoration ecology. The term "restoration" has been used in a variety of ways, justifiably generating concern and confusion. To address this, MacMahon (1997) classified the *"five Rs" of restoration ecology:* restoration, rehabilitation, reclamation, re-creation, and recovery. Each implies a different level of intervention and a different goal. Restoration means to return a landscape to an earlier condition. Rehabilitation implies restoring elements of system structure or function, but not necessarily complete restoration to a former state. Reclamation refers to rehabilitation of the most degraded sites. Re-creation is aimed at reconstructing an entire ecosystem when disturbance has been so severe that almost nothing remains of the original. Recovery means allowing the system to rebuild through natural processes. There are many situations where full restoration is not possible or appropriate. The five Rs can serve as a toolbox of approaches with full "restoration" as one option along a continuum of possibilities. Moreover, although they were developed in relation to principally natural systems, there is no reason why they cannot be applied to more cultural landscapes as well. Some such applications will require integrating ecological concerns with other human goals, such as improving ecological function on degraded lands to produce a sustainable economic return (Johnson and Campbell 1999).

In a related vein, *green infrastructure* (Chapter 11) is a relatively new term that has arisen with the growing attention to ecology in urban planning. It is used to prioritize features in existing or developing urban areas that can serve to protect or restore living systems within the urban fabric, and to reduce impacts on areas outside of it. *Urban infrastructure,* the source for the term, means roads, sewers, electrical and gas utilities—all the landscape features that people build to convey themselves, materials, resources, and energy. Green infrastructure is a mirror image—features that serve important ecological functions such as native species habitat or movement corridors; wetlands, floodplains, and stormwater swales for water purification, groundwater recharge, and flood control; and natural areas for human enjoyment. In general, landscape features derived from long-term geomorphic processes provide a more reliable framework for green infrastructure than those derived from recent land-use practices.

Green infrastructure represents a turn from ideas of nature as wilderness to a recognition that nature is everywhere (Chapter 2). Locally and regionally important ecological features are interwoven in urban fabrics, including wetlands, riverine systems, and habitats for rare and endangered species. One of the most dramatic events in this respect has been the listing of Pacific Northwest salmon runs under the Endangered Species Act—some of whose survival may depend on land-use reform in urban areas.

Together, ideas of reserve design and matrix management, ecosystem management and urban design, and different types of restoration and green infrastructure offer a strategic set of approaches that may be adapted and combined in a palette of design responses. A key element of all recent ecological frameworks is the recognition that change within limits is essential. Understanding how concepts of disturbance can serve as a common currency for considering changes caused by human and nonhuman agency is an important complement to strategies for spatial organization in design and planning (Chapter 3).

As plans and proposals are formulated, the next stage is to consider a comprehensive picture of their potential consequences.

What Impacts Might Result from Implementing a Design or Plan?

Impacts can propagate through landscapes in unexpected ways. Exploring a range of potential physical, biological, and cultural effects through conceptual or quantitative models is essential to address system complexity. Inherent uncertainties of living systems limit our ability to anticipate effects, however. In particular, it becomes increasingly difficult to predict effects over large spatial and temporal scales.

We have stressed the importance of respecting the integrity of a landscape as an interconnected, dynamic living system no matter what its context or scale. How can designers and planners responsibly address the potential impacts of a design or plan in ways that reflect that understanding and intention? First, they must explore a range of possible social and ecological consequences that extend beyond the intended effects of a design or plan. Rather than delving into specific metrics and models that may be employed to assess potential impacts, we present a framework for exploring how impacts may occur. Specifically, we focus on (1) how effects propagate through living systems, (2) key classes of impacts from human activities and system responses to them, and (3) uncertainties inherent to landscape change. While the concepts presented are drawn to a great extent from ecology, we feel they provide a broad framework for both cultural and ecological concerns, and offer insight for evaluating cultural concerns.

Propagation of effects.

Direct effects occur with no intermediaries between the causal agent and the result in question. *Indirect effects* are those that occur through secondary causes following the original impact, such as the *cascading effects* on trophic webs when a keystone species is removed (Chapter 3). Indirect effects are extremely important in ecosystems and may in fact dominate food web interactions (Patten 1985, 1991). Moreover, secondary effects from human impacts are considered to be one of the four major causes of extinction (Diamond 1989). *Cumulative effects* that accrue from multiple impacts of the same type or from different kinds of impacts may be greater than the sum of their parts. This can occur when the same disturbance occurs so frequently that the system cannot recover between events. For instance, too-frequent fires can degrade even fire-dependent communities (Carroll and Meffe 1997). Alternatively, the effects of different types of stress can be synergistic (Myers 1997). Multiple types of *anthropogenic stressors* must be addressed in landscape planning projects (Johnson and Campbell 1999),[6] and their cumulative effects should be a concern for human well-being (Chapter 8) as well as for their impacts on other species.

Cumulative effects may push a system toward a *threshold* beyond which change can suddenly and unexpectedly accelerate. Results from spatially explicit mathematical modeling suggest that the rate of local species extinctions following sequential rounds of habitat loss and fragmentation initially may be low, leading even astute observers to conclude that further impacts will be small. After crossing a threshold, however, species numbers may begin to plummet. Furthermore, a lingering extinction debt may cause continuing but delayed extinctions long after habitat destruction has ceased (Tilman et al. 1994). Other spatially explicit ecological modeling efforts (Kareiva and Wennergren 1995) offer further cautions for land managers, designers, and planners.

Positive and negative feedback can be seen in many types of processes. Positive feedback feeds upon itself. Negative feedback is stabilizing. When hikers take a shortcut across a switchback on a steep trail, they often initiate a cycle of positive feedback. Other hikers begin to follow the signs of their passage, initiating erosion that deepens the trail. This invites more use by hikers and continued erosion. With sufficient fuel, fire can feed upon itself, creating its own weather that sweeps the fire across the landscape. Repeated applications of nonselective pesticides may necessitate increasingly heavier applications or the development of more toxic pesticides. Many predator-prey cycles, such as Canadian Lynx–Snowshoe Hare dynamics, have been considered classic examples of negative feedback loops. Recent empirical evidence supports this phenomenon, albeit with more complex dynamics than previously understood, involving three trophic levels rather than two (Krebs et al. 2001; Krebs, Boutin, and Boostra 2001). In the realm of human affairs, many regulatory and policy frameworks are based on providing social and economic feedback that attempts to stabilize system dynamics or to reinforce desired trends.

Simplification, fragmentation, and system responses.

One of the key consequences of many contemporary human interventions in ecosystems is *simplification* in both space and time. Simplification of habitat, processes, and species diversity all are important. Cutting off the supply of large wood in a stream or straightening its channel reduces the number and variety of niches, refuges, and resources for diverse species that make up complex aquatic food webs. Suppression of historic disturbances such as fires and flooding can simplify landscape structure at both broad and fine scales. For instance, fire can create a mosaic of vegetation patches of different types and ages that are important to many wildlife species in terrestrial habitats (Agee 1993; DeBano, Neary, and Ffolliott 1998). Bratton (1976) showed that habitat complexity at even very fine spatial scales fostered higher herb diversity in southern Appalachian forests. Another major way that simplification occurs is through the introduction of exotic species, including pests, diseases, and invasive species. Exotic species can reduce or eliminate existing species, as well as alter trophic pathways, and nutrient cycling.

Many of the keys to an ecological approach to design lie in the ability of designers to conceive of creative ways to restore complexity at both broad and fine scales, and to cultivate an appreciation for "messy" ecosystems (Nassauer 1995) among the public and other clients. Loss of human cultural and social diversity is another kind of simplification designers should guard against. The same can be said for reduced opportunities for rich and diverse experiences by individuals or for people of different backgrounds and abilities.

Another key way that humans impact systems is habitat *fragmentation* (see Noss and Csuti 1997 for a good summary). Fragmentation is comprised of two connected processes: the reduction of total habitat and the division of the remainder into smaller, more isolated portions of the original. The latter process is important because of physical and biological edge effects as well as disruption of biotic and abiotic flows across the landscape. The effects of fragmentation on different species depend on their movement and dispersal abilities, their needs for movement at the level of individuals or populations, and their level of habitat sensitivity. The concept of fragmentation is applicable to humans as well, as when new freeways fragment urban neighborhoods.

Ideas of community or ecosystem *stability* that implied landscapes were static over time have been largely overthrown (Chapter 3). The stability of an ecological system, to the extent it exists, is only relative, bounded by different degrees of variability as it responds to disturbances. *Resistance* is the ability to withstand or resist alteration when subjected to environmental change or disturbance. *Resilience,* on the other hand, is the tendency to return to the original state following disturbance, as well as the speed of recovery. A lawn is fairly resistant to trampling by people's footsteps, but it does have limits. A temperate forest has little resistance to chain saws but often is fairly resilient, growing back rapidly to a similar community. On the other hand, fragile desert environments may have little resistance or resilience to mining or off-road vehicle use. Many tropical forests may recover if cut, but if turned into pasture soon may pass a threshold of soil degradation that makes it unlikely they will return to forest in the foreseeable future. Social systems also have limits to their resistance and resilience in the face of disturbances. When living systems are pushed beyond certain limits of resistance and resilience, some changes may be for all practical purposes *irreversible.* Once prime farmland has been converted to a shopping mall, both economic and ecological processes make it unlikely it will ever again be farmed. The more irreversible a potential impact, the more carefully it should be considered.

Uncertainty.

Uncertainty is an ever-present partner of environmental design. Will pedestrians use this new walkway? Can a warbler population survive a planned subdivision? Just as designing for people faces uncertainty because of the complexities and variety of human responses, design and planning for ecosystems also faces intrinsic uncertainties. Some uncertainties are related to *incomplete knowledge* of how ecosystems may react to perturbations, while others are a consequence of the inherent uncertainties of living systems (Meffe and Carroll 1997b). A close parallel between ecological and social concerns is the focus on rare, endangered, or otherwise sensitive species and the increasing attention paid to the effects of designs for public landscapes on vulnerable social groups, including children, the elderly, the poor, minorities, and the disabled.

The consequences of specific perturbations are only partially predictable due to (1) *system complexity, individualistic contexts,* and *nonlinear responses;* (2) the possibility of *multiple responses* to a certain stimulus; and (3) the introduction of *novel disturbances,* especially by humans, whose consequences are essentially unforeseeable for organisms. *Chance events* (Chapter

6) and, in particular, environmental, demographic, and genetic stochasticity (Chapter 3) play a major role in structuring ecosystems and further confound predictions. Extrapolating from incomplete knowledge into an uncertain future is difficult to say the least. The further ahead one attempts to predict, the greater the uncertainty. Uncertainty leads to the precautionary principle in decision making (discussed on page 339). At the same time, many impacts are predictable with high probability. The key to how predictions of impacts are weighed against other priorities lies in how decisions are made.

How Should Decisions Be Made?

Designers and planners apply ecological understanding in diverse decision-making forums and policy frameworks. In all cases we believe designers and planners bear a professional responsibility to ensure that the interests of the broader land community, both human and nonhuman, are adequately considered and represented.

In this section, we briefly examine the roles of designers and planners in decision making. We next consider how principles emerging in collaborative planning may provide an ethical framework for decisions regardless of a project's specific legal or policy requirements. We then consider how the precautionary principle, and adaptive monitoring, can guide sound decision making in the light of uncertainties. We focus on the role of the designer and cover many important issues briefly, if at all, such as those of *cultural differences*, and *power and its expression* (Lynch 1981) in decision making.

Forums and frameworks.

Designers and planners may perceive themselves as experts, facilitators, advocates, teachers, or informers (Lynch 1981). Fischer (1993) argues for reconceptualizing the expert-citizen relationship to one where the professional serves citizens by interpreting and mediating between theoretical knowledge and competing practical arguments. The key contribution of planners to decision making may lie in their ability to bring attention to "long-term effects, the interests of an absent client, the construction of new possibilities, the explicit use of values and the ways of informing and opening up the decision process" (Lynch 1981, p. 47). Ecological principles are most likely to affect land-use decisions if they are framed to be relevant in the context of markets, policy, laws, and politics (Hulse and Ribe 2000)—a further reason why collaboration among ecologists, designers, and planners is critical.

How decisions are made and who is involved in making them varies dramatically depending on the site and its scale and context. On private properties, decisions may only require agreement between designer and client, contingent on complying with public policies and regulations. In public sector work, designers and planners may be responsible to a public agency and called upon to act within regulatory frameworks, such as the National Environmental Policy Act (NEPA). In watershed planning, planners may be involved in public forums that address issues on hundreds and even thousands of square miles. At these broader scales, the strategic challenges include not only the large spatial extent, but also planning and management across a mosaic of land-use types and a diverse set of public and private ownerships. As Ahern et al. (Chapter 15) point out, this scale and complexity of planning require the involvement of a wide range of disciplines. In addition, participation from a broad array of affected parties becomes an ethical and practical imperative.

Collaborative decision making increasingly has been considered an essential component of planning for large landscapes (Yaffee 1996), including by the federal government (Thomas 1996; Dombeck 1996). In particular, the success of such projects may depend on the meaningful involvement of a wide variety of stakeholders, or affected parties (Slocombe 1993; Grumbine 1997). They may include local landowners and citizen coalitions; county, state, and federal agencies; corporations and national environmental organizations; and the public at local, regional, or national levels. Collaboration, however, involves more than simply soliciting opinions from stakeholders or holding common forums. Rather, it requires attention to social and economic dynamics, providing full and equal opportunity for all parties to participate in and affect decisions, and commitment and accountability by all participants (Gamman 1994; McGinnis, Woolley, and Gamman 1999). Truly collaborative processes must address imbalances in power and support equitable power sharing (Grumbine 1997). Toward these ends, all affected parties should be involved throughout the planning process. Even the first stage of problem definition establishes who may or may not benefit from the project (Fischer 1993).

Despite the strong ecological priorities of most landscape-based planning projects, ecological science and public involvement rarely appear to be effectively linked, although promising strategies are emerging (Johnson and Campbell 1999). In a parallel look at the social dynamics of community participation in urban planning, Hester (1996) observes that although local participation is at an all-time high, its capacity to address environmental racism and poverty has diminished. He posits that participation needs revitalization through a renewed focus on a sense of community, environmental injustice, and empowerment of the disenfranchised.

Collaborative decision making and attending to the needs of a wide variety of stakeholders are not required in many projects. Its principles, however, offer a basis for sound, ethical practice, even if they affect only the thinking or recommendations of the designer. Furthermore, for many of the reasons that Stone (1974) argued that trees and other natural objects should have legal rights, designers bear an ethical imperative to examine decisions from the perspective of a full complement of human and nonhuman stakeholders.

Adapting to uncertainty.

How can environmental decision makers respond to risk and uncertainty? One approach is the *precautionary principle* (Chapters 8 and 12). The basic premise is that when there are uncertainties, one should *guard against potential harm.* The rapid changes imposed on the biosphere from expanding human populations and development, along with a sense of humility based on human history, should make the precautionary principle a priority when decisions have the potential for widespread harm to people or the environment. One way to implement the precautionary principle is to *shift the burden of proof* so that those who propose such changes must demonstrate that their actions will not cause significant harm (Shrader-Frechette and McCoy 1993; Lemons 1998). Another possible approach emphasizes weighing risks and benefits against a range of potential outcomes to *develop plans that are robust to competing models* of how the landscape works. As Ludwig, Hilborn, and Walters (1993, p. 551) offer: "The most effective action is often one that is robust to a variety of possible states of the world. It is seldom achieved by optimizing after fixing upon a single best estimate or most likely state of the world." Speaking in a complementary vein from a design perspective, Lynch (1958) examined multiple ways to *incorporate adaptability* to uncertainties and future contingencies in built works.

A third way toward the precautionary principle is to *institutionalize learning from experience.* Neither the development of a design nor its implementation should be seen as the final stage of the design process. Uncertainties, as well as poor execution of the design or management plan, can lead to unintended consequences. The concepts of *adaptive management* (Holling 1978; Walters 1986) and *postoccupancy evaluations* (Cooper Marcus and Sarkissian 1986; Cooper Marcus and Francis 1998) both emphasize that design and planning knowledge is provisional. By including a feedback loop from implementation to monitoring and back to design and management, not only can case-specific results be modified but the process can build deeper knowledge bases for design and management theory. Neither of these two practices, one initiated in ecological resource management and one in urban design, is as yet widely implemented. Analysis of built works in teaching and academic research as well as consideration of how to institutionalize monitoring in practice, where cost is a major consideration, both deserve close attention. Together, the ideas of protecting against broad harm, shifting the burden of proof, developing plans that are robust against uncertainties, building adaptability into physical design, and institutionalizing monitoring offer an array of methods for design and planning professions to engage the precautionary principle.

Strategies to Enhance Collaboration in Education and Practice

So far we have focused on identifying a framework for dialogue among designers and ecologists. We now shift to examining strategies to apply dialogue to collaboration. We begin by comparing modes of landscape representation and communication in the two disciplines. Next we consider a strategic alignment of the processes of design and ecological science by considering designs as hypotheses. Finally, we examine several ways in which dialogue and collaboration can be enhanced within educational settings.

Modes of Representation and Communication

Designers and ecologists have different ways of representing landscapes and communicating their analyses and proposals. Each should become familiar with the other's methods and expand their repertoire of skills to include new ways of representing and communicating ideas.

Landscape representation is a means to understanding—whether through analysis, expression, or communication. Designers and ecologists each employ a variety of methods to represent landscapes that may not be fully understood by those in the other field. To enhance collaboration, students and practitioners of each discipline need to become adept at interpreting and integrating these different types of representation. This includes those that are quantitative or qualitative; spatially explicit or nonspatial; of historic, existing, or projected conditions; and those used for landscape analysis and design development as well as for communication to stakeholders and clients. Some of the most commonly used methods in one or the other discipline are numeric data and results, classificatory schema, written and oral representations, and graphic representations. In the long run, a creative and thoughtful melding of the two discipline's approaches can strengthen their joint abilities to describe landscapes in ways that support decision making.

While there is substantial overlap in the use of *numeric representations* of landscapes, it is in ecology that they are most highly developed. Design students can benefit from learning to use and interpret statistical summaries or estimates of selected parameters, including their variability and confidence intervals, as part of site analysis and setting design objectives. Applied ecology students, in turn, can benefit from learning to frame quantitative investigations in ways most useful to design and planning applications.

Both disciplines use *classificatory schemas* as the basis for landscape assessments. A common system used in both fields is landcover types, but other useful qualities may be employed, such as disturbance frequency, buildable lands, biodiversity or cultural "hot spots," restoration potential, or divergence of biota from minimally disturbed sites. Classification schemes that build integrated understandings of cultural and ecological phenomena are particularly needed.

The *written and oral representations* of both disciplines typically include fact-based evaluations of existing or potential conditions. The design disciplines often incorporate normative characterizations, either those of people who use or otherwise know the site or those of the designers themselves. These characterizations can be important in determining what qualities of the landscape should be conserved or altered.

Graphic representations may include conceptual diagrams, graphs, drawings, photographs, and maps. Ecologists have long used conceptual diagrams to represent landscape processes, and although designers often use bubble diagrams and flowcharts to similar ends, ecologists have a broader array of graphic and analytical methods used in conceptual modeling. Likewise, the use of graphs based on numeric data has been more the domain of scientists than of designers. Showing design students creative scientific graphs and conceptual models as a prelude to developing their own can develop skills in applying and adapting scientific modes of representation to design applications. On the other hand, sketches, sections, perspectives, and plan-view drawings are used extensively by designers for site analysis, design development, and design communication, and less so by ecologists. Such visual representations may capture or synthesize spatial and experiential landscape qualities in ways that numeric results may not. Students from both disciplines can not only learn respect for one another's approaches when asked to employ them collaboratively, they also deepen their abilities to understand landscapes from a variety of standpoints (Box 13-1). Edward Tufte has authored three books on the visual display of information (Tufte 1983, 1990, 1997) that can be very useful in this respect. Finally, designers develop site drawings at multiple scales to focus on the spatial qualities of existing and proposed conditions at different levels of resolution. Directing this toward formal understandings of issues related to hierarchy and scale can be a bridge between ecologists' and designers' investigations.

The rapidly growing use of GIS (Chapter 9) and associated computer-mapping technologies are revolutionizing the representation of large landscapes in both design and ecology. It is becoming easier to overlay multiple classifications on a single map, to georeference descriptive data, and to drape a plan view over a digital elevation model. In one such emerging application, designer David Hulse has spearheaded an effort to link ecological research and stakeholder involvement through the development and analysis of alter-

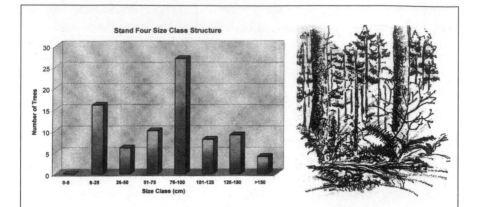

BOX 13-1. Combining Modes of Representation

At the University of Oregon, students in Bart Johnson's Principles of Applied Ecology class are given rapid-fire exercises in interpreting different kinds of scientific graphs. Later they are asked to develop creative graphic approaches to reporting the results of class field data while working in interdisciplinary teams. In the course of this project they also are required to construct and interpret graphs of tree size–class distribution alongside sketches of different forest stands as part of reconstructing stand history, thus comparing two modes of representation. Students from each discipline find themselves challenged by being beginners in new activities; at the same time they learn to appreciate the different skills of their colleagues.

native landscape future scenarios, which are then presented as three-dimensional computer simulations (Hulse et al. 1997; <http://ise.uoregon.edu/muddy/muddy-abstract.html>). Combining traditional maps with such visualization tools provides a common frame of reference for experts and lay people with different types of expertise and experience.

The ability to portray and analyze a landscape from multiple points of view also is useful—and sometimes essential. When terrestrial and aquatic ecologists began working together on the President's Forest Plan in the Pacific Northwest, each group initially came up with different kinds of maps. The aquatic experts made network maps derived from the importance of stream-system connectivity for aquatic species and processes. The terrestrial experts made patchwork maps, reflecting the distribution of landcover types and their importance as habitats for terrestrial species. Only by combining the two perspectives into one map of a combined patchwork/network system could they capture the ecological character of the landscape (Fred Swanson, USFS, pers. comm.).

Design as Hypothesis

When designs are considered as hypotheses, design process and scientific inquiry can be more strategically aligned. At the same time, it raises substantial questions about the feasibility of treating designs as hypotheses in practice.

Part of the gap between ecological science and design arises because the design, or its installation, is often considered the endpoint of the design process. As such, designers often omit a formal feedback loop from implementation to new knowledge that can be gained through post-implementation monitoring. If, however, a design is considered to embody a set of implicit and explicit hypotheses about how it would change a site or the way it functions, the roles science and practice become more strategically aligned. In this sense, design development is a process of hypothesis formation. The initial "test" of the design hypothesis is whether it holds up to rigorous evaluative critique in much the way that a scientific model examines phenomena through logical probing and simulation. Once a design passes this stage, the installation itself becomes the real-world test, a quasi-experiment in progress, with monitoring to determine whether the design "works" as predicted.

True design experiments seem unlikely to be feasible in professional practice. Formal experiments are not always practical in ecology, either, even under the best of circumstances. Design theorists such as Bruce Archer advanced the idea of design as hypothesis in the 1960s. However, they encountered difficulties identifying suitable measurable tests for such hypotheses (Simon Swaffield, pers. comm.). Furthermore, true experiments would require much greater control over design and installation than can realistically be expected, as well as replicates of experimental controls and treatments. The concept, however, has two main advantages. The first is that even if true experiments are not conducted in practice, the idea is a solid heuristic device to teach students to critically consider the consequences of their designs and the logic and evidence that support their predictions. Second, whether or not designs are implemented as experiments, post-implementation monitoring is a useful case study approach. There are many analytical techniques used by ecologists when true experiments are not feasible that could be applied. They range from statistical surveys of existing conditions under different design "treatments" to actual quasi-experiments that are missing one or more experimental attributes, such as randomization or replication, but were carefully planned to include baseline data on the design site and a paired reference site prior to implementation.

Elements of such an approach are emerging in practice through the analysis of built works (Cooper Marcus and Francis 1998), the monitoring of wetland mitigation projects, proposals for adaptive management (Holling 1978; Walters 1986), and the use of ecological restoration as a tool of scientific inquiry (Jordan, Gilpin, and Aber 1987). To make such procedures common in design practice would require deep changes in how the profession operates, including addressing the issue of who bears the added costs. But it is an idea that makes deep sense, albeit requiring designers trained to think in such ways and a professional or academic infrastructure for monitoring project outcomes over time. We examine a possible framework in our discussion of collaborative research, below.

To bring the concept of designs as hypotheses into design education requires introducing students to the logic required for strong inference. As a start, students must learn to articulate the cultural and ecological hypotheses represented in their designs and to rigorously evaluate the potential consequences of their proposals. To evaluate success, there must also be measurable criteria to assess the achievement of specific objectives. To the extent that design students are not taught that the examination of explicit (or implicit) hypotheses is part of a design generation process, their work will be less relevant to the emerging practice of ecological design and planning. In fact, we argue that such critical thinking will make them better designers, no matter what the context.

Forums for Collaboration

The authors of the other working-group chapters in this volume each discuss ways to integrate ecology into design and planning education, and through many of them, to enhance dialogue among designers and ecologists. Their proposals range from teaching techniques to course content recommendations, to curricular strategies, and to lifelong learning. In our final section, we consider where and when dialogue and collaboration can occur within educational settings and identify three principal forums: collaborative teaching, collaborative research, and institutional support systems.

Collaboratively taught classes can provide students with experience in applying interdisciplinary dialogue to problem solving. Essentially, there is a continuum of possibilities that range from building sequential course content to linked but individual courses, to classes taught jointly by designers and natural scientists. Science and design students can collaborate in problem-solving teams in any of these situations. The most suitable approaches will depend on both the content and context of individual academic programs.

Collaborative research is another way to enhance dialogue. The benefits for students may be direct, such as research assistantships, or indirect through publications that become the backbone for classroom discussions and debate.

The first step to fostering collaborative research is to establish common goals and agendas that can link the two fields of study. One such connection is the two new urban, long-term ecological research (LTER) sites in Phoenix and Baltimore (Parlange 1998) and for which a merging of the natural and social sciences is essential (Pickett et al. 1997). Collaborative opportunities abound for ecologists, city planners, and landscape architects in urban areas. For example, the scale-dependent effects of urban landscape designs on animal communities provided a research topic in which ecologists and landscape architects could collaborate (Hostetler 1999). Cross-disciplinary studies are coming to the forefront of science, but collaborative work among designers and ecologists is far from common. Although some theory exists for cross-disciplinary work (e.g., Pickett et al. 1997), putting theory into practice may be the largest hurdle.

Shrader-Frechette and McCoy (1993) argue that for practical problem solving, bottom-up approaches to ecological research, particularly case studies, are a necessary complement to top-down approaches to theory building and application. The strengths of case studies lie in their practical approach to problem solving and, in particular, their systematic, organized, and detailed attention to a wide range of evidence. They allow for rough generalizations and may provide insight into similar situations. Case study approaches have their weaknesses, including the increased potential for researcher bias and the difficulties of generalization, but there are ways to address these issues (Shrader-Frechette and McCoy 1993).

Collaborative research needs professional forums where it can be nurtured, disseminated, and debated among peers. The International Association for Landscape Ecology (IALE) and the Society for Restoration Ecology (SER) are notable for their explicit inclusion of ecologists, designers, and planners from their inception (Chapter 1). Along with other professional organizations and associated publications, they can serve key roles in fostering interdisciplinary research.

Institutional infrastructures that distance designers and ecologists conceptually and physically are common impediments to dialogue and collaboration between the two disciplines. While some design programs are allied with natural resources programs or related environmental disciplines, many are principally aligned with architecture and the arts with little historic or current connection to the natural sciences. Making root changes in campus affiliations is not easily undertaken, but neither is it impossible. The University of Georgia has recently embarked on a funding initiative to secure an endowment that would house ecology and environmental design/urban planning under one roof in a new College of the Environment (University of Georgia 2000).

The express goal of the new college would be to restructure the university to support environmental problem solving through interdisciplinary collaboration. Specific objectives would be to support innovative teaching and curricular development; redevelop professional and academic degree tracks; provide seed grants for interdisciplinary research; establish graduate and postdoctoral fellowships, new faculty positions, and student internships; and serve as a clearinghouse for environmental information at the university. The initiative includes a cross-campus Academy of Environmental Studies that would extend the college's roots and reach to other academic units. Should it occur, such a merger may require deep adjustments for some designers and ecologists. For instance, will designers consider quantitative research to play as central a role in joint efforts as ecologists do? Will ecologists accept that qualitative research, rigorously done, is a useful complement to quantitative research?

Not all solutions require such deep restructuring. For example, faculty at the University of Washington have spent the last four years developing proposals for a curriculum in urban ecology in which students would frame research issues through interactions with government agencies (Kristina Hill, pers. comm.). Creating graduate degree programs in this way may achieve similar goals for collaboration without attempting to create a new college or disband current academic alliances with the arts.

Institutions can also work to convert their physical infrastructure into a living classroom where faculty and students analyze and redesign the campus itself along environmental standards (Orr Foreword, this volume). Such initiatives may have seemed unlikely just a few years ago, but the realization of an ambitious project for the Oberlin College Environmental Studies building and grounds (Orr 1997) demonstrates how such efforts may blossom given sufficient will, imagination, and financial support. Other institutions may not be far behind. The Evergreen State College in Washington has developed a "zero-impact development plan" that focuses on creative stormwater management to maintain a healthy watershed with no measurable downstream impacts through the integrated planning of site and structures (SCA Engineering 2000). One of the goals is to involve students in seminars and projects, including experimental installations and monitoring. Moreover, the campus is developing stronger environmental standards for new buildings as well as planning for how to cost-effectively renovate existing buildings to reduce negative impacts on the environment and on users (Michel George, Director of Facilities, The Evergreen State College, pers. comm.). Meanwhile, the University of Georgia plans a state-of-the-art "green architecture" building to house the proposed College of the Environment. Such steps seem as conservative as they are radical, extending learning beyond the confines of academic disciplines to utilize the campus environment as a basis for scholarly inquiry.

Conclusions

The pressing need for new ways to integrate ecology and human culture within landscapes underscores the importance of creative collaboration among designers and ecologists. We began this essay by considering how the different training and goals of designers and ecologists could be harnessed toward designing and planning landscapes that continue to express Earth's rich evolutionary and cultural heritages. We illustrated how key concepts from both disciplines might be juxtaposed and integrated in the course of asking five key questions about landscapes that serve the need of decision-making processes. In doing so, we stressed the following:

- Understanding how landscapes work over broad spatial and temporal frames by systematically exploring them as individualistic, living systems with interactive physical, biological, and cultural attributes.
- Developing an evaluative framework of inclusive design for both humans and nonhumans in which livability, sense of place, and aesthetics for people are aligned to support ecological health and biological integrity as well as social justice and cultural diversity.
- Ways to critically and imaginatively conceive interventions in space and over time that respect the integrity of a landscape as an interconnected, dynamic, living system, whatever its context or scale.
- The need to understand key classes of impacts from human activities in light of the importance of secondary effects and the inherent uncertainties of living systems.
- How principles from collaborative planning in conjunction with the precautionary principle may provide an ethical framework for decision making, including a professional responsibility to ensure that the interests of the land community, human and nonhuman, are always adequately considered and represented.

We then examined strategies for enhancing dialogue and collaboration in education. We began with a look at how design and ecology students could expand their personal skills repertoires by learning each other's ways of representing and communicating ideas. Next, we suggested a strategic approach to collaborative case studies that aligns science and practice by treating designs as hypotheses of ecological and cultural outcomes. Finally, we described a range of ways to foster joint ecological-cultural understanding in education, including institutional frameworks in which designers and applied ecologists may be able to more fully integrate their teaching, research, and practice.

The solution cannot be for designers and applied ecologists to meet, exchange ideas, and go their own ways. According to designer Anne Whiston Spirn, designers in the 1960s and 1970s were inspired by new and emerging ideas from ecologists such as Robert MacArthur and Eugene and Howard Odum. These ideas became the foundation for a profoundly different approach to professional practice and teaching. But in too many cases, these ideas became frozen in the minds of teachers and students. They held onto the concepts they had learned long after ecology had moved on (Anne W. Spirn, pers. comm.). Ongoing dialogue and collaboration are essential if we are to forge a dynamic relationship between the two disciplines. Both designers and ecologists offer unique perspectives essential to landscape decision making that respects, celebrates, and perpetuates a rich heritage of ecological and cultural diversity.

Citations

Agee, J. K. 1993. Fire ecology of Pacific Northwest forests. Island Press, Washington, D.C., USA.

Ahl, V., and T. F. H. Allen. 1996. Hierarchy theory: a vision, vocabulary, and epistemology. Columbia University Press, New York, New York, USA.

Allen, J. F., and M. Burns. 1986. Cataclysms on the Columbia. Timber Press, Portland, Oregon, USA.

Allen, T. F. H., and T. W. Hoekstra. 1992. Toward a unified ecology: complexity in ecological systems series. Columbia University Press, New York, New York, USA.

Allendorf, F. W., and R. F. Leary. 1986. Heterozygosity and fitness in natural populations of animals. Pages 57–76 in M. E. Soulé, editor. Conservation biology: the science of scarcity and diversity. Sinauer Associates, Sunderland, Massachusetts, USA.

Angermeier, P. L., and J. R. Karr. 1994. Biological integrity versus biological diversity as policy directives. BioScience 44: 690–697.

Appleton, J. 1975. The experience of landscape. John Wiley and Sons, London, UK.

———. 1984. Prospects and refuges revisited. Landscape Journal 3: 91–103.

Banerjee, T., and M. Southworth. 1995. City sense and city design, writings and projects of Kevin Lynch. MIT Press, Cambridge, Massachusetts, USA.

Bourassa, S. C. 1991. The aesthetics of landscape. Belhaven Press, London, UK.

Bratton, S. P. 1976. Resource division in an understory herb community: responses to temporal and microtopographic gradients. American Naturalist 110: 679–693.

Callicott, J. B. 1995. Some problems with the concept of ecosystem health. Ecosystem Health 1: 101–112.

Callicott, J. B., and K. Mumford. 1997. Ecological sustainability as a conservation concept. Conservation Biology 11: 32–40.

Carr, S., M. Francis, L. G. Rivlin, and A. M. Stone. 1992. Public space. Cambridge University Press, New York, New York, USA.

Carroll, C. R., and G. K. Meffe. 1997. Management to meet conservation goals: general principles. Pages 347–383 in G. K. Meffe and C. R. Carroll, editors. Principles of conservation biology, second edition. Sinauer Associates, Sunderland, Massachusetts, USA.

Christensen, N. L., et al. 1996. The report of the Ecological Society of America committee on the scientific basis for ecosystem management. Ecological Applications 6: 665–691.

Coley, R. L., F. E. Kuo, and W. C. Sullivan. 1997. Where does community grow? The social context created by nature in urban public housing. Environment and Behavior 29: 468–492.

Colinvaux, P. 1993. Ecology 2. John Wiley and Sons, New York, New York, USA.

Costanza, R., B. G. Norton, and B. G. Haskell. 1992. Ecosystem health: new goals for environmental management. Island Press, Washington, D.C., USA

Costanza, R., and H. E. Daly. 1992. Natural capital and sustainable development. Conservation Biology 6: 37–46.

Cooper Marcus, C., and C. Francis. 1998. Post-occupancy evaluation. Pages 345–356 in C. Cooper Marcus and C. Francis, editors. People places: design guidelines for urban open space, second edition. Van Nostrand Reinhold, New York, New York, USA.

Cooper Marcus, C., and W. Sarkissian. 1986. Housing as if people mattered: site design guidelines for medium-density family housing. University of California Press, Berkeley, California, USA.

Cottam, G. 1987. Community dynamics on an artificial prairie. Pages 257–270 in W. R. Jordan, M. E. Gilpin and J. D. Aber, editors. Restoration ecology, a synthetic approach to ecological research. Cambridge University Press, New York, New York, USA.

Dale, V. H., S. Brown, R. A. Haeuber, N. T. Hobbs, N. Huntly, R. J. Naiman, W. E. Riebsame, M. G. Turner, and T. J. Valone. 2000. Ecological principles and guidelines for managing the use of land. Ecological Applications 10: 639–670.

DeBano, L. F., D. G. Neary and P. F. Ffolliott. 1998. Fire's effects on ecosystems. John Wiley und Sons, New York, New York, USA.

Denslow J. S. 1985. Disturbance-mediated coexistence of species. Pages 307–323 in S. T. A. Pickett and P. S. White, editors. The ecology of natural disturbance and patch dynamics. Academic Press, Orlando, Florida, USA.

Diamond, J. M. 1989. The present, past, and future of human-caused extinctions. Philosophical Transactions of the Royal Society of London B 325: 469–477.

Dombeck, M. P. 1996. Thinking like a mountain: BLM's approach to ecosystem management. Ecological applications 6: 699–702.

Falk, D. A., and K. E. Holsinger. 1991. Genetics and conservation of rare plants. Oxford University Press, New York, New York, USA.

Fischee, M., and J. Stocklin. 1997. Local extinctions of plants in remnants of extensively used calcareous grasslands, 1950–1985. Conservation Biology 11: 727–737.

Fischer, F. 1993. Citizen participation and the democratization of policy expertise: from theoretical inquiry to practical cases. Policy Sciences 26: 165–187.

Folke, C., A. Jansson, J. Larsson, and R. Costanza. 1997. Ecosystem appropriation by cities. Ambio 26: 167–172.

Forman, Richard T. T. 1995. Land mosaics: the ecology of landscapes and regions. Cambridge University Press, New York, New York, USA.

Forman, R. T. T., and L. E. Alexander. 1998. Roads and their major ecological effects. Annual Review of Ecology and Systematics 29: 207–231.

Forman, R. T. T., and M. Godron. 1986. Landscape ecology. John Wiley and Sons, New York, New York, USA.

Frampton, K. 1982. Modern architecture and the critical present. Architectural Design Profile, New York, New York, USA.

Franklin, J. F. 1993. Preserving biodiversity: species, ecosystems or landscapes? Ecological Applications 3: 202–205.

Franklin, J. F., K. Cromack, W. Denison et al. 1981. Ecological characteristics of old-growth Douglas-fir forests. USDA Forest Service General Technical Report PNW-118. Pacific Northwest Forest and Range Experiment Station, Portland, Oregon, USA.

Gamman, J. K. 1994. Overcoming obstacles in environmental policymaking: creating partnerships through mediation. State University of New York Press, Albany, New York, USA.

Gilpin, M. E., and M. E. Soulé. 1986. Minimum viable populations: processes of species extinction. Pages 19–34 in M. E. Soulé, editor. Conservation biology: the science of scarcity and diversity. Sinauer Associates, Sunderland, Massachusetts, USA.

Gottlieb, R. 1993. Forcing the spring: the transformation of the American environmental movement. Island Press, Washington, D.C., USA.

Grime, J. P. 1979. Plant strategies and vegetation processes. John Wiley and Sons, New York, New York, USA.

Grumbine, R. E. 1994. What is ecosystem management? Conservation Biology 8: 27–38.

———. 1997. Reflections on "What is ecosystem management?" Conservation Biology 11: 41–47.

Hartig, T., M. Mang, and G. W. Evans. 1991. Restorative effects of natural environment experience. Environment and Behavior 23: 3–26.

Harvey, D. 1973. Social justice and the city. Johns Hopkins University Press, Baltimore, Maryland, USA.

Helphand, K. 1997. Defiant gardens. Journal of Garden History 17: 101–121.

Hester, R. 1996. Wanted: local participation with a view. Pages 42–52 in J. L. Nasar and B. B. Brown, editors. Public and private places. Proceedings of the 27th annual conference of the Environmental Design Research Association, June 1996, Salt Lake City, Utah. EDRA, Edmond, Oklahoma, USA.

Hobbs, R. J., and D. A. Norton. 1996. Towards a conceptual framework for restoration ecology. Restoration Ecology 4: 93–110.

Holling, C. S. 1978. Adaptive environmental assessment and management. John Wiley and Sons, New York, New York, USA.

Hostetler, M. E. 1999. Scale, birds, and human decisions: a potential for integrative research in urban ecosystems. Landscape and Urban Planning 45: 15–19.

Hulse, D., J. Eilers, K. Freemark, C. Hummon, and D. White. 2000. Planning alternative future landscape in Oregon: evaluating effects on water quality and biodiversity. Landscape Journal 19: 1–19.

Hulse, D., L. Goorjian, D. Richey, M. Flaxman, C. Hummon, D. White, K. Freemark, J. Eilers, J. Bernert, K. Vache, J. Kaytes, and D. Diethelm. 1997. Possible futures for the Muddy Creek Watershed, Benton County, Oregon. Institute for a Sustainable Environment, University of Oregon, Eugene, Oregon, USA.

Hulse, D., and R. Ribe. 2000. Land conversion and the production of wealth. Ecological Applications 10: 679–682.

Hutchinson, G. E. 1965. The ecological theater and the evolutionary play. Yale University Press, New Haven, Connecticut, USA.

Johnson, B. R., and R. Campbell. 1999. Ecology and participation in landscape-based planning within the Pacific Northwest. Policy Studies Journal 27: 502–529.

Jones, J. A., F. J. Swanson, B. C. Wemple, and K. U. Snyder. 2000. Road effects on hydrology, geomorphology, and disturbance patches in stream networks. Conservation Biology 14: 76–85.

Jones, S., and P. Welch. 1999. Evolving visions: segregation, integration, and inclusion in the design of built places. Pages 106–116 in T. Mann, editor. The power of imagination. Proceedings of the 30th annual conference of the Environmental Design Research Association, June 1999, Orlando, Florida. EDRA, Edmond, Oklahoma, USA.

Jordan, W. R., M. E. Gilpin, and J. D. Aber. 1987. Restoration ecology: ecological restoration as a technique for basic research. Pages 3–21 in W. R. Jordan, M. E. Gilpin and J. D. Aber, editors. Restoration ecology: a synthetic approach to ecological research. Cambridge University Press, New York, New York, USA.

Kaplan, R., and S. Kaplan. 1989. The experience of nature: a psychological perspective. Cambridge University Press, New York, New York, USA.

Kareiva, P., and U. Wennergren. 1995. Connecting landscape patterns to ecosystem and population processes. Nature 373: 299–302.

Karr, J. R. 2000. Health, integrity and biological assessment: the importance of measuring whole things. Pages 209–225 in D. Pimental, L. Westra and R. F. Noss, editors. Ecological integrity: integrating environment, conservation, and health. Island Press, Washington, D.C., USA.

Krebs, C. J., R. Boonstra, S. Boutin, and A. R. E. Sinclair. 2001. What drives the 10-year cycle of snowshoe hares? Bioscience 51: 25–35.

Krebs, C. J., S. Boutin, and R. Boonstra. 2001. Ecosystem dynamics of the Boreal Forest: the Kluane Project. Oxford University Press, New York, New York, USA.

Krummel J. R., P. R. Coleman, R. H. Gardner, G. Sugihara, and R. V. O'Neill. 1987. Landscape patterns in a disturbed environment. Oikos 48: 321–324.

Kuo, F. E., W. C. Sullivan, R. L. Coley, and L. Brunson. 1998. Fertile ground for community: inner city neighborhood common spaces. American Journal of Community Psychology 26: 823–851.

Landres, P., P. Morgan, and F. Swanson. 1999. Overview of the use of natural variability concepts in managing ecological systems. Ecological Applications 9: 1179–1188.

Lemons, J. 1998. Burden of proof requirements and environmental sustainability: science, public policy, and ethics. Pages 75–103 in J. Lemons, L. Westra, and R. Goodland, editors. Ecological sustainability and integrity: concepts and approaches. Kluwer Academic Press, Boston, Massachusetts, USA.

Ludwig, D., R. Hilborn, and C. Walters. 1993. Uncertainty, resource exploitation and conservation: lessons from history. Science 260: 17, 36.

Lyle, J. T. 1985. The alternating current of design process. Landscape Journal 4: 7–13.

Lynch, K. 1958. Environmental adaptability. Journal of the American Institute of Planners 24: 16–24.

————. 1960. The image of the city. Technology Press, Cambridge, Massachusetts, USA.

————. 1981. A theory of good city form. MIT Press, Cambridge, Massachusetts, USA.

Lynch, K. and L. Rodwin. 1958. A theory of urban form. Journal of the American Institute of Planners **24**: 201–214.

MacMahon, J. A. 1997. Ecological restoration. Pages 479–511 in G. K. Meffe and C. R. Carroll, editors. Principles of conservation biology, second edition. Sinauer Associates, Sunderland, Massachusetts, USA.

May, R. M. 1992. How many species inhabit the earth? Scientific American **267**: 42–48.

McGinnis, M. V., J. T. Woolley, and J. K. Gamman. 1999. Bioregional conflict resolution: rebuilding community in watershed planning and organizing. Environmental Management **24**: 1–12.

Meffe, G. K., and C. R. Carroll, editors. 1997a. Principles of conservation biology, second edition. Sinauer Associates, Sunderland, Massachusetts, USA.

————. 1997b. What is conservation biology? Pages 3–27 in G. K. Meffe and C. R. Carroll, editors. Principles of conservation biology, second edition. Sinauer Associates, Sunderland, Massachusetts, USA.

————. 1997c. Genetics: conservation of diversity within species? Pages 161–201 in G. K. Meffe and C. R. Carroll, editors. Principles of conservation biology, second edition. Sinauer Associates, Sunderland, Massachusetts, USA.

————. 1997d. Sustainable development case studies. Pages 599–642 in G. K. Meffe and C. R. Carroll, editors. Principles of conservation biology, second edition. Sinauer Associates, Sunderland, Massachusetts, USA.

————. 1997e. The species in conservation. Pages 57–86 in G. K. Meffe and C. R. Carroll, editors. Principles of conservation biology, second edition. Sinauer Associates, Sunderland, Massachusetts, USA.

Montgomery, D. R., G. E. Grant, and K. Sullivan. 1995. Watershed analysis as a framework for implementing ecosystem management. Water Resources Bulletin **31**: 369–386.

Mozingo, L. A. 1997. The aesthetics of ecological design: seeing science as culture. Landscape Journal **16**: 46–59.

Myers, N. 1997. Global biodiversity II: losses and threats. Pages 123–158 in G. K. Meffe and C. R. Carroll, editors. Principles of conservation biology, second edition. Sinauer Associates, Sunderland, Massachusetts, USA.

Nassauer, J. I. 1995. Messy ecosystems, orderly frames. Landscape Journal **14**: 161–170.

Norton, B. G. 1991. Toward unity among environmentalists. Oxford University Press, New York, New York, USA.

Norton, B. G., and R. E. Ulanowicz. 1992. Scale and biodiversity policy: a hierarchical approach. Ambio **21**(4): 244–249.

Noss, R. F. 1990. Indicators for monitoring biodiversity: a hierarchical approach. Conservation Biology **4**: 355–364.

————. 1992. The Wildlands project: Land conservation strategy. Wild Earth (Special Issue): 10–25.

Noss, R. F., and B. Csuti. 1997. Habitat fragmentation. Pages 269–304 in G. K. Meffe and C. R. Carroll, editors. Principles of conservation biology, second edition. Sinauer Associates, Sunderland, Massachusetts, USA.

O'Neill, R. V. 1989. Perspectives in hierarchy and scale. Pages 140–156 in J. Roughgarden, R. M. May, and S. A. Levin, editors. Theoretical ecology. Princeton University Press, Princeton, New Jersey, USA.

Orr, D. W. 1997. Architecture as pedagogy II. Conservation Biology **11**: 597–600.

Parlange, M. 1998. The city as an ecosystem. Bioscience **48**: 581–585.

Patten, B. C. 1985. Energy cycling, length of food chains, and direct vs. indirect effects in ecosystems. Canadian Bulletin of Fisheries and Aquatic Sciences **213**: 119–138.

————. 1991. Network ecology: Indirect determination of the life-environment relationship in ecosystems. Pages 288–351 in M. Higashi and T. P. Burns, editors. Theoretical ecosystem ecology: the network perspective. Cambridge University Press, London, UK.

Peters, R. L., and J. D. S. Darling. 1985. The greenhouse effect and nature reserves. Bioscience **35**: 707–717.

Peters, R. L., and T. E. Lovejoy. 1992. Global warming and biological diversity. Yale University Press, New Haven, Connecticut, USA.

Pickett, S. T. A., W. R. Burch Jr., S. E. Dalton, T. W. Foresman, J. M. Grove, and R. Rowntree. 1997. A conceptual framework for the study of human ecosystems in urban areas. Urban Ecosystem **1**: 185–199.

Pickett, S. T. A., V. T. Parker, and P. L. Fiedler. 1992. The new paradigm in ecology: implications for conservation biology above the species level. Pages 65–88 in P. L. Fiedler and S. K. Jain, editors. Conservation biology: the theory and practice of nature conservation preservation and management. Chapman and Hall, New York, New York, USA.

Rappaport, D. R., Costanza, P. R. Epstein, C. Gaudet and R. Levins. 1998. Ecosystem health. Blackwell Science, Malden, Massachusetts, USA.

Rees, W. E., and M. Wackernagel. 1994. Ecological footprints and appropriated carrying capacity: Measuring the natural capital requirements of the human economy. Chapter 20 in A. M. Jansson, M. Hammer, C. Folke, and R. Costanza, editors. Investing in natural capital: the ecological economics approach to sustainability. Island Press, Washington, D.C., USA.

Romme, W. H. 1997. Creating pseudo-rural landscapes in the mountain West. Pages 139–161 in J. I. Nassauer, editor. Placing nature: culture and landscape ecology. Island Press, Washington, D.C., USA.

SCA Engineering. November, 2000. Toward zero impact: Evergreen State College campus opportunities for zero impact development and redevelopment. Facilities Department, Evergreen State College, Olympia, WA, USA. <http://www.evergreen.edu/user/facilities/projects_and_reports /epa_zero_impact.htm>

Shrader-Frechette, K. S. and E. D. McCoy. 1993. Method in ecology: strategies for conservation. Cambridge University Press, New York, New York, USA.

Sierra Nevada Ecosystem Project. 1996. Final report to Congress, vol. 1, assessment summaries and management strategies. Wildland Resources Center Report No. 36. Centers for Water and Wildland Resources, University of California, Davis, California, USA.

Silbernagel, J. 1997. Scale perception: from cartography to ecology. Bulletin of the Ecological Society of America **78**: 166–169.

Slocombe, D. S. 1993. Implementing ecosystem-based management. BioScience **43**: 612–622.

Soulé, M. E., and B. A. Wilcox. 1980. Conservation biology: its scope and its challenge. Pages 1–8 in M. E. Soulé and B. A. Wilcox, editors. Conservation biology: an evolutionary-ecological perspective. Sinauer Associates, Sunderland, Massachusetts, USA.

Sousa, W. P. 1984. The role of disturbance in natural communities. Annual Reviews of Ecology and Systematics **15**: 353–391.

Steinitz, C. 1990. A framework for theory applicable to the education of landscape architects (and other environmental design professionals). Landscape Journal **9**: 136–143.

Stone, C. D. 1974. Should trees have standing? Toward legal rights for natural objects. W. Kaufmann, Los Altos, California, USA.

Sullivan, W. C., F. E. Kuo, and S. Deppoter. In preparation. Tree cover and social activities in inner-city neighborhood common spaces.

Swanson, F. J., J. A. Jones, and G. E. Grant. 1997. The physical environment as a basis for man-aging ecosystems. Pages 229–238 in K. A. Kohm and J. F. Franklin, editors. Creating a forestry for the 21st century: the science of ecosystem management. Island Press, Washington D.C., USA.

Thomas, J. W. 1996. Forest Service perspective on ecosystem management. Ecological Applications **6**: 703–705.

Thompson, D. W. 1961. On growth and form. Cambridge University Press, Cambridge, UK.

Tilman, D., R. M. May, C. L. Lehman, and M. A. Nowak. 1994. Habitat destruction and the extinction debt. Nature **371**: 65–66.

Tuan, Yi-fu. 1997. Space and place: the perspective of experience. University of Minnesota Press, Minneapolis, Minnesota, USA.

Tufte, E. R. 1983. The visual display of quantitative information. Graphics Press, Cheshire, Connecticut, USA.

———. 1990. Envisioning information. Graphics Press, Cheshire, Connecticut, USA.

———. 1997. Visual explanations: images and quantities, evidence and narrative. Graphics Press, Cheshire, Connecticut, USA.

Turner, M. G., and R. H. Gardner. 1991. Quantitative methods in landscape ecology. Springer-Verlag, New York, New York, USA.

Ulrich, R. S., R. F. Simons, B. D. Losito, E. Fioriteo, M. A. Miles, and M. Zelson. 1991. Stress recovery during exposure to natural and urban environments. Journal of Environmental Psychology **11**: 201–230.

University of Georgia. 2000. Meeting the environmental challenge, the role of the University of Georgia. A report of the Environmental Programs Enhancement Committee. University of Georgia, Athens, Georgia, USA.

Urban, D. L., R. V. O'Neill, and H. H. Shugart. 1987. Landscape ecology: a hierarchical perspective can help scientists understand spatial patterns. BioScience **37**: 119–127.

Vrijenhoek, R. C. 1989. Population genetics and conservation. Pages 89–98 in D. Western and M. C. Pearl, editors. Conservation for the twenty-first century. Oxford University Press, New York, New York, USA.

Wackernagel, M., and W. E. Rees 1996. Our ecological footprint: reducing human impact on the Earth. New Society Publishers, Gabriola Island, British Columbia, Canada.

Walters, C. J. 1986. Adaptive management of renewable resources. McGraw-Hill, New York, New York, USA.

Wilson, E. O. 1992. The diversity of life. Belknap Press of Harvard University Press, Cambridge, Massachusetts, USA.

Woodley, S., J. Kay, and G. Francis. 1993. Ecological integrity and the management of ecosystems. St. Lucie Press, Delray Beach, Florida, USA.

Wright, S. 1931. Evolution in Mendelian populations. Genetics 16: 97–159.

Yaffee, S. L. 1996. Ecosystem management in practice: the importance of human institutions. Ecological Applications 6: 724–727.

Zube, E. 1987. Perceived land use patterns and landscape values. Landscape Ecology 1: 37–45.

———. 1990. Landscape research: planned and serendipitous. Human Behavior and Environment: Advances in Theory and Research 11: 291–334.

Notes

1. We will use "designers" as a shorthand to refer to designers and physical planners as a group and "ecologists" to include biophysical scientists in environmental disciplines often allied with ecology. Distinctions between the terms "design" and "planning" in the landscape professions typically relate to spatial scale, the degree of spatial resolution, and how change is implemented. While there are no hard-and-fast rules, design tends to involve smaller areas with a high degree of spatial resolution and is implemented through direct intervention on the land by a discrete number of individuals. Planning tends to involve larger areas and typically guides the overall process and outcomes of development by facilitating or inhibiting certain types of changes within a legal policy framework. For example, the use of the terms "urban design" and "urban planning" typically adheres to these distinctions. While "ecologist" refers to someone with a specific training and academic degree, many scientists with training in biology, botany, zoology, forestry, geography, geology, and other fields play similar roles in land management and conservation practice. In this arena, the term "plan" is commonly used in the sense of a "plan of action" or a "management plan," emphasizing a document that offers criteria and a process for decisions about the landscape. It may include components of both a design and a plan in the terminology of landscape designers and planners.

2. Pickett, Parker, and Fielder (1992) describe a discipline's worldview as composed of its "beliefs, values, and techniques," and more specifically that it could be thought of as a "family of theories that undergird a discipline."

3. The choice of scales for study requires consideration of both *grain* and *extent*. Ahl and Allen (1996) describe grain as analogous to the size of the holes in a

fishing net, whereas extent would be analogous to the size of the net, the area fished, or the length of the fishing expedition. To avoid confusion among ecologists and designers, it is best to refer to scales as broad or fine. While ecologists use "large scale" and "small scale" to indicate spatial extent, designers use the cartographic system, which refers to the proportional representation of an object— the opposite of that used by ecologists (Allen and Hoekstra 1992; Forman 1995; Silbernagel 1997; Turner and Gardner 1991).

4. Some authors writing on landscape ecology use the term "structure" to indicate spatial organization or pattern not including composition. Franklin et al. (1981) employed the term as one of three system attributes: composition, structure, and function. Forman and Godron (1986) promoted using the term to indicate both composition and pattern, and the word continues to be used in multiple ways. Structure also has other usages in ecology. It is used to describe the vertical and horizontal structure of vegetation, or in the sense of population structure, especially as it relates to the numbers of individuals of different ages, genders, or sizes in relation to reproductive abilities. These multiple meanings can be a particular source of confusion for design students and others outside the field. For consistency we will use structure to denote landscape composition and pattern, unless otherwise distinguished.

5. At its core, the term "sustainability" refers to the ability to manage a system (social, ecological, economic) so as to perpetuate it indefinitely without compromising the ability to continue to do so in the future. The idea has been receiving substantial attention in ecology, design, and other disciplines, yet its meaning remains controversial (Costanza and Daly 1992; Ludwig, Hilborn, and Walters 1993). A number of authors have attempted to define sustainability so as to explicitly link biological and cultural criteria (for instance, see Meffe and Carroll 1997d). It is hard to imagine environmental reform without some concept of sustainability, albeit one guided by a keen awareness of ecological limits. For the latter reason we have used the concept only within the context of health as defined here and do not discuss it independent of such biological or social criteria.

6. In a survey of 140 Pacific Northwest landscape planning projects that ranged in size from 1 square mile to 30,000 square miles, Johnson and Campbell (1999) found that projects were addressing an average of six major stressors. Ranked by percent of projects addressing, the principal stressors were non-point-source pollution (75%), infrastructure (e.g., roads, power lines, rights-of-way) (75%), timber management (73%), recreational uses (66%), land conversion to suburban or rural development (56%), agricultural practices (56%), point-source pollution (52%), grazing and range management (50%), introduction of exotic species (46%), mining (38%), and land conversion to agriculture (28%).

CHAPTER 14

Interweaving Ecology in Design and Planning Curricula

Ken Tamminga, Louise Mozingo,
Donna Erickson, and John Harrington

The inaugural Shire Conference presented a remarkable opportunity to collectively explore how landscape planning and design curricula inadequately incorporate ecological science. Our group's assessment led to two broad queries, the first self-evident to conference attendees, the second a series of related strategic questions. First, should ecology be bolstered throughout the curricula of landscape design and planning programs? Second (and if so), is all landscape design and planning "ecological," or are we concerned with ecology only as a distinct subject under certain conditions?[1] Should ecological critique apply to all landscape design and planning? What circumstances and precepts determine an improved ecological curriculum? What epistemology is at work; how is ecological knowledge acquired and applied? What hurdles exist?

As the group charged with linking ecology and curriculum, we established an operating framework to discuss specific and ongoing elements of an ecology-infused curriculum. Landscape architects and planners seem poised to embrace ecological tenets as central to their practice. The Shire Conference signals that we seek a more rigorous and cohesive understanding of the essence and horizons of our profession and, by extension, what is taught and how it is taught. We also addressed specific issues, realities, and approaches that various programs might consider in remedying the tenuous connections between ecological science and landscape planning and design (and the ecologist and designer/planner).

We premised our discussions on the conviction—addressed by Johnson,

Hill, Forman, and many others in this volume—that ecological science is essential to learning in landscape design and planning. We recognized the importance of a choreographed, carefully timed introduction of ecology's conceptual models, relevant principles, modes of investigation, and use of hypothesis testing. We further agreed that, from both curricular and professional perspectives, ecologists and landscape designers and planners are indispensable allies in addressing the imbalances and contradictions that mark the contemporary North American landscape: ecologically healthy versus sick, culturally inspired versus mundane. Our present curricula mimic these dichotomies, some inevitable through our status as a microcosm of the university and society.

Retrospection on the curriculum is in order. Jaroslav Pelikan (1992) reminds us of the intimate, problematic connection to the larger issues of the century "that makes a reexamination of the idea of the university so essential and yet so complicated" (p. 14). The shadows of looming social and environmental problems compel us to more vigorous interaction between science and design in the academy. As education visionary Ernest Boyer (Boyer and Mitgang 1996) writes, "Beyond question, the attitudes and behavior fostered in school will profoundly influence professional behavior later on" (p. 108).

Telltale Issues

The Shire Conference allowed an exchange of institutional perspectives, anecdotes, course lists, and syllabi. Over several intensive days focused on the curriculum, we noticed a pattern of small triumphs in a few academic programs overshadowed by a status quo of curricula troubled by polarities and incongruities. Telltale concerns expressed included the following:

- Course sequences that place ecology as a single isolated subject either very early (the introductory course-from-hell) or very late (the token gesture) in the curriculum.
- Studios where "ecological design," "sustainability," and similar buzzwords are promoted, but where ecological science *content* barely surpasses the symbolic or rhetorical, and where ecologically informed interventions are more serendipitous than commonplace.[2]
- Faculties that lack bona fide ecologists and environmental scientists and that maintain insubstantial relationships with ecologists beyond departmental walls, sometimes resulting in project settings that operate with outdated or inaccurate ecological concepts and techniques.
- Assignments (and, reportedly, entire curricula) that employ abstracted ecological models set apart from the complexities, contingencies, and constituencies of real ecosystems.

- A widespread and seemingly entrenched institutional distance between design and natural science disciplines, people, and endeavors.

Although the design and planning professions are particularly obligated to be informed by a spectrum of liberal arts and sciences, these problems are not entirely of our own making. They appear symptomatic of a broader malaise that affects the entire academy, briefly addressed below.[3]

First Principles for an Interwoven Curriculum

We present no one-size-fits-all parcel of problems and remedies, but rather highlight key issues and a field of responses for university programs seeking positive change. These discussions can only suggest the range and diversity of program adaptations and "best" practices.

As an overture, the first principles describe appropriate tenets for constructive curriculum change. Since we are unable to address the individual contingencies of each program, these tenets provide a *context* for rethinking curriculum and teaching. Our intent is to meld these ideals with practical strategies. Core concepts can address common challenges and coalesce in a robust concept of ecology in the curriculum. That, in itself, is our significant goal.

The University: Learning in a Connected Community

On more than one occasion the conference raised suggestions for a radical dissociation between landscape architecture programs and the university. These voices conjured remote, Taliesan-like, live-work cloisters, stand-alone nonprofits built on service-learning in the inner city, or the like. The tone reminded us of the common campus bumper sticker exhorting motorists to "subvert the dominant paradigm." While extracting some of the effective elements of these models, we see the "dominant paradigm" not as something to be subverted, but something to be fixed. So, before we examine curricular strategies and precedents, we must acknowledge our larger institutional context.

Landscape design and planning as academic pursuits are intimately tied to the university as an idea; they partake of its traditions and core values— learning for learning's sake, albeit integrated with select pragmatic principles. When we contemplate severing our ties either physically or through curricular detachment, we remove ourselves from a unique concentration of knowledge. Scholars such as Jaroslav Pelikan, Ernest Boyer, David Orr, and earlier luminaries such as John Newman remind us that, as for all professional curricula in the university, landscape architecture and planning programs are strategically but precariously positioned between real-world utilitarian and traditional liberal education. This reflects the principle that knowledge and application be intimately linked. To mold our curriculum, educators in design and planning must cull the "Arts-Sci" continuum for knowledge directed to application, a difficult and consuming task.

This task is complicated by the tendency of the Western university to fragment knowledge, as is portrayed in Asa Briggs's metaphoric protest:

> In the modern map of learning within the universities, students and teachers . . . all too often figure as inhabitants of separate continents. A few boats pass between them, fewer still on regular service; there are a number of distinguished travelers and a diminished number of visitors; there is little long-distance migration, either temporary or permanent. Inhabitants know a little of their adjacent territories, but their ideas of what happens in the more distant regions are usually imprecise, frequently prejudiced, and often wrong. (quoted in Eckhardt 1978, p. 38)

David Orr (1992) goes further:

> Disconnectedness in the form of excessive specialization is fatal to comprehension because it removes knowledge from its larger context. Collection of data supersedes understanding of connecting patterns, which is, I believe, the essence of wisdom. (p. 101)

Whatever the downside of the contemporary university education, there are few alternatives that offer such opportunity for a network of knowledge. Considering the potential pooling of insights from the natural sciences, humanities, and social sciences, perhaps no other setting can provide the stage for connected learning. The singular, eclectic forum of the university can build stronger ties between designer/planner and environmental scientists.

Ecological Science and Other Connections

Ultimately, the university binds the liberal arts (which historically included the sciences), ethical discourse, professional design and planning programs, and the landscape. In discussing the relationship between the university and the state of the environment, Pelikan (1992, pp. 20–21) states: "Anyone who cares simultaneously about the environment and about the university must address the question whether the university has the capacity to meet a crisis that is not only ecological and technological, but ultimately educational and moral." Boyer (1997) provides direction in how this can be addressed: "To be truly educated means going beyond the isolated facts, putting learning in larger contexts, and, above all, it means discovering the connectedness of things" (p. 108). And, as Boyer more optimistically writes, "The connectedness of things is what the great university is all about" (p. 80).

If any discipline engages rigorously in the notion of connectedness, it is ecology. As landscape architecture's "native" science, ecology is a primary way of understanding how landscapes function. But, more profoundly, "Its goal is not just a comprehension of how the world works, but, in the light of that knowledge, a life lived accordingly" (Orr 1992, pp. 87–88). Our programs will be increasingly called to shape curricula that teach healing, reconnection, and design in stressed and dysfunctional places. They must engender in our students a desire for ecological literacy, which "forces us to reckon with the roots of our ailments, not just the symptoms" (Orr 1992, pp. 87–88).

The essence of ecological science demands attention to an expansive design continuum, from urban brownfield to suburb to old-growth forest. Since ecology is neither spatially nor temporally bounded, neither should its application in and evaluation of landscape design and planning be so bounded. Design and planning programs—and by extension their curricula—should serve as centripetal forces, pulling natural and social science into a more synergistic organization. The planning and design curriculum necessitates holism; ecology brings the gift of a particularly connected form of inquiry. The pattern-form focus of the former enriches the functional focus of the latter, and vice-versa.

The Interwoven Curriculum

We advocate more than mere juxtaposition, but to *integrate* ecological science into design and planning curricula is a questionable challenge. Such integration is a laudable but probably unattainable goal. A rigorous yet tenable curriculum must distinguish among integrated curricula, integrated knowledge, and integrated responses.

An aggressively "ecologized" curriculum could presume to matriculate "design-ecologists." The biosphere is an immensely complex, interconnected, and stochastic web of relationships, cycles, and flows. We trivialize this foundation when we formulate curricula that claim "enough" integration and then dispense with the ecologist or assure students (if only implicitly) that the knowledge base they develop by the end of their residency remains viable for their entire career. By so doing, we instill in our students an unwarranted sense of empowerment and do them a disservice.[4]

Regarding our attempt to teach "all" of ecology, Nassauer writes, we "exhaust students' creative energy on an impossible task." We distract students from "the important work of making the landscape while they ceaselessly tread the slope of preparation to act" (Chapter 9, p. 222). As we detail below, students of design and ecological sciences are at their best when they are strategic partners in knowledge and real-world learning. This approach encourages each to draw from different or converging heritages, stimulating a common vocabulary and sparking an exploration of roles and responsibilities.

Integration also raises the question of timing. The intensity, scope, and degree of understanding of ecological principles and application should reflect first, the level of the program, and second, the mission and flavor of the program. In progressive, stepwise fashion, the teaching of ecological basics and principles as "robust rules of thumb" (Steinitz, Chapter 10, p. 236) should be followed up by ascending ecological complexity as learning intensifies.

We suggest, then, that the process of *interweaving* ecological science into design and planning curricula gives rise to a more resilient and relevant substance. Integration requires the formation of a cohesive whole, or, pedagogically, a complete incorporation of knowledges, traditions, techniques, and values drawn from two or more disciplines. The terminology of interweaving ecological science into the curriculum seems more appropriate in describing a close-knit threading of interdependent, yet still distinct, parts.[5]

While we question the feasibility of integration of ecological science in its full sense within the curriculum, we aspire to the ideal of an integrated and holistic *outcome,* for that is ultimately what matters in the landscape. And between curriculum and outcome, the process of achieving an integrated understanding lies primarily in the mind and heart of the student.

Engendering Interrelationships

A persistent theme of the Shire Conference was the fundamental role of the ecologist in studios and courses covering basic and advanced ecology. We shared accounts of faculty relying on outdated ecological snippets tucked away since their days as undergraduates in the 1960s. Ecologists and allied scientists admitted that like the object of their research, scientific principles shift and facts come and go. Several eminent educators called for curricula that thoroughly reform the basics, and a continuous gleaning of contemporary principles of the type laid out in Chapter 13. Of the new basics, Boyer (1997, p. 109) writes, "If I were reshaping the school curriculum to help students see connections, I would have, at the very core of common learning, one major strand of study called the 'Life Cycle' . . . Being truly educated means reflecting sensitively on the mystery of birth and growth and death." Others proposed fundamental scientific canons—the laws of thermodynamics, ecological cause and effect, for example—as equally essential. Participants widely concurred that, as applied professions, landscape design and planning rely on ever-changing ecological discoveries and sensibilities. These raised stakes call for an arrangement that ensures the close and continual proximity of ecologists.

However, there are considerable disincentives to a workable, consistent dialogue between ecology and design. As discussed in more detail below, some are endemic to our profession, while some we've inherited as institutional hurdles. Beyond the technical hurdles of re-formed curricula, Schneekloth and Shibley (1995) expose the disciplinary biases and power assumptions of professional practice that are no less intimidating for academics considering curricular change: "There are real professional risks for practitioners who enter the dialogic space of placemaking and choose to work with the messy conception that many knowledges are legitimate... The condition of being open and vulnerable... calls for commitment to a process that continually makes, takes, and remakes places through the practice of dialogue and environmental action" (pp. 199–200).

The sensitive reflection on, and application of, ecological knowledge (from ecological foundations to in situ observations) requires the joint efforts of both ecology and nonecology faculty. Nassauer writes that we may bring ecology into an appropriate relationship with design place by place, looking carefully together (Chapter 9). Frequent interactions with field ecologists and practicing designers and planners are also vital to grounding the curriculum. The dynamic mix and interchange of people and their activities—students and faculty in design/planning and environmental sciences, professionals of various stripes—is no less a part of an effective curriculum than its slate of courses. When curricula "farm out" ecological learning somewhere else on campus, or isolate ecology faculty to a single lecture class, they lose the occasion to interweave ecology into design and planning education.

As both design and ecological science seek to understand and creatively intervene in the landscape, the potential for mutual learning is higher than ever. Most landscape design and planning programs possess inherent strengths and employ pedagogical models that many other disciplines are only just discovering. The young synthetic sciences of landscape ecology, restoration ecology, ecosystems management, and the like are becoming increasingly visible on campuses and increasingly responsive to the public interest. Finding common ground across disciplinary boundaries requires reciprocity and humility (Luymes, Nadenicek, and Tamminga 1995). While the design and planning student will benefit greatly from interaction with ecological science faculty and peers, the scientist will be exposed to the designer's ability to contribute "through the reflective act of design" (Boyer and Mitgang 1996, p. xv).

Real Ecosystems, Real Places

There is no substitute for diligent co-inquiry between designer and ecologist, and the best place for this relationship to occur is on-site. As a profession concerned with landscape, our curriculum and its pedagogy must engage an ecology that is not remote from its subject. In developing an epistemology for ecological sustainability, Gregory Bateson promotes the concept of interrelatedness that is understood through "patterns that connect," asserting that "we are not outside the ecology for which we plan . . . we are always and inevitably part of it" (in Orr 1992, p. 37). This "partness" implies an obligation to comprehend one's context. Sustainable practices begin with a thorough knowledge of ecological factors that are confirmed and inspired through the action of getting to know and appreciate a *place*. A number of esteemed educator-practitioners have long enacted a pedagogy that links strong understandings of site and region. In Chapter 11, Michael Hough writes: "The emphasis on a fieldwork approach to ecological design has, in my experience, contributed largely to my students' understanding of natural processes and their links with the urban community. It is a dynamic interaction that reinforces the realities of human and natural evolution. And it reflects a fundamental principle: that landscapes, to have meaning, must be tied to their regional geography, their climate and vegetation, their political and social contexts, and their local urban environment" (p. 264).

One significant and common effect of on-site learning is multidirectional invigoration: Life is breathed into the data of ecology, and life is breathed into design through ecological vitality. Our curricula demand cross-pollination between campus and field/community, while generating a symbiotic discourse between designer and scientist.

Envisioning Curricular Change

The Shire Conference was a first, constructive step in forming a collective vision for a new type of curriculum. The several dozen programs represented at the conference encompassed a variety of departmental agendas and geographical situations. We had different stories to tell, different frustrations and accomplishments. But despite our distinctions, rarely was there a narrative to which we could not relate. These encounters reveal that landscape design and planning programs are intimately connected by heritage, overarching goals, shortcomings, and expectations. Our commonalities enabled a vision of ecology-infused curricula. Therefore, we see a need for curricula that

- facilitate a co-mingling of design and ecology knowledge bases and underlying value sets;
- both respond to and anticipate needs of communities and environments;

- instill a sense of the merit of perpetual interdisciplinary alliances and cooperation in studio, lab, field, and community; and
- stimulate a love for continual learning about the inhabitants, processes, and patterns of the living landscapes we affect.

In *Building Community*, Boyer concludes that design education is about "fostering the learning habits needed for the discovery, integration, application, and sharing of knowledge over a lifetime" (Boyer and Mitgang 1996, p. xvi). Can we achieve some consensus on directions for interweaving ecology with the design and planning curriculum? The realities and strategies discussed below suggest that we can.

Challenging Realities

Several tensions, even dichotomies, impede the enthusiastic acceptance of ecological science as a cornerstone of landscape design and planning curricula. While these tensions may be felt more or less acutely depending on program and faculty, they are nonetheless identifiable and familiar to those engaged in landscape design and planning education. Most basic among these is the perception of ecological science as constraining rather than expanding creative design and planning. Particularly during the last half century, professional design culture has emphasized the designer's singular creativity in the face of the design task, the weight of imagination resting mightily on the designer's shoulders. This certainly formed a resistance to seeing the bearer of ecological science, the scientist, as a peer in the design process, as a participant who explicates creative possibilities rather than sets tiresome limits. Conversely, the scientist, steeped in the analysis of the "what is," not the "what can be" of design, can resist design's speculative thrust.

Beyond False Oppositions

The prevailing oppositional stance reflects a competition between rational science and intuitive imagination for superiority in the intellectual process of landscape design and planning. Landscape design demands the integration of both. Perhaps most destructive to the welcome interweaving of ecological science in the design and planning curricula is a falsely oppositional stance between ecology and culture; design does not require a choice between sound science and design invention. Instead, the ecologically constructive and the culturally resonant should not be seen as mutually exclusive.

The creativity of scientific insight and the creativity of design are not only similar but also potentially allied. To see ecological science as part of the creative design process requires a shift in design and planning education that presents ecological science parameters as opportunities, not constraints, to the creativity of landscape design. This requires a collaborative process that moves beyond the utilization of ecological data as instrumentalist information to make design more acceptable or convincing. In this collaboration, science and design fuse in the creative leap that produces the solutions that characterize the best landscape design and planning.

The Challenge of Collaboration

Moving beyond outdated and counterproductive mutual suspicions requires an emphasis on the collaborative process of design that, in turn, infuses the curriculum. Of course, the collaborative process is much more the reality of design practice, but university education builds from individual achievement and evaluation, and this institutional environment is not usually conducive to collaboration.

Planning and design students learn in at least two ways: by the doing of design and planning projects and by the observation and analysis of existing projects, especially as presented by various forms of professional communication. The implications of the former are that more opportunity needs to be given to students to participate in collaborative projects across disciplines, including ecological science. The implications of the latter are that as educators or professionals or both, we need to present and underscore our projects as collaborative endeavors.

However, given that the issues here deal with the interweaving of ecological science into design programs and not vice versa, educators need to remain cognizant that form making is a positive force in the design of ecological landscapes. Form making, with all its ecological and cultural implications, attracts and motivates the design and planning student. In conceiving curricula, educators should hold steadfast that design requires both convincing aesthetic and ecological strategy.

Faculty Expertise

Currently, design and planning curricula use four approaches to incorporate environmental scientists. First, many landscape design and planning faculties have no scientists within their ranks and rely on other departments to impart ecological science to their students. Most programs, particularly undergraduate ones, require students to take a general ecology course. Understanding of basic ecological principles, without specifically addressing the studio work at the core of landscape architecture education, produces a tenuous and superficial understanding of the ecological implications in planning and design. Design and planning faculty themselves may teach specialized

courses or structure studio work so that students generate and analyze ecological data as they apply to design sites. The advanced insight that the
trained scientist provides is absent. More subtly, the scientist, or for that matter any academic, who is engaged in active research generally provides a more
honed intellectual approach to that material. Programs confronted with the
obvious weaknesses that these alternatives provide can forge personal relationships with environmental scientists in their universities. Required courses
with high enrollments of design and planning students can include specifically relevant course material. A creative solution in departments that cannot
accommodate scientists, the latter relies on personal relationships and collegiality built up between particular faculty and can be vulnerable to the
vagaries of faculty turnovers.

Second, joint appointments can generally provide departments with environmental science courses specifically tailored to design students and taught
by a trained scientist. Students usually have a greater opportunity to collect,
review, and interpret environmental data as particularly applied to design and
planning. Difficulties arise since joint-appointment scientists are likely to be
more engaged in their science department than with design and planning
faculty, where they are inevitably in a minority. Environmental science
courses often stand apart from the studio, and the scientists may rarely critique or review design projects.

Third, as detailed in the following section, a small minority of departments has designated positions for environmental science faculty. This provides an opportunity for scientists to be fully integrated in lecture and studio
courses. Most important, this allows the environmental scientist to steer or
influence the intellectual course of the department, at the very least as a
minority voice.

The caveat to the inclusion of ecological science faculty within departments is that each department cannot be a university in microcosm. Understandably, academic design and planning departments that are both applied
and interdisciplinary will seek to tailor courses and faculty from both the
social and physical sciences to their particular needs. But even with ample
university funding, the comparatively small number of positions in most
design departments imposes choices between design faculty and specialized
faculty. This risks the dilution of departmental capacity to teach design. The
design and planning student well trained in ecological science but unable to
carry that through to a project design undermines the particular expertise and
contribution of the profession.

Disciplinary challenges also arise in placing scientists within design
and planning faculty. It requires a meeting of the minds between
rational and intuitive types of disciplines and intellectual traditions.
Though an atmosphere of open intellectual inquiry most often is

dependent on collegiality and personality, there are nonetheless some structural and institutional parameters that can facilitate scientists' productive inclusion in the faculty. Foremost, a curriculum approach, which stabilizes the ecological content in the curriculum, assures its centrality in the departmental pedagogy. When inevitable inconsistencies and conflicts arise in departmental decisions, the discussion cannot challenge whether ecological science should be included in the curriculum, only how. Without this acknowledgment of ecological science as one of the intellectual bases of the department, retaining scientists with whatever strategy would be very difficult.

Fourth, interdepartmental lecture courses can present both science and design and planning material within the same venue, giving both students and faculty the opportunity to see each as part of the same academic and professional endeavor. This requires institutional support through interdisciplinary course listing, joint appointments, or scientists within the department. Interdisciplinary courses are also essential fora for explicating the inconsistencies and, it is hoped, the resulting inspirations generally between science and design traditions. The least effectual presentation of ecological science material is in a separate lecture course taught solely by the environmental scientists. Unintegrated in the design studio, it relies too heavily on student initiative and insight to move ecological science into the design realm.

The preferred strategy is to move the environmental scientist into the design studio, both as intellectual resource and critic. But obstacles to this are significant. It requires either joint-appointment faculty or full-time departmental faculty. In departments with scarce resources, the high number of contact hours required in studio courses may not be feasible for a scientist to accommodate, especially if the scientist is also teaching a laboratory course and conducting research. Intellectually, it compels the designer to open the studio situation to rational inquiry and critique, and the scientist to concede to the cultural implications of manipulation of the landscape, not just the measurable scientific ones. This can be a challenge for both designer and scientist.

Ideally, the injection of science into the design curriculum would have its basis not only in joint pedagogy, but in joint research. Acknowledging all the onerous university demands of narrowly defined research for tenure and promotion, integrated research between designers and scientists is presently rare in landscape planning and design. Maximizing the interweaving of ecological science in the curriculum will ultimately need to be based on discipline-defining research results that advance both design and ecological science within the same intellectual construct.

Curricular Structure

Making science faculty, ecological content, and scientific inquiry stable within the curriculum begs the question of achievable goals. Our experience is that students frequently conflate environmentalism and ecological science; a goal within the department curriculum must be to distinguish between them. The curriculum should nurture environmental conscience while developing critical ecological rigor.

At the same time, ecological science within the curriculum should train design and planning students, not produce ecologists. To misrepresent the designer-planner as ecologist, rather than the designer-planner armed with ecological thinking, dilutes the specific contribution of designers and planners to both the environmental science and environmental design fields. Students need to emerge from programs with a thorough "talking knowledge"— an intellectual and experiential understanding—though not necessarily a "working knowledge" of ecological science.

Curricular goals to introduce ecological science beyond classroom theory must by necessity also be based in fieldwork. This implies that the geographic location of the program will define the emphasis of ecological science concepts, data gathering, and analysis methodologies. The student population and the probable location of their professional careers must influence teaching strategies. Schools in urban areas that draw students with urban interests might be most effective focusing on urban ecology; those schools in primarily agricultural states might have greatest success focusing on rural ecology. Not only does this increase the vividness of presenting ecological science materials, it expands the possibilities for greater concordance with socioeconomic learning.

Each of the course types within landscape design and planning curricula provide different capacities to present ecological science material. Lecture courses are invaluable for the concentrated transference of knowledge, yet if divorced from relevant studio activity will only vaguely influence design tasks. Field courses, a fundamental part of any environmental science curriculum, literally bring planning and design students into the landscape where they will work. The experience of seeing and observing ecological process across the landscape is irreplaceable. However, the emphasis in ecological science field courses on existing conditions—data evaluation as an end in and of itself—may leave design and planning students hanging as to its relevance. Predictive data in ecological science prove harder to measure and evaluate, hence ecological scientists tackle it less frequently, and field courses can unwittingly display ecological evaluation as a blind alley for designers.

Studio courses tend to culminate in the final design plan and presentation of a project. In studio, the capacity of engaging aesthetics to supercede the ecological science content of a project can suppress tough questions regarding ecological impacts and exasperate participating scientists. Verbal and graphic presentation skills, certainly necessary in design and planning curricula, emphasize the design aspects of projects, not their ecological content. Studio instructors are challenged to find new ways to communicate these new project parameters. While GIS computing promises much in this direction, its cutting-edge techniques are with specialized faculty in the department, not in the hands of most design studio leaders.

Extramural Concerns

The challenges of providing both skill building and scholarship are also affected by professional accreditation. Accreditation bodies' priority on servicing existing practice rather than anticipating future needs tends to guide accreditation processes. Though many academics, as well as professionals, see ecological design as an exciting opportunity for the future of landscape planning and design, the current accreditation review process favors more traditional landscape architectural and planning practice. The inclusion of ecological science in landscape architecture programs implies a shift or expansion of traditional practice. The willingness of departments to lead rather than follow this shift varies, with many departments concerned with risking the employability of their graduates if they venture too far beyond known professional parameters. This is a familiar discussion among faculty regarding the degree to which academic programs can or should anticipate or serve the profession.

Ultimately, departments must evaluate their own success in interweaving ecological science in their curricula. While valuable and relevant, the immediate gauge of favorable course evaluations and facility displayed in student projects may not be reflected in integrated professional work over the long term. The professional outcome of graduates' influence on the profession, the foundation and expansion of scholarship, and the appearance of built projects that are informed by sound ecological science all need to be part of evaluating the success of regularizing ecological science within landscape design and planning curricula. In the final analysis, whether or not changes in the landscape wrought by designers and planners are ecologically constructive, and whether new graduates can credibly respond to the market for ecologically sound design, will determine the success of this curriculum shift.

Key Strategies

The discussion of the principles and realities of an interwoven ecological curriculum generated specific questions:

- What ensures the consistent and continuous interweaving of ecology into the studio and classroom?
- When is ecology introduced into the curriculum?
- What works in the actual teaching of ecology in the curriculum?
- Where does the infusion of ecology best occur in a curriculum?
- What are effective frameworks for introducing ecology to design and planning students?
- How does a department gain faculty expertise, and who is involved?
- Is research a component of the injection of ecology into design and planning education?

In this section, we review these questions and propose general strategies as a response. Many of the strategies described here are well established; others are experimental. Along with these strategies, we provide brief case studies from several schools participating in the Shire Conference. They exemplify how programs attempt to instill ecological thinking into their curricula. Our intent is not to be all-inclusive, but to establish a dialogue for future strategic planning to bring ecology into landscape design and planning curricula.

The context in which ecology and design interact and are presented at any given institution will vary according to the emphases and goals of a particular program. Recognizing this, not all of the strategies discussed below are possible for any given program. Some have urban applications; others stress student fieldwork, or concern a particular aspect of ecology such as ecological restoration or landscape ecology. Understanding these differences and crafting an ecology-design interaction accordingly is imperative to fulfilling any department's unique mission. Tied to the strategies, case studies from Arizona State University, University of California-Berkeley, University of Michigan, University of Oregon, Penn State University, and University of Wisconsin-Madison provide examples of departmental curricula that interweave the science of ecology.

What Ensures the Consistent and Continuous Interweaving of Ecology into the Studio and Classroom?

Clearly, ecology provides theories, concepts, and language that are important to the disciplines of landscape design and planning, and consequently to the education of designers and planners.

Strategies

To interweave it with design teaching, educators must conceive of ecology in its broadest context. Ecology can be integrated at all scales and in all phases of the design process, regardless of landscape context. Ecology can be explored as a scientific and as an aesthetic framework. Such frameworks help to engage students and faculty in discussions on theories and applications of ecology, as well as the role of ethics and environmentalism.

Ecological theories, concepts, and language become an automatic and central part of design thinking and the design process. As such, they begin to challenge the manner in which one approaches problem solving. This explores interrelationships among culture, art, and ecology—ingraining habits of critical and integrative thinking. Interweaving ecological theory, applications, and literacy into the curriculum requires a process that neither trivializes nor blindly promotes it as dogma or rhetoric. A major goal of such curricula is to convey to students that design and ecology are not mutually exclusive.

CASE STUDY: UNIVERSITY OF MICHIGAN

The University of Michigan's interwoven curriculum is strengthened by an institutional structure where landscape architecture is one of three concentration areas, rather than a discreet department within the School of Natural Resources and Environment (SNRE). This permeable structure, grounded in an ecological approach to problem solving, allows Master of Landscape Architecture (MLA) students to interact with faculty and students in landscape ecology, ecosystem management, aquatic ecosystems, and conservation biology. Students are required to take two courses from Resource Ecology faculty. Subsequently, MLA students typically enroll in a number of other ecology courses on an elective basis—for example, Forest Ecology, Wetlands Ecology, and/or Forest Hydrology. Conversely, non-MLA graduate students in SNRE typically take coursework from LA faculty, in courses such as Ecological Restoration, Planning in Rural Environments, Land-Use Planning, and Landscape Ecology.

When Is Ecology Introduced into the Curriculum?

Ecological science can be interwoven throughout the fabric of the design curriculum, or sewn in as distinct threads. It can be introduced at an early level or in more advanced coursework. It can be taught in stand-alone courses, or merged with design or analysis material. Which of these choices is best?

Strategies

Earlier is better than later. Many schools find it advantageous to link ecology and landscape design quite early in a student's education. Consequently, ecology becomes a central feature in design thinking, alongside aesthetic principles, social tenets, and other core bodies of knowledge. In addressing ecology early, students and faculty do not perceive or treat ecology as an add-on—a discretionary specialty that one may or may not pursue in the profession.

Often is better than occasionally; rarely is unacceptable. Student fear of, and reluctance to engage, ecology can be minimized through early and frequent interactions attuned to the students' level. Returning to ecological principles repeatedly throughout the program reinforces these ideas and their importance, especially through a variety of landscape contexts.

Deepen ecological understanding midway. Stand-alone ecology courses need to occur in the early to middle parts of the curriculum. Students at that middle stage have sufficient experience, both in the sciences and in design and planning, to explore the relevance of such courses to their studies. During the remaining time in their academic career, students can be encouraged to apply and deepen their ecological understanding.

CASE STUDY: UNIVERSITY OF WISCONSIN-MADISON

Ecology is integrated into the curriculum from year one, beginning in the program's introductory studios. The department requires a stand-alone ecology course and continues to weave ecology into design and planning studies as well as in special workshops (see Figure 14-1). Relevant BSLA and MSLA courses include basic ecology (external lecture), soils (external lecture), geography (external lecture/lab), Site Inventory and Analysis (studio, second year), Site Planning and Design (studio, second year), urban ecology (new lecture, third year), Plants and Environment (studio, third year), Open Space Design (studio, third year), Regional Analysis and Design (studio, fourth year), senior capstone project (studio, fourth year), Restoration Ecology (studio, fourth year/graduate), Field Course in Wisconsin Native Plant Communities (fourth year/graduate), Restoration and Management in Natural Systems (workshop, fourth year/graduate), Seminar in Restoration Ecology (graduate), and Theory of Landscape Change (graduate). Undergraduate students must also enroll in a three-course professional sequence of which one option is Conservation and Ecosystems.

FIGURE 14-1.
Reading the Door County landscape is an optional student/faculty field trip open to students at all class levels. University of Wisconsin-Madison (J. Harrington photo).

CASE STUDY: PENN STATE UNIVERSITY

Relevant BLA and MLA curricula include Site Ecology (lecture/field studies, second year); Ecosystems Biology (external lecture, second year); Regional Analysis and Design, including basic landscape ecology (studio/seminar, third year); sustainability in site design (design and implementation studios/lectures, both third year); regional ecosystems transect (field course, second year); two physical geographies (external lecture, timing varies); senior capstone project (studio/seminar, fifth year option); and the intercollege watershed stewardship option (19 of 44 MLA graduate credits).

CASE STUDY: UNIVERSITY OF CALIFORNIA-BERKELEY

Relevant MLA curricula include Ecological Analysis (second year), Ecological Factors in Urban Design (second year studio), Environmental

Geology for Planners (lecture/field), Hydrology for Planners (lecture/field), Introduction to California Landscapes (seminar/field), Vegetation Analysis and Management (lecture/studio), Urban Forest Planning and Management (lecture/studio), Restoration of Rivers and Streams (lecture/field/research), and The Sustainable Landscape (seminar/studio).

What Works in the Actual Teaching of Ecology in the Curriculum?

Devising methods to interweave ecology into the curriculum can stymie even dedicated faculty. Students need clear, substantive applications of ecological science in the curriculum.

Strategies

Approach design as hypothesis-testing. Landscape design, as a profession and as an undergraduate curriculum, does not often approach design in terms of testing cause-and effect relationships. According to several instructors, students seem to fear this approach, believing it runs counter to creativity. Exposing students to case study examples where professional designers and planners have used a hypothesis-testing approach—and have documented it—has been effective in counteracting student apprehensions.

Draw from the methodologies of science in the design curriculum. Explore the development of questions and hypotheses with students. Work with students to determine the necessary data and the means for collecting them. Assist students in analysis and application of data sets as part of problem solving. Ecological methods include monitoring and evaluation of ecosystem change or success in obtaining goals; such methods should be adopted into the design process and into landscape management over time. Such approaches can be phased into the curriculum until they become second nature. The general process model is familiar to the profession—from site inventory/analysis to postevaluative studies—but the specific context, objectives, and data collection methods for ecological studies are not typical to our curricula. Field courses are an excellent mechanism for covering data collection and understanding the relationship of data to questions posed and the analysis needed.

Differentiate between ecological design and naturalistic design. Environmentally sound design is often confused with a naturalistic style. A design with an informal or naturalistic style is not necessarily ecologically sound and, conversely, an ecologically sound design can take many forms. This distinction needs to be made clear and often in students' design education. Some instructors drive this point home by assigning ecologically appropriate design problems where more formal design styles must be used.

CASE STUDY: UNIVERSITY OF WISCONSIN-MADISON

Restoration Ecology and Restoration and Management of Native Ecosystems, two courses taken by seniors and graduate students, use an hypothesis-testing format under the terminology of "adaptive management." Teams are assigned projects and clients to determine best use, land cover, and management schemes for specific sites or regions. Teams develop objectives and determine data needs and data collection methods (see Figure 14-2). Most teams work with a series of test solutions and monitoring schemes using an if-then approach. Solutions may be implemented with monitoring thresholds, and strategies are developed to modify solutions if the targeted goals are not met. Objectives are allowed to be fluid as long as broader overall goals are being met, such as those involving improved water quality, sediment reduction, vegetation structure, and compatible site use. Not only do these courses expose students to hypothesis testing and mechanisms for design evaluation and modification, but they also provide for interactions between graduate students (many with more science than design background) and undergraduates (who are much more design-oriented).

CASE STUDY: UNIVERSITY OF CALIFORNIA-BERKELEY

As part of the core design studio sequence, all MLA students must take the studio "Ecological Factors in Urban Design" while simultaneously taking an ecological analysis lecture and field laboratory course. These courses share instructors, as the two environmental scientists who teach the ecological analysis course also co-teach the studio course with design faculty and fully participate in studio design critiques. The field data gathering and analysis techniques developed in the ecological analysis course concern the design project of the studio. Design projects must quantify the positive ecological effects of their designs, based on previous data gathering. The connection between ecological science and site design is not hoped for, but a requirement of the studio.

FIGURE 14-2.
Graduate and undergraduate students in Landscape Architecture and Environmental Studies learn to collect data to address questions on the relationship between vegetation structure and patterns and microtopography of a wet mesic prairie. University of Wisconsin-Madison (photo by J. Harrington).

Where Does the Infusion of Ecology Best Occur in the Curriculum?

The course context in which students are exposed to ecological ideas can profoundly expand the application of ecological science in design and planning projects. Problem solving and place making in the traditional studio format can consider ecology and ecosystems as well as the social sciences and humanities. Ecology can be linked to form-giving both descriptively and analytically. This links ecology in quite a different way from separate coursework that does not involve design.

Strategies

Use real places for course projects and immerse students in the site. For ecological material to be understood, instructors must link ecology and design together explicitly in real places. For studio projects, real sites grounded in tangible landscape systems are indispensable. Courses that emphasize the site's ecology must allow for intensive field time, including exposure to ecological field techniques. Teaching observation skills in the field gives students opportunities to record diverse types of data. Although difficult and time consuming, assisting students to integrate ecological data with social and behavioral data results in an environmentally and socially balanced—and more convincing—project.

Conduct special-topic seminars. Seminars provide opportunities for students to increase awareness of literature and issues in ecology that pertain to their interests and future goals. Seminars often allow for guest lecturers, literature reviews, and discussions on specific topics rather than general broad-based concepts. Students are then provided the opportunity to lead the exploration of a concept and its applicability to design.

Expose students to successful case studies in which designers and ecologists have collaborated. Lectures, seminars, and field courses can use case study methods to illustrate problem solving that integrates the science of ecology and ecosystems into design.

CASE STUDY: UNIVERSITY OF CALIFORNIA-BERKELEY

Courses that now fulfill breadth requirements, such as Restoration of Rivers and Streams and Vegetation Analysis and Management, began as topical seminars that over time built credibility with students and fellow faculty. These courses typically provide a series of lectures and require students to either develop their own project or conduct original case study research on projects that integrate design and environmental science.

CASE STUDY: UNIVERSITY OF WISCONSIN-MADISON

Graduate students can enroll in an interactive seminar for Restoration Ecology. Topics change yearly but focus on the exploration of ecosystem processes and their application to planning and management within natural systems. Guest lecturers (academics or practitioners) who are experts in their fields participate in exploring developing issues and engaging in discussion with students. Students are responsible for some presentations and leading discussions. Wisconsin also requires each fourth-year undergraduate to participate in a semester capstone seminar followed by a semester capstone studio. Each student is set up with a complex project that involves multiple clients: communities, industry, and public agencies. One requirement of that project is the consideration and incorporation of ecological concepts and environmental protection measures. Students are required to explore and report on theory and case studies relevant to their situation and in support of their solutions.

CASE STUDY: PENN STATE UNIVERSITY

Penn State's undergraduate studio sequence is "shadowed" by one-credit seminars on design theory and precedent. Initially conceived of primarily in aesthetic and historical terms, the seminars are increasingly making room for ecological content that directly informs studio projects. The seminars and colloquia that parallel the fifth-year ecological design capstone studio (usually one section) are focused on ecology and its application in design and planning.

Make use of alternative course formats. Special topic, continuing education, and distance education courses are all useful in testing new courses and bolstering faculty confidence in the challenge of introducing ecological material into the curriculum.

Field courses in the summer may provide similar benefits by immersing students in the landscape—for example, comparing restored and natural systems. Team instruction involving faculty from several disciplines in ecology can be highly effective.

CASE STUDY: UNIVERSITY OF OREGON

The University of Oregon program holds a one-week optional field course, Reading the Landscape of the Oregon High Cascades, prior to the start of the fall semester. Graduate students in the Landscape Ecology concentration spend seven days backpacking and exploring forest dynamics, soils, geology, hydrology, vegetation, and climate. In addition, students meet with professionals on-site to discuss wilderness perceptions, recreational impacts, and landscape history. Prior to the trip, students select a topic in which they are to become the "resident expert" throughout the course. Several meetings that provide background and prepare students are held during the semester and summer prior to entering the field.

FIGURE 14-3.
Wetland specialist Andy Cole, assistant professor of Ecology and Landscape Architecture, discusses the function of an automated test well at a reference wetland. The optional summer course for professionals and students links hydrogeomorphic processes with appropriate wetland interventions. Penn State University (photo by K. Tamminga).

CASE STUDY: PENN STATE UNIVERSITY

Principles of Wetland Design is a summer continuing education course that is team-taught by a wetland ecologist and a landscape architect, both faculty in the Department of Landscape Architecture. It emphasizes principles of wetland ecology and hydrology, with field excursions to compare natural and created wetlands in the Ridge and Valley physiographic region. Current field techniques in site ecology and hydrology are also demonstrated (see Figure 14-3). The confidence generated by the interweaving of science and design perspectives since 1996 has helped inspired similar initiatives in the regular undergraduate site planning, regional design, and capstone studios, as well as the new graduate Watershed Stewardship option.

CASE STUDY: UNIVERSITY OF WISCONSIN-MADISON

Field Studies of Wisconsin Native Plant Communities is taught as an intensive three-week field course during the early summer. Jointly taught by faculty trained in plant ecology and landscape design and planning, the course meets daily. The first two weeks explore southern Wisconsin grassland and wetland communities; the third week occurs

FIGURE 14-4.
Students collect data to determine the vegetation structure of a floodplain forest during LA 667: Field Studies of Wisconsin Native Plant Communities. University of Wisconsin-Madison (photo by J. Harrington).

in northern Wisconsin, immersing students into the ecology of bogs, swamps, and coniferous and hardwood forests. The course concentrates on the ecology and aesthetics of state natural areas representative of a variety of ecosystem types. Students learn sampling techniques (see Figure 14-4) for the specific vegetation structure of the day as well as observational techniques of the community's structure, functions, and aesthetics. Students visit one or two restoration sites for comparisons with community models. They then respond to a series of questions that help to translate their experiences and data to projects in which they may be involved in the future.

What Are Effective Frameworks for Introducing Ecology to Design and Planning Students?

Deciding what aspects of ecology to introduce to design students is daunting. Should ecologists introduce food chains, population models, hydrology, wildlife management, plant community ecology, landscape ecology, or genetic studies? The list goes on. Perhaps each of us knows what should be discussed in terms of ecology or ecosystem structure and function, but there is an enormous range of what that might include. Are we talking about the breadth of ecology or selected areas within the discipline of ecology? (For one scenario on what constitutes the breadth, see Chapter 6).

Two pedagogical questions identify the horizons of ecological breadth and scope. First, instructors can ask this fundamental question: What are the lessons from ecology that explain linkages and connections of landscape patterns, and that tie energy flows to landscape function and biodiversity? Relatedly, ecological principles provide powerful criteria on which to test design models that predict change against current realities. The second question is, What are the cause-and-effect relationships inherent in landscape change?

Strategies

Show explicitly that ecological concepts apply at all scales and incorporate projects that demonstrate the shifting scales of design and planning. The best order to work on various scales responds to the focus and student cohort of a particular program. For some programs with strong natural resource and regional design foci, this might mean working from system to site. However, for site- or urban-oriented design students, working from smaller to larger scales may be more effective—from site to system. These students may start by examining postindustrial urban sites, urban stream corridors, and a continuum of urban impacts, thereby engaging students in places relevant to their own social background and interests. In other cases, to work from system to site and back again is appropriate, shifting scales throughout the curriculum and emphasizing multiscalar relationships.

CASE STUDY: UNIVERSITY OF CALIFORNIA-BERKELEY

At UC-Berkeley the ecological design studio met pedagogical success only when it moved the project site from large-scale urban-edge projects to urban restoration and infill projects. This reflected the predominate goals of students who hoped to work within urbanized contexts. The shift in project context focused on site and then moved outward to system, rather than from system to site in the classic McHargian manner. This was essential in engaging design students who understandably see site as the fundamental medium of landscape design. The crucial task for the instructor is to guide students outward to system and impel their understanding of system as a collection of sites over which they have influence.

Express landscape context and pattern variation over space. Students often perceive that ecology is relevant only to nonurbanized landscapes, or only to particular land uses—parks, conservation areas, or river corridors, for example. To dispel this myth, course content must illustrate the linkage of ecology and design in a variety of landscape contexts, in all land uses, from urban to pristine. Recognizing that landscapes vary over time, as well as in scale, can help students understand a site's pattern and function as part of its larger systems context. Including courses that present concepts and applications of landscape ecology is one mechanism to encourage such understanding.

CASE STUDY: UNIVERSITY OF MICHIGAN

Courses consider ecology across a variety of landscape types and patterns: brownfields to wilderness. The city of Detroit, with many brownfields, vast amounts of open and abandoned land, and diverse ethnic and historic neighborhoods, provides a vivid teaching laboratory. As Detroit rebuilds, there is an incredible opportunity to do ecological design and planning within an urban fabric; the MLA program makes use of, and contributes to, this opportunity in studio courses. In addition, the older and newer suburbs of the metropolitan area contain fascinating project sites, where issues of water use, open space planning (see Figure 14-5), and "greenfields" protection comprise MLA projects at the design/ecology interface.

FIGURE 14-5.
Students take notes in an old-field context, where ecologist Mat Heumann explains impacts and issues associated with human use in public open space systems. University of Michigan (photo by D. Erickson).

FIGURE 14-6.
An Allegheny National Forest ecologist and Penn State University students inspect a salamander pitfall trap during the Ecological Transect field course. (photo by K. Tamminga).

CASE STUDY: PENN STATE UNIVERSITY

Penn State's second-year Ecological Transect field course extends from New Jersey's barrier islands to Lake Erie's Presque Isle. It includes stops at a variety of ecologically contrasting sites, including a native vegetation "pocket park" in downtown Philadelphia and an old-growth forest on the Allegheny Plateau (see Figure 14-6). The fifth-year capstone studio option in ecological design and restoration periodically addresses postindustrial sites and brownfields in inner-city Pittsburgh. These and a variety of other courses through the curriculum help students experience the ubiquitous nature of ecology.

Express landscape change over time as well as in space. Each of us recognizes that landscapes are dynamic, yet few of us demonstrate that understanding in the final products of design studios. Design plans seldom recognize or illustrate the evolution of that landscape over time. The ability to recognize and predict potential changes in light patterns, vegetation structure, hydrology, and neighborhood context (stormwater runoff, site use and management) is vital to functioning ecological systems. Strategies to develop and portray design solutions that allow for change, and yet maintain site goals, can permeate the curriculum.

How Does a Department Gain Faculty Expertise, and Who Is Involved?

THE ROLE OF ECOLOGISTS.

Clearly, one way to bring ecology into the design curriculum is to bring ecologists into the studio. This sounds simple. However, there are a number of questions inherent in this seemingly obvious solution. What is the role of ecologists in design? How are they involved? What's in it for them? How is their interest in studio collaborations maintained, where the emphasis is often on design and their expertise may be misunderstood or casually treated over time? Conversely, design faculty may find it difficult to welcome non-designers and their methods into studios.

At some institutions, the infrastructure is not set up for easy interdisciplinary interaction among faculty. In response, some faculty members involve ecologists outside academia with their classes. As previously discussed, these other excellent arrangements are sometimes vulnerable or unstable based on personalities, workloads, goodwill, and retirements.

Involving an in-house faculty ecologist can increase continuity. Well over a dozen schools have landscape architects with strong ecological science expertise on staff, as well, or have strong relationships with ecologists in external departments.[6]

Ecologists come in a variety of "costumes" and disciplinary affiliations, similar to designers. Many individuals who can potentially enrich our design pedagogy don't often interact with the academy. Many do not necessarily call themselves ecologists. Most programs can benefit from experts who have field experience, particularly those who have engaged in a consulting role with designers and planners. Similarly, programs may introduce designers who have participated in consulting roles with ecologists and natural scientists.

Strategies

Include an ecologist on the department faculty or participate in joint-appointment hiring with another department.

CASE STUDY: UNIVERSITY OF WISCONSIN-MADISON

The department has a tenured plant ecologist on the faculty who has participated as a co-instructor in undergraduate courses covering site inventory and analysis, plants and design, open space design, and the capstone studio at the undergraduate level. At the graduate level, the plant ecologist teaches a workshop in vegetation management in natural systems and co-teaches restoration ecology with a faculty member trained in landscape architecture, and theories of landscape change with a geographer. The plant ecologist also participates in the advising of students and is heavily involved with graduate student research. One means that helped the department integrate ecology into the curriculum was to enlist the ecologist as an active member on the curriculum committee.

Develop strong associations with ecologists from area agencies; create adjunct positions with benefits to the holder.

CASE STUDY: UNIVERSITY OF MICHIGAN

The landscape architecture program has a roster of adjunct professors who work in both public and private practice and interact with design students in a number of ways. Many of these adjuncts provide specific information in ecological design and planning, in direct ways for students (see Figure 14-7). For instance, one adjunct faculty is a wetlands ecologist who regularly lectures in landscape architecture courses, participates in critiques, provides internships for students, and advises opus projects. In another case, an adjunct faculty is a practicing landscape architect who specializes in ecologically sound stormwater management and design. Her involvement takes a number of forms, including design critiques. She is a great example of a "bridger"; her work is grounded in ecological processes and experimentation, while at the same time stressing urban forms and creativity.

Heighten opportunities for interactions between students and ecologists.

CASE STUDY: UNIVERSITY OF WISCONSIN-MADISON

UW-Madison's restoration ecology course focuses on reclaiming and recreating natural systems and processes in disturbed landscapes. Projects involve clients such as The Nature Conservancy, U.S. Fish and Wildlife Service, U.S. Army, and the International Crane Foundation, as well as more urban-oriented clients, including the UW-Arboretum, the UW-Campus Natural Areas, Madison parks, and schools. Most projects serve the above clients and are often implemented or adopted into planning strategies. Student enrollment comes from several majors

FIGURE 14-7.
Ecologist Mat Heumann introduces University of Michigan students to an interpretive nature trail in a Washtenaw County park (D. Erickson photo).

besides landscape architecture, including land resources, water resources, forest ecology, and conservation biology. Under this guise, faculty from the Departments of Botany and Civil Engineering (hydrologists), professional designers, ecologists, and students interact frequently and with depth.

Bring multiple ecology-related disciplines into the design studio. Landscape ecology, restoration ecology, plant and community ecology, wildlife ecology, hydrologic modeling, soils ecology, conservation biology, geomorphology, physical geography, and disciplines from other natural sciences have relevance to planning and design curricula. Programs must continue to embrace design as a multidisciplinary and interdisciplinary experience involving experts in expanding fields to the greatest extent possible.

CASE STUDY: UNIVERSITY OF CALIFORNIA-BERKELEY

Environmental scientists participate not only in final reviews but also in weekly desk reviews. They require design students to quantify rather than speculate on the positive ecological effects of their designs. This also requires design students to be able to convincingly complete hydrologic calculation and vegetation assessment. The purpose behind this is not to make scientists of designers but for designers to be able to use and critique environmental data effectively. Conversely, design faculty are invited to review the project and case study presentations in the applied environmental science courses, and are expected to bring their concerns, especially cultural and aesthetic concerns, to the discussion.

THE ROLE OF DESIGNERS AND PLANNERS.

What is the responsibility and role of the professor in facilitating ecological thought and practice in planning and design education? Given that any one instructor may not have both depth and breadth in the ecological sciences, this inherently means reaching out to others, as we have discussed in previous sections.

We view one role of the design professor in the design-ecology integration as that of a choreographer—positioning a range of teachers and learners to interact in the studio and in the field. The design faculty member maintains the critical connection to design throughout, while other professionals help deepen the scientific understanding. One of the challenges to this approach has to do with buy-in across design and planning programs. One's colleagues need to accept and embrace this heightened interweaving of ecology, and the extra resources that it entails. In some cases, this may seem an insurmountable challenge.

Strategies

Encourage participation of design/planning-trained faculty directly in students' ecology education. Whether ecology is taught to design and planning students as a traditional course with a lab, in the field, or in the studio, the design faculty should find a way to participate in the class. Depending on one's training, this might happen through participation in lectures, designing and leading lab exercises, or sitting in class as a student.

CASE STUDY: ARIZONA STATE UNIVERSITY

Frequently, a landscape architect and planner teach the introductory graduate-level environmental planning studio. The students come from diverse backgrounds, including the environmental design arts and sciences as well as the social sciences, humanities, engineering, law, and business. The instructors have found it is useful to immerse the students in real places. Often these places are parts of "opus" projects that are the focus of actual planning discussions and faculty research. In recent years, the watershed has been the level of inquiry for several projects. For example, professors at ASU have been studying the upper Gila River basin (in Arizona and New Mexico) for the Arizona Department of Environmental Quality and the upper San Pedro basin (in Arizona and Sonora, Mexico) for the U.S. Environmental Protection Agency. The student reports of the upper Gila and San Pedro basins have influenced planning and decision making in both these large, mostly rural, regions.

Encourage joint instructional responsibility for courses with both a design and an ecology component.

CASE STUDY: PENN STATE UNIVERSITY

The Landscape Architecture Department and the School of Forest Resources jointly offer the Watershed Stewardship option. Ten to fifteen graduate MLA and MS students co-mingle in a program of breadth and depth courses in natural and social sciences that culminates in an integrated second-year practicum based at the Center for Watershed Stewardship. A nonprofit conservancy usually sponsors this Keystone Project. Watershed assessment and design are blended during the project, including collaborative stream and riparian assessments in the field (see Figure 14-8). The two Center directors hold cross-appointments, assisted by two Faculty Fellows representing environmental planning and science.

Promote advanced education of landscape design and planning faculty in the ecological sciences. Academic-track landscape architects can pursue advanced degrees in an ecology-related discipline. For example, one member of the faculty at the University of Wisconsin has an undergraduate degree in landscape architecture and a Ph.D. in forest ecology with an emphasis in landscape ecology. The University of Michigan's Ph.D. program in landscape architecture attracts landscape architects who want to conduct scholarly research at the interface of design, ecology, and the social sciences.

STUDENT ROLES AND INTERACTIONS.

Students need to accept that ecological knowledge is as much a part of design as aesthetics and social science. They need to recognize through

FIGURE 14-8.
Faculty biologist Lysle Sherwin and an MLA student conduct D-net sampling for stream macroinvertebrates. Graduate Watershed Stewardship option, Penn State University (photo by K. Tamminga).

first-hand experience that the incorporation of ecological science (landscape ecology, restoration ecology, etc.) in design need not impinge upon creativity, but rather can serve to expand and enhance it. Students may learn this best from each other—particularly by working with students from diverse design and ecology backgrounds. The student-teacher relationship is, of course, critical; the potential relationships between students and mentors and advisors with experience in ecological science can have great benefits.

Strategies

Link undergraduate and graduate learning in ecology. Invite graduate students to help make ecology more relevant to undergraduates.

Make opus projects interdisciplinary and "real." To the extent possible, create team projects that use real places, real clients, and real problems. Create teams that incorporate design and planning students with other students, for instance, in resource management or in aquatic ecology.

Combine design and planning students and ecology students in coursework. For instance, landscape design and planning students may participate in an ecological analysis course with ecology students, using design problem sets. Arizona State University, University of Oregon, University of Wisconsin-Madison, Penn State University, University of Michigan, and Utah State University each have in-house courses that are advertised to and attract students from disciplines in ecology.

CASE STUDY: UNIVERSITY OF MICHIGAN

The opus experience for MLA students at the University of Michigan is an interdisciplinary problem-solving master's project. Graduate students interact with Resource Ecology and Management students, as well as Resource Policy and Behavior students, over a one-and-one-half-year period, starting midway through the second year of the three-year curriculum. These four- to eight-member projects integrate graduate students from across the School of Natural Resources and Environment to work with outside "clients" on various landscapes and resource issues. Outside clients help advise, with one or more faculty advisors. Students typically work with federal and state agencies, local jurisdictions, nonprofit groups, neighborhood organizations, or other project sponsors.

CASE STUDY: PENN STATE UNIVERSITY

Institutional limitations often demand alternative and opportunistic means to ensuring student-ecologist interactions. At Penn State, this consists of several tactics: appointment of an adjunct ecologist (now an assistant professor), periodic forging of links between field-based ecologists and students on real projects, inviting visiting ecologists as Bracken Lecturers, and arranging for summer internships in ecologically oriented research centers and consulting firms. Even student advising becomes important, for instance, in guiding students to take the Geography Department's landscape ecology course or the School of Forestry's conservation biology course as electives. Recently established centers are providing a new venue for interdisciplinary learning, particularly through joint appointments to the Center for Watershed Stewardship.

CASE STUDY: UNIVERSITY OF CALIFORNIA-BERKELEY

The core design studio, Ecological Factors in Urban Design, is required for both MLA design (second year) and MS environmental planning students (first year). They must work in integrated groups in data gathering and analysis and at the master plan level of project development. Then each student, whether planner or designer, must present a site design that constitutes part of the master plan. Inevitably, students learn to appreciate each other's strength and expertise (generally, the environmental planning students have extensive environmental science backgrounds). This ecological design course usually serves to introduce design students to the importance of environmental science and most go on to take the advanced specialized applied geology, hydrology, and vegetation science courses offered in the department.

Is Research a Component of the Injection of Ecology into Landscape Design and Planning Education?

Incorporate research at the undergraduate level.

CASE STUDY: PENN STATE UNIVERSITY

Penn State's fifth-year capstone studio encourages students to engage in projects that incorporate ecological science. The ecological design and planning option requires submission of a study proposal that includes a literature review of ecological science journals, field analyses (site or system scale), a compilation of ecological principles and techniques, peer review sessions arranged in "guilds" reflecting student project contexts, and interaction with external ecologist-mentors (agency, academic, or practitioner). For several students this course has served as a stepping-stone to graduate studies and practice in ecological planning and design. Relatedly, Penn State's Inter-College Graduate Program in Ecology provides a forum for interdisciplinary research and dialogue. Since the election of several landscape architecture faculty to the program, a number of useful collaborations between scientist and designer have taken place. A "trickle-down" effect has been felt in the undergraduate curriculum in terms of problem statement content and the presence of ecologists as regular studio guests.

Incorporate ecology into research at the graduate level.

CASE STUDY: UNIVERSITY OF WISCONSIN-MADISON

The graduate program requires a research thesis in several areas of concentration, including restoration ecology and ecological design. Students are required to develop and defend questions and hypotheses, propose and carry out methodologies, and defend results to a committee composed of two departmental faculty and at least one faculty member external to the department. Students also present their study in a colloquium that is open to all students. Students often have adjunct committee members from public and private agencies such as the Wisconsin Department of Natural Resources and The Nature Conservancy.

CASE STUDY: UNIVERSITY OF CALIFORNIA-BERKELEY

Since the introduction of the graduate ecological design studio, many students have gone on to complete theses explicitly integrating design and environmental science with integrated committee members. These thesis projects all concern the ecological process of urban sites. For example, constructed wetlands for wastewater treatment, urban creek restoration and design, urban stormwater treatment in street redesign, integrating ecological design in urban parks, and phytoremediation of an urban industrial site. Although these projects obviously require the development of valid science-based strategies, they are executed and critiqued as landscape projects as well, with the full complement of design concerns. Succinctly, they must be aesthetic and ecologically sound.

CASE STUDY: UNIVERSITY OF MICHIGAN

Faculty members address ecological questions in their own research. Graduate seminars are an excellent medium to convey this work with students, especially when these seminars include not only MLA students but also those from other fields. The benefits go both ways, in that faculty learn a tremendous amount from offering a research-based seminar of this type. Faculty research is also conveyed through short studio modules that deal with cutting-edge issues in design and planning. Again, ecology and resource policy students are invited to enroll in these courses. Typically, the interaction of LA and non-LA graduate students is beneficial and dynamic.

Conclusion: Advancing the Vision

Which direction will lead to a curriculum interwoven with ecological science? What is the right degree of emphasis to give to curriculum change? We have described how changes in curricula emphases and relationships are beginning to occur where they should—at the departmental and university levels—accommodating institutional and disciplinary differences that could never be captured in any attempt at a single, prescriptive continent-wide agenda. Each program must make its own choices, tempered by specific contexts.

Evolving an educational metaphor that is more ecological than Cartesian—where our relationships with our colleagues in ecology are immediate, seeking integrated, creative solutions—is not easy. Yet the state and prospects of environments all around us impel action, not passivity.

The only responsible posture is that of academic leadership and vision resulting in intelligent and inclusive curriculum change, the seminal means to achieve informed, inspired, and relevant landscape planning and design. Let us intensify our conviction to continue the interdisciplinary discourse on this matter, resolving destructive tensions and highlighting commonalities.

Citations

Boyer, E. L. 1997. Selected speeches. The Carnegie Foundation for the Advancement of Teaching. Princeton University, Princeton, New Jersey, USA.

Boyer, E. L., and Lee D. Mitgang. 1996. Building community: a new future for architectural education and practice. The Carnegie Foundation for the Advancement of Teaching. Princeton University, Princeton, New Jersey, USA.

Eckhardt, C. D. 1978. Interdisciplinary programs and administrative structure: problems and prospects for the 1980s. Center for the Study of Higher Education, Pennsylvania State University, University Park, Pennsylvania, USA.

Luymes, D., D. Nadenicek, and K. Tamminga. 1995. Across the Great Divide: landscape architecture, ecology and the city. Pages 187–196 in K. L. Niles, editor. Proceedings of the 1995 conference of the American Society of Landscape Architects. ASLA, Washington, D.C., USA.

Nadenicek, D., and C. M. Hastings. 2000. Pages 131–161 in Environmental rhetoric, environmental sophism: the words and work of landscape architecture. Environmentalism in landscape architecture, volume 22 of Dumbarton Oaks history of landscape architecture colloquium series. Dumbarton Oaks, Washington, D.C., USA.

Orr, D. W. 1992. Ecological literacy: education and the transition to a postmodern world. State University of New York Press, Albany, New York, USA.

Pelikan, J. 1992. The idea of the university: a reexamination. Yale University Press, New Haven, Connecticut, USA.

Schneekloth, L., and R. Shibley. 1995. Placemaking: the art and practice of building communities. John Wiley and Sons, New York, New York, USA.

Notes

1. We agree with Carl Steinitz (Chapter 10, p. 232): "Regardless of whether design is directed toward intentional change or intentional conservation, it has the pri-

mary social objective of changing peoples' lives by changing their environment and its processes, including its ecological processes."

2. For a reading on the tendency to invoke "ecology" in the promotion of planning and design schemes, see Nadenicek and Hastings (2000).

3. The tradition of inquiry into the mission and function of the Western university is a venerable one and of value in tracing our places as academic professions; refer, for instance, to the work of John Henry Newman (1840s–1870s), Alfred N. Whitehead (1920s–1940s), William C. DeVane (1940s–1960s), Robert M. Hutchins (1950s–1960s), Jaroslav Pelikan (1960s–1990s), and Ernest L. Boyer (1970s–1990s).

4. This is reminiscent of more traditional foibles common to design schools blindly striving for interdisciplinarity, such as the "teaching" of architecture to landscape architecture students and vice versa. Long-allied and mature programs pay attention to the distinction between siphoning, dabbling, and respectful, collaborative exploration.

5. An arguably more rigorous but esoteric term is "interfusion"—the root "fusion" suggesting something more strongly bound than through weaving, which can unravel.

6. The University of Wisconsin-Madison, Rutgers University, Harvard University, University of Oregon, Penn State University, University of Washington, University of Michigan, and University of Minnesota, among others, have ecologists on their faculty.

CHAPTER 15

Integrating Ecology "across" the Curriculum of Landscape Architecture

Jack Ahern, Robert France, Michael Hough, Jon Burley,
Wood Turner, Stephan Schmidt, David Hulse,
Julia Badenhope, and Grant Jones

As landscape architecture continues to evolve from a profession to a discipline, lively debates engage numerous issues, including the role and importance of research in the field of design (Riley 1990; Chenoweth 1992; Lloyd and Scott 1995; Selman 1998). Most of the debated issues need to be considered in the context of landscape design and planning pedagogy (Rodiek 1998, Zube 1998). In particular, some have expressed concern that if the bias of design as a largely site-based, problem-solving activity in landscape education persists, then "the wider intellectual development of the discipline may be hindered" (Thwaites 1998, p. 197). Through education then, "the landscape academy needs to continue to work vigorously to develop its research sensibilities, cultures, traditions and agendas" (Benson 1998, p. 203). This chapter explores concepts and models for the integration of ecological thinking and learning, according to an integrated "across the curriculum" pedagogy for landscape architecture.

The Dimensions of Ecology

"Ecology" in the profession of landscape architecture and planning cannot be understood solely as meaning the relationship between nonhuman life-forms and their environment. The term "ecology" is traditionally used as shorthand for the sum of the biophysical forces that have shaped and continue to shape the physical world. Thus there are other dimensions to be recognized if we are to understand the key nature of ecology: that of process, integration, and humanity. In the following discussion, ecology is understood as an intellectual and professional endeavor that includes physical, biological, and cultural dimensions, and occurs across multiple spatial and temporal scales, in contexts ranging from urban to pristine.

Humans and their activities and impacts are explicitly included in this concept of ecology. This reflects Barry Commoner's adage that "everything is connected to everything else," and is implicit in viewing an ecosystem as a home rather than just a house (Christie et al. 1986). As Michael Hough (1995) asserts, understanding any local place requires an understanding of its larger context. Consider the example of an urban waterfront. To many, a waterfront is simply a narrow band of land next to a river, lakeshore, or sea. When understood in its larger context, however, a waterfront is understood to be hydrologically linked to an entire watershed by rivers, aquifers, water mains, and storm and sanitary sewers. What goes down the sewer in the local residential area has an effect on the watershed, including its rivers and lakes and waterfronts many hundreds of kilometers away. Human uses of the land—transportation, housing, industry, business, and recreation—tie waterfronts to the larger region. Understanding these broad contextual issues is fundamental to applying ecology in landscape planning and design.

Flaws in Existing Pedagogy

The landscape architecture studio setting has the potential to provide an environment in which students can explore the implications of a broad conception of ecology, drawing from numerous sources and promoting group participation among students from varied backgrounds. However, whereas there is much promise, there are also many flaws in much of the existing studio pedagogy. More often than not, there seems to be only a minimal influence in the studio from non-landscape-architecture disciplines. Currently, many studios fail to draw students from other disciplines and backgrounds, in particular from the sciences. This results in a feeling of otherworldliness about studio projects in which, for the most part, the only issues addressed are those involving design elements. Other important considerations (i.e., financial, regulatory, even scientific) are glossed over or deemed irrelevant. While this sort of focus is understandable, especially in introductory studios, it does seem somewhat limited and limiting. Further, such an isolationist perspective is dangerous as it supports the erroneous contention that design professionals can work by themselves on complex projects keeping "all the balls up in the air." In short, we need to instill the idea that landscape design and planning is best approached with feelings of humility, not hubris. Understanding the merits and challenges of transdisciplinary research and practice is the key.

The most common pedagogical approach in landscape architecture education in the United States is to develop parallel paths of studio and core-subject courses. This approach often leads to a balkanization of the curriculum, one in which ecology is marginalized or trivialized. In the extreme, ecological concerns can be misunderstood or ignored outright.

In response to these perceived flaws, we offer three broad pedagogical goals for a more complete and successful integration of ecology across the landscape architecture curriculum: Instill an informed "sense of place"; integrate ethics into the design curriculum; and learn to think and work in a transdisciplinary manner. These goals are discussed below, followed by ten teaching and learning models that address the goals in an integrated manner.

Goals for Integrating Ecology "across" the Curriculum

Instill an Informed "Sense of Place"

In 1994, a presidential memorandum on "environmentally and economically beneficial practices on federal landscaped grounds" was released and subsequently endorsed by the ASLA on the grounds that "implementation of the proposal provides an opportunity, by the example it sets, to recreate a 'sense of place' in the American landscape, instead of the *place-lessness* of so much of the urban fabric" (ASLA 1995, p. 1). How do we instill *place-rootedness* in a group of students, many of whom have traveled far to take our courses, and who will immediately leave the region upon graduation, perhaps never to return? Even students from the immediate bioregion may become dissociated from the geographic location in which they find themselves immersed for study: "Suffice it to say, that though bodily I have been a member of Harvard University, heart and soul I have been far away among the scenes of my boyhood, those hours that should have been devoted to study have been spent in scouring the woods and exploring the lakes and streams of my native village" (Thoreau 1837).

Objectives for Developing an Informed Sense of Place

- Be able to understand a place in its larger context of space and time through detailed inventory, assessment, and synthesis of information.
- Be able to recognize a place's intrinsic resources essential for self-sustenance,

potential resources for human use, and whether any of these are exceptional examples in comparison with other places.

- Be able to articulate impressions of the place in both technical and nontechnical language and through skillful use of words, images, and spatial information systems.
- Understand a place's robustness/sensitivity to change—that is, its *resilience* and where that place is on a trajectory of ecological change.
- Learn to experience the places that are about to change as a basis for deciding whether they should change at all, and if so, what kind of change is needed.
- Learn to design "in the field" without paper and ink or computers; to engage the media of the landscape intimately; to walk it, to smell it, and to know it.

Integrate Ethics into the Design Curriculum

Motivated by fear that "the future of the profession is at stake," the ASLA adopted a declaration on the environment and development. This action was an attempt to encourage landscape architecture to play a "key role in shaping an ecologically healthy and regenerative world in the 21st century" rather than degrading into "little more than a minor decorative art" (ASLA 1993a). Despite the assertion by Ian McHarg (among many others) that "the study of environmental ethics, with its roots in ecology, is absolutely crucial to landscape architecture, very few design education programs have incorporated environmental ethics into their curricula" (ASLA 1993b). A 1992 survey revealed that only three of forty-three programs had ever offered a full course on environmental ethics taught by a landscape architecture faculty member! This is not only extremely embarrassing, it is outright dangerous! Landscape architecture claims to promote wise stewardship of the land. Exercising informed judgments in this regard demands education that explicitly addresses ethics, or else the profession stands to jeopardize its stewardship role. For example, Mozingo (1997) begins her paper on the aesthetics of ecological design with the following alarming anecdote: "In a recent awards issue of *Landscape Architecture* magazine, a jury member made the statement: 'We award the projects that are really beautiful and a little irresponsible, but never those that are environmentally responsible but a little ugly'" (p. 46). In order to alter this trend, ethics needs to be addressed more overtly and explicitly in the landscape architecture curriculum.

Objectives for Integrating Ethics into the Design Curriculum

- Be able to make morally reasoned judgments from a defensible ethical platform through becoming familiar with the writings and actions of philosophers, practitioners, and educators.

- Be able to deal with competing value systems and understand the inherent conflicts with professional practice (e.g., client versus site), and understand that ultimately the place is the real client.
- Avoid the objectification of nature; rather, develop ethically informed values upon which decisions can be based and defended.
- Don't begin with a mandate to intervene; rather, start with questions as to the appropriate level of intervention in terms of considering the "null model" of zero development.

Learn to Think and Work in a Transdisciplinary Manner

Most landscape architecture faculty and students use terms such as "multidisciplinary" and "interdisciplinary" when discussing the collaboration between specialists in climate, hydrology, geomorphology, ecology, heritage, human and animal behavior, and economics in design and planning projects. Such terms may be perceived to be interchangeable. They imply a collection of experts, each having a role to play in the resolution of complex landscape problems, but who have little direct connection or dialogue with other members of the team, or knowledge of their areas of expertise. The term "transdisciplinary" implies a true collaboration, requiring a general working knowledge of the different areas of expertise, recognition of and respect for each contribution to the problem, and implicit understanding that dialogue is paramount. While this definition may be perceived as arcane hairsplitting, transdisciplinary collaboration means that everyone works together in every sense of the word. Since almost every project in real life requires a unified integration of disciplines, it becomes an essential component of landscape-based problem solving. When true transdisciplinarity is achieved, landscape planners and designers will become equal partners with scientists, having a role not only in the intelligent application of knowledge, but being integral to its discovery (Ahern 1999).

Supporting Objectives

- Recognize that the role of the landscape designer/planner often lies in coordinating and integrating the various components of a project, or that such an individual frequently acts as a specialist subconsultant.
- Understand what questions to ask of the different disciplines and where they fit into the overall team structure.
- Be able to function effectively in team projects either as the coordinator or as a specialist.

Selected Teaching and Learning Models

This section presents ten teaching and learning models, organized into place, ethics, and transdisciplinary approaches. Although these models are classified into three categories, they are meant to be understood within an integrated pedagogical model.

Place-Based Models

Ecology, restricted to being taught only in the classroom, is a far from perfect approach to fostering the understanding about how the natural world works (Orr 1992), which is an essential requisite for successful, environmentally conscious, design. One circumventing approach associated with teaching ecology for planners and designers is to utilize the environment as the classroom. This approach facilitates extended opportunities for analysis and thoughtful integration of ecological knowledge into planning and design applications, while minimizing the recurring need to become acquainted with and learn about yet another new site. We have identified three models that merit discussion concerning the use of the environment as the classroom.

Site field station.

Traditionally, the field station has been affiliated with academic studies in biology, natural resources, forestry, geology, recreation, and studio arts (Orlich et al. 1980). Nevertheless, the field station setting has also proven useful to other academic disciplines and professions such as engineering. Many higher education schools have field stations (a location not on the main campus and usually within special natural resource surroundings), but not all planning and design programs make use of this resource. Several landscape architectural programs, however, do require their students to attend the field station, taking specific courses integrated into the landscape architecture curriculum, some taught by an ecologist associated with the program (Burley 1998). These are very specific place-based settings where the faculty develop, refine, and build upon their knowledge of these environments to prepare focused, meaningful, and substantial exercises. The advantage of this focused approach is that the instructors can attempt to provide guidance and incorporate planning and design significance into the field exercises.

Several landscape architecture/planning programs have traditionally utilized their field stations by advising students with a strong ecology interest to attend courses held at these locations. This approach is very helpful for the mature student who is able to synthesize and integrate ecological exercises without much guidance. However, this approach can backfire for students who are unable to broadly consider the significance of the assignments. For example, field station courses in entomology, limnology, or wildlife ecology may stress sampling procedures and statistical analysis without much application to planning and design. For the young landscape architecture/planning student who wishes to become a "designer," exercises that emphasize population estimation techniques or memorization of more nomenclature may seem irrelevant. In contrast, this approach is highly flexible and allows graduate students who wish to have their theses focus on wetlands or some other natural resource project to learn more about specific ecologically related topics germane to their research without forcing all students to comply.

Site revisitation.

Feelings of land connectivity develop through deep and continued familiarity with one's surroundings, both abiotic and cultural (Chapter 11). One of the easiest ways of fostering this in design curricula is through repeated exposure to a particular site in courses and studios. At some institutions the very first course students take is a week-long field immersion into the ecological and cultural mosaic of their bioregion before the formal fall term begins. There the students engage in detailed site analyses of patterns in terrestrial vegetation and processes in riparian zones, followed by a glimpse of how these have been influenced since European colonization. During their second academic year, the students revisit this area for a core studio. This time the emphasis is on design on both regional and local scales, moving through a rigorous exercise in environmental impact assessment. Having previously developed some ecological understanding and environmental sensitivity for the area, the students are much less likely to impose irresponsible designs upon the landscape, thereby moving toward the "less is often best" aspiration for development.

Therefore, in contrast to optional studio experiences where a new site for a development project may be chosen every year, we believe that when conducting ecological planning and design studies, it is often beneficial to employ the same site each year, especially where field surveys and longitudinal information may be helpful. This approach facilitates understanding change, making predictions, and appreciating the biosphere as a continually changing environment.

Site historical perspective.

Important ecological insights can be gained from the study of past events that may have influenced a site (Steedman et al. 1996). It is important for students to realize that ecology is not solely limited to something "out there" in the wilderness, but is an important issue in urban settings as well (Chapter 11). Repeated field trips to a nearby site with a rich cultural heritage are important for instilling a temporal view of human-environment relations. These outings illuminate the cultural history of the site—for example, why the factory was first located there, the environmental degradation that occurred as a consequence of this and other industries, the changing technologies that can be traced in the various buildings that became sequentially redundant as new manufacturing processes were introduced, the industry's impact on the valley during operation, and the natural regeneration that has emerged after the site was abandoned—as well as its much longer geological history: examples of overburden removal to expose sedimentary deposits revealing different ice ages, changing climates, and flora and fauna that have appeared and vanished over tens of thousands of years.

Engaging in such a process allows students to recognize the relationship between the city and the industry that helped build it, the symbolic interdependence of nature and urbanism, and strategies for cultural heritage protection that focus on the evolutionary processes that have shaped both early industry and the site's geomorphology. It's a dynamic interaction that reinforces the realities of human and natural evolution, and it reflects a fundamental principle that landscapes, to have meaning, must be tied to their regional geography and urban context, vegetation and climate, political and social contexts, and local environment. Such landscapes then become, in the students' eyes, rooted to the notions of continuity and place.

Ethics-Based Models

Ecology is a life philosophy (Chapter 6), and the teaching of "deep ecology" (sensu Devall 1988) is as much about instilling a code of conduct, or ethics, as it is about simply explaining technical scientific knowledge (Chapter 11). There are at least three different approaches concerning how environmental ethics might be taught and become established in existing design curricula. These approaches differ whether either a professional ethicist or an ethically sensitive landscape architect teaches the course, whether the course should be an elective or mandatory, and finally as to whether there should actually be a separate ethics course or rather have ethical issues firmly integrated into all studios.

Within the department.

An in-house course has the advantage of being able to be tooled specifically to the needs of landscape designers and planners, but may be limited by the knowledge base accumulated by the faculty member teaching outside her/his particular field of study. The latter problem can be alleviated somewhat by selection of texts that specifically focus on issues and concerns of land ethics and land development (e.g., Beatley 1994; Freyfogle 1998).

Another option that combines aspects of both "place" and land ethics involves a comprehensive reading of classics and recent books pertaining to specific locations. From such a background, students can "springboard" into discussions about landscape, memory, and topophilia from their own personal backgrounds. By understanding and explaining what it is about the area they grew up in and now possibly cherish, they can begin to grasp the impact of their future development projects on someone else's childhood neighborhood.

A final approach is one based on case-based instruction wherein students are presented with a series of real-world, increasingly complex, and morally engaging examples that require reasoning through problem solving and constructing value-engaging scenarios selected and conceived to bring ethical issues into focus.

Infusing real-world opportunities into the curriculum.

Another approach with the potential to be both informative and fun, yet at the same time challenging, is to engage environmental advocates (e.g., Greenpeace, environmental lawyers, tribal elders, NGOs, The Nature Conservancy) for desk critiques, initial design reviews, and even final juries. In such a pedagogical context, the environmental advocate can review design work from a more ethical perspective, complementing a jury consisting of artistic, biological, and technical panelists. This approach ensures that ethics doesn't remain as a tangential or optional concern. Rather, its centrality to design will be clearly reinforced in the presence of other jurists.

Separate and outside the department.

Although the focus pertaining to landscape architecture/planning in such a course may be limited, a specialized instructor of environmental ethics may imbue a broader understanding of the issue through exposure to other disciplines' ethical views and perspectives.

Transdisciplinary-Based Models

Landscape architecture students bring a different perspective to environmental problem solving, with both descriptive and prescriptive dimensions. This perspective promises useful insight and reward (Ryder and Swoope 1997). There is recognition, therefore, that a tighter interlinking of ecology, culture, and design needs to be developed in our curricula (Chapter 9). We identified and examined the merits of four nontraditional approaches for fostering transdisciplinary relationships in design/planning curricula.

"Back to the lab."

There is a need to experiment with teaching programs in a problem-solving framework, best accomplished by opening them up and "airing them out"

(Apostol 1998). There is often a wide, sometimes alarmingly so, gap between ecology courses and studio projects. One way to bridge this gap is through a tight integration of courses with studios, each building upon the other. Another approach is to do away with the false compartmentalization to begin with and design a new type of course—a "research workshop" or "lab" course that offers elements of lecture series and studios in addition to both seminars and independent theses. To be successful, such a research workshop needs to address sites and ecological issues at multiple scales, for example, site, landscape, watershed, and region. Understanding how and when landscape architects fit into a team composed of engineers, environmental scientists, policy analysts, government regulators, economists, citizen activist groups, and the like, is crucial to their education.

In such a course, student projects could provide written, oral, and visual elements integrated toward formulating practical solutions to issues of local or regional development. Site visits and lectures by practicing professionals help to scope project definition, foster background research, explore varied design options, and focus project resolution. Most important, each selected problem necessitates the exciting (and no doubt sometimes frustrating) opportunity for landscape architecture students to participate in multidisciplinary groups involving students in civil engineering, hydrology, limnology, ecology, and environmental policy from other departments and even other local universities. In some cases, these other, non-landscape-architecture students will work on a project both before and after the landscape students do.

Spatial integration.

Issues of scale are paramount to understanding ecology as a mature discipline (Allen and Hoeskstra 1992). Likewise, there is a cardinal need to educate about the importance of considering off-site effects and causes when examining any development project (Chapter 5). Another teaching approach that addresses the environment as a classroom involves the "sliding scale" or "nested scale" series of studies that allows students to be exposed to planning and design issues across spatial scales. We prefer the model where students intimately know a somewhat small site first, possibly with several assignments associated with this first site. However, as the student moves through the curriculum, the smaller, somewhat intensely studied site is set within a larger context such as a habitat design project or a transportation project, and eventually a land-use regional planning project. Notwithstanding, changing scales from fine to broad may be only one type of multiscale experience for the student. Students can participate in projects carrying an ecological thread by studying as a team various scales at once, where some students are conducting design projects at the site level and other students may be preparing regional land-use conservation plans. Thereby, ecological knowledge can be shared as a team project. In addition, a series of projects may start at the regional level and then proceed until the student has completed a project at the site level. In this way the student would gain experience at design development and bring a sense of the relationships between the appropriate significance of various types of ecological information at different scales.

In such a strategy the instructor must weigh the benefits of many exercises versus a few lengthy exercises that engage a specific environment more deeply. For example, many planners and designers may prefer giving their students many projects and experiences (breadth). In opposition, many ecologists may prefer that their students pursue one or two projects at a highly intimate level (depth). There are merits for both approaches. Achieving a healthy balance between the two is important for developing a successful curriculum containing place-based ecological studies.

Inclusive-integrative.

This approach uses "alternate futures analysis and planning" and is oriented primarily at understanding the relationships among human activities and changes occurring in natural systems (Steinitz et al. 1996; Hulse et al. 1997; Schoonenboom 1995). A principal activity is the development of spatially explicit landscape analyses and plans for plausible future configurations of human uses of land and water. Using geographical information systems (GIS) and related tools, the efforts produce digital and paper representations showing the past, present, and potential future conditions of particular watersheds, landscapes, and river basins. These are used to identify trends over space and time in human occupancy, in the nature of land and water use, in the composition and structure of vegetation, in hydrology, in the quality and quantity of sensitive habitats and threatened species, and in other natural resources. From these trajectories of change, correlations can be developed and applied to resolving environmental problems and managing population growth.

Such situations often benefit from the involvement of representative citizen groups interested in exploring plausible options for the future of places they live in and care about. Based on a set of values and desired future conditions developed through working with the citizen groups, faculty and students produce a series of maps and landscape visualizations depicting a range of alternative ways of achieving the intended future conditions. These scenarios are then revised and refined in a process of consensus building. The final alternative future landscapes are evaluated by computerized models to analyze important biophysical and socioeconomic resources and processes such as water availability and quality, biodiversity, and agricultural productivity. Through these analyses, the planner/designer can determine the effects that the different alternatives may have on those resources and processes. This approach allows future consequences of present choices to better inform critical decisions and provides an integrative, replicable framework for combining descriptive and prescriptive modes of working.

Inclusive-catholic.

This approach provides for the selection of a public project that is both highly controversial and complex relative to the number of disciplines involved. This allows students to cast a critical eye on the issues and illustrates the numerous facets of a project that must be considered when evaluating their environmental, social, and economic impact on society. Topics and questions that could be examined include the following:

- Politics and the decision-making process. How were the decisions made, and who made them at the government level and in the private sector to proceed with this project? What were the political implications of job creation in a region of high unemployment, for example?
- Energy-related issues. What were the energy needs relative to the regional and provincial demands? What energy alternatives were considered? What natural resources could be tapped? What would be the environmental and economic implications?
- Economic issues. How was the project to be funded? What were the roles of the federal and state/provincial governments and the private sector? What were the implications of borrowing, debt, and taxation?
- Social issues. What are the social implications of the development on existing industries, agriculture, cottage industries, local communities, and native people?
- Environmental issues. What would be the consequences of the development on wildlife, fisheries, forestry, or mining and their dependent local industries?
- Scenic and design factors. How would the development alter the famed scenic and experiential quality of the area? What might be the alternative technical design approaches that might meet the same objectives while providing similar benefits in other ways?

These and many other factors can be discussed in such a course, not as separate issues, but as to how they all interrelate. Such a strategy allows students to understand that one avenue of action has implications on many other factors. Courses in this vein vary considerably depending on the issue, but in general, the terms and principles of interrelationships remain the same.

Summary and Conclusion

The teaching of ecology is best conceived as an integrated approach, taught not in isolated courses, but across the curriculum. The essence of this approach is contained in the following notions:

- At present, the ecological dimensions that need the attention of landscape architects have not been adequately identified or addressed.
- Shortcomings in the way ecology is taught in landscape architecture programs tend to be persistent problems that are difficult to solve within the typical landscape architecture curriculum.
- Efforts to work collaboratively across disciplines—particularly the design and scientific disciplines—demand new approaches that are more rigorous, more interactive, and transdisciplinary.

The goal of landscape architecture programs should be to facilitate—aggressively and comprehensively—place-based learning among students, focusing primarily on landscapes and ecosystems in close proximity to their institution. The intent of this type of learning should be to instill in students a dynamic working knowledge of the ecological processes in natural and urban systems characteristic of their particular geographical area, and to foster an understanding of the interconnectedness of those processes and systems. This place-based education should do the following:

- Connect students to the landscape where they are receiving their training.
- Provide students with tools for conceptualizing and understanding the future landscapes that they plan and design.
- Foster a stewardship-oriented land ethic, with a foundation in at least one required environmental ethics course.
- Serve as a tool for integrating the overall curriculum and establishing the landscape architect as a critical and effective collaborator (or even coordinator) in the use and communication of ecological knowledge.

There are opportunities within the current pedagogical framework of landscape architecture programs to move toward addressing such goals. Many of these opportunities lie in reconceiving the relationship between the discipline of landscape architecture and the fields of ecological science that influence landscape design and planning projects. One approach to redefining these relationships is to move away from an *inter*disciplinary approach toward a more *trans*disciplinary approach, whereby landscape architecture programs do not simply rely on the unidirectional infusion of ecological information and research into their studio projects by science professionals and faculty, but instead engage those in the ecological sciences in a dialogue about the continual interaction between design and ecology. In effect, landscape architecture programs should commit as much to teaching those in ecology about the discipline of landscape architecture as they should to facilitating a comprehensive, ecological understanding of "place" among their students.

Strategies for integrating a more effective approach to ecological education into the typical landscape architecture curriculum are at least threefold. They can be place-based, ethics-based, or transdisciplinary-based.

Place-based models assume that much ecological learning should occur outside of the traditional classroom. The following are three particular approaches for this type of training:

- *The use of a site field station:* Site field stations can be used as broadly adaptable programmatic tools to provide both fundamental ecological training for students with limited science background and extensive opportunities for advanced learning among those students with a deeper understanding of and interest in ecology.
- *The process of site revisitation:* The site revisitation approach is, in some ways, simply an extension of the site field station idea. The basic idea is that for landscape architecture students to get the most out of their ecological training, they should work on varied projects in the same or similar settings throughout a given academic year or even throughout their time in the program. This type of immersion can give students an opportunity to elevate their understanding of one particular site, as opposed to getting a more watered down exposure to ecological processes and issues by constantly changing sites.
- *The assessment of a site historical perspective:* The site historical perspective approach is intended to present students with opportunities to assess the processes of temporal landscape change both conceptually (through field visits) and substantively (through the use of technologies and various research methods). Comprehensive studies of the forces of historical change to landscapes can be powerful tools for increasing both the depth and the breadth of ecological knowledge and the perception that landscape architects in training will be able to offer as professionals.

Landscape architecture programs might pursue ethics-based models for increasing ecological understanding in one of at least three ways:

- *Within the department:* This provides the opportunity to tailor the ethics course to the specific needs and issues facing landscape architects.
- *Infusing real-world opportunities into the curriculum:* This provides the opportunity to expose students to a variety of philosophical and political perspectives relative to the ethical training they have received.
- *Separate and outside the department:* The opportunity exists for valued partnerships between the landscape architecture department and a variety of humanities departments throughout the university.

Finally, transdisciplinary relationships can be structured in a number of different ways to support increased ecological knowledge among landscape architects in training. Some of the approaches might include the following:

- *"Back to the lab":* A "research workshop" model whereby studio and non-studio courses—such as seminars and lectures—are better integrated so that all work done outside of the studio is ultimately geared toward informing the project-based work occurring within.
- *Spatial integration:* "Nested scale" design and planning studies, whereby the studio focuses on a singular site throughout the course but explores that site's ecological processes at a number of different scales at different stages in the course. All students could study different scales in progression, or different students could address different scales and be charged with their overall integration.
- *Inclusive-integrative:* A comprehensive assessment of the ecological and human forces of landscape change from the perspectives of diverse stakeholders in order to foster critical analysis and thinking.
- *Inclusive-catholic:* Somewhat similar to the inclusive-integrative approach—that ecologically oriented studio projects be selected with controversy in mind so that landscape architecture students are pressed to understand in more depth the range of issues that influence the way development decisions get made that have both beneficial and negative effects on landscapes. It assumes that landscape architecture students are not trained to have only a cursory understanding of ecological processes in a vacuum, shielded from other nonscientific pressures or disturbances.

Citations

Ahern, J. 1999. Integration of landscape ecology and landscape design: an evolutionary perspective. Pages 119–123 in Issues in landscape ecology. J. A. Weins and M. R. Moss, editors. International Association for Landscape Ecology, University of Guelph, Canada.

Allen, T. F. H. and, T. W. Hoeskstra. 1992. Toward a unified ecology. Columbia Univ. Press. New York, New York, USA.

Apostol, D. 1998. An open letter to meeting participants. Shire Conference, Unpublished.

ASLA. 1993a. Taking up the challenge. Land 35: 5

———. 1993b. Environmental ethics: elective only? Land 35: 2.

———. 1995. Raising the standard: VP Gore's national historic performance review addresses landscape. Land 37: 1.

Beatley, T. 1994. Ethical land use, principles of policy and planning. Island Press. Washington, D.C., USA.

Benson, J. F. 1998. On research, scholarship and design in landscape architecture. Landscape Research 23: 198–204.

Burley, J. 1998. Determining when contents of space make a difference: examining ecological landscape treatments: a biostation experience—training tomorrow's landscape scientists. Unpubl. manuscript. Michigan State University, East Lansing, Michigan, USA.

Chenoweth, R. 1992. Research: hype and reality. Landscape Architecture 82:47–48.

Christie, W. J., M. Becker, J. W. Cowden, and J. R. Vallentyne. 1986. Managing the Great Lakes Basin as a home. Journal of Great Lakes Research 12:2–17.

Devall, B. 1988. Simple in means, rich in ends: practicing deep ecology. Gibbs Smith Publishing. Salt Lake City, Utah, USA.

Freyfogle, E. T. 1998. Bounded people, boundless lands: envisioning a new land ethic. Island Press. Washington, D.C., USA.

Hough, M. 1995. Cities and natural process. Routledge Press, London, UK.

Hulse, D., L. Goorjian, D. Richey, M. Flaxman, C. Hummon, D. White, K. Freemark, J. Eilers, J. Bernert, K. Vache, J. Kaytes, and D. Diethelm. 1997. Possible futures for the Muddy Creek Watershed, Benton County, Oregon. Institute for a Sustainable Environment. University of Oregon, Eugene, Oregon, USA.

Lloyd, P., and P. Scott. 1995. Difference in similarity: interpreting the architectural design process. Environment and Planning B: Planning and Design 22:383–406.

Mozingo, L. A. 1997. The aesthetics of ecological design: seeing science as culture. Landscape Journal 16: 46–59.

Orlich, D. C., and others. 1980. Teaching strategies: a guide to better instruction. D.C. Heath and Co., Lexington, Massachusetts, USA.

Orr, D. W. 1992. Ecological literacy: education and the transition to a post-modern world. State University of New York Press. New York, New York, USA.

Riley, R. 1990. Editorial commentary: some thoughts on scholarship and publication. Landscape Journal 9:47–50.

Rodiek, J. E. 1998. Special issue: landscape architecture research and education. Landscape and Urban Planning 42:73–75.

Ryder, B. A., and K. S. F. Swoope. 1997. Learning about riparian rehabilitation: assessing natural resource and landscape architecture student teams. Journal of Natural Resources Life Science Education 26:115–119.

Schoonenboom, I. J. 1995. Overview and state of the art of scenario studies for the rural environment. Pages 15–24 in Scenario studies for the rural environment, Proceedings of the symposium Scenario Studies for the Rural Environment, Wageningen, The Netherlands, September 1994, J. Schoute, P. A. Finke, F. R. Veenenklaas and H. P. Wolfert, editors. Kluwer Academic Publishers, Dordrecht, The Netherlands.

Selman, P. 1998. Landscape design as research: an emerging debate. Landscape Research 23:195–196.

Steedman, R. J., and others. 1996. Use of retrospective information for aquatic habitat conservation and restoration. Canadian Journal of Fisheries and Aquatic. Sciences 53 (Suppl. 1):415–423.

Steinitz, C., M. Binford, P. Cote, T. Edwards Jr., S. Ervin, R. T .T. Forman, C. Johnson, R. Kiester, D. Mouat, D. Olson, A. Shearer, R. Toth, and R. Wills. 1996. Biodiversity and landscape planning: alternative futures for the region of Camp

Pendleton, California. Harvard University Graduate School of Design. Cambridge, Massachusetts, USA.

Thoreau, H. D. 1837. Harvard classbook.

Thwaites, K. 1998. Landscape design is research: an exploration. Landscape Research 23:196–198.

Zube, E. H. 1998. The evolution of a profession. Landscape and Urban Planning 42:75–81.

CHAPTER 16

Building Ecological Understandings in Design Studio: A Repertoire for a Well-Crafted Learning Experience

Kathy Poole, Susan Galatowitsch, Robert Grese, Douglas Johnston, J. Timothy Keller, David Richey, Lee R. Skabelund, Carl Steinitz, and Joan Woodward

Situating Ecology within Studio

Ecology is the scientific study of the interrelationships among organisms and between organisms, and between them and their living and nonliving environments.

Thus, by analogy, design studio is a unique kind of "ecosystem." It is a discrete entity made of living parts (students and instructors) and nonliving parts (drawings tables, computers, reams of paper, and scores of physical models), all of which interact to form a system in (it is hoped) some sort of dynamic equilibrium. It does not cycle nutrients, but studio repeatedly cycles ideas and forms, constantly reworking them into new processes and forms. And the flow of energy makes studio an extraordinarily productive learning environment—and a model that is distinct within the academy.

Studio as Learning Environment

Studio is similar to a science or engineering lab that accompanies a lecture course because it is time-intense and an effort-intense work session that allows students to explore issues and phenomena in depth. Yet, studio is distinct in the comprehensiveness of subjects that it addresses, a range that includes many environmental sciences, engineering, sociology, history, architecture, geography, urban studies, anthropology, and art. And the study of these subjects is not partitioned. In fact, the primary objective of a design is to arrive at a superlative, unique, creative *synthesis* of the many subjects—to sieve the significant relationships from the peripheral and to transform them into something that is truly wonderful in the world.

What is most distinctive about design studio is that students produce ideas and drawings and three-dimensional models (in miniature) about how the world *should be*. While their designs are, like scientists' experiments, *observations* about the world and, like many disciplines, *hypotheses* about how the world might operate, designs project possible *realities*. They are full of ideas (like all disciplines produce) but also more than that. They are physical, tangible built landscapes that "work" (like engineering) but also more than that. They are visceral, aesthetic *experiences*—places—that are full of meaning on many levels for the people who experience them.

All of this makes design studio a unique model as a learning environment, in this case an environment for building ecological understandings. The primary role of studio is not one of introducing ecological information. Instead studio is a unique synthetic environment for demonstrating to students how to apply those concepts, for teaching them how to critique ecological knowledge, and for helping them communicate information clearly to themselves and their clients. It is a place for students to make the relationships between ecological content and all of the other cultural, political, social, personal, and experiential content that coalesce to form wonderful projects.[1]

The Place of Ecology

Ecological concepts are potentially pertinent to all design because all designs and design studios engage natural systems. They are all hypotheses about the biophysical environment. And natural systems are one of landscape design's fundamental knowledge bases. They are neither more nor less important than the discipline's other principal vocabularies. Consequently, ecology should not be privileged content. It should be addressed in much the same way as any other design aspect, as one set of relationships that must be integrated with every other set, creating a new synthetic expression whose unique relationships continue to reveal new aspects about each of its individual vocabularies.

The problem is that there is often confusion about ecology's content and place in design. On the one hand, the emotional fervor that often surrounds it tends to displace it from the realm of science into a philosophical or even pseudoreligious realm. Unfortunately, this causes some designers to consider ecology as content without rigor or objectivity and, therefore, to dismiss its potential. Others dismiss ecology in reaction to dogmatic applications of it, an attitude antithetical to design creativity.

On the other hand, both scientists and designers often forget that ecology, ultimately, is a social construct, framed within larger cultural values.[2] Science holds no moral authority, only systematized study, testable hypotheses, and statistical evidence—all very valuable. And science itself continues to change its pursuits and conclusions along with changing cultures, sparking a reciprocal relationship between cultural expression and science that is healthy and creative. Yet, this dynamic relationship makes it all the more difficult to balance scientific knowledge with cultural values, often leaving designers with the impression that there is no dependable knowledge, that all ecology is good ecology.

Whichever the case, we maintain that ecology is still a new content area relative to other knowledge bases. Its critical mass and level of expertise within the landscape design and planning disciplines are still not on par with other areas like history and landscape construction. Instructors and professionals alike often fear stepping into the abyss guarded by the beast of "Science," afraid that such an empirical endeavor will strip them of their artistic talents. Add to this the lack of published precedents on how to *apply* ecological knowledge to design, and we approach design studios unsure of ecology's relevant content. The territory is not fully explored, let alone codified, leaving us questioning how to integrate ecology in design efforts, how to fill the gaps between what we know and what we do not know, and how to involve ecologists, whom we always hope (and trust) can give us guidance.

This chapter focuses on landscape planning and design studios that specifically include in their intentions helping students to build and apply ecological knowledge. The authors' collaboration has reaffirmed our beliefs that there are positive things happening all over the country and that sharing them will help us all to build upon these efforts. Our explorations have also reaffirmed that there are clear, shared, and articulated aspects that give footing to the shifting territory. Our collating of these efforts is far from constituting a method. Just as each ecological question must be framed within particular cultural values, we avoid proposing "ideal" models. Each landscape architecture program is unique and requires a unique approach. (To illustrate the range of curricular realities, we have included in three curricular "types." Appendix 16A, page 464) In this chapter, we offer a *repertoire* of successful strategies and tactics from which studio instructors in various institutional and curricular situations can craft their own best relationships.[3]

Key Relationships between Ecology and Design

Specifically addressing ecology highlights some inherent differences between ecology and traditionally administered planning and design studios that warrant special attention. These six key differences span design program types, site scales, and ecosystem types. At present these differences tend to be impediments, but this need not be the case. Ecological concepts can be addressed in almost every design project to various levels of sophistication. And the resolution of differences holds exciting potentials for design.

General social constructs + specific methods.

In all designs, decision criteria are necessary for assessing a design's success. The problems with the current pursuit of ecological decision criteria are ironically two facets of the same issue of considering ecology in all its fullness: (1) neglecting to situate ecology within general social constructs, and (2) lack of precision and specificity.

First, we often fail to acknowledge that ecology is a social construct and subject to the same dynamic changes in theory, cultural needs, and emphasis as any other issue within design. The observable "facts" of ecology may not change in a revolutionary way, but each cultural group's theoretical framing and implementation of those facts, like all design issues, is based on cultural models that must be understood, evaluated, and creatively engaged—for *each* place and, in time, for *each* project. What one culture or interest group considers ecologically "good" will rarely be coincident with the judgment of another group. Unless we properly situate ecology within its appropriate social construct, we risk misperceiving facts and assuming values that have been placed by others upon particular ecological issues that we might be considering.

Plus, analytic exercises tend to occupy an increasing and significant amount of studio time and energy. The result has been the treatment of "analysis" as a notably different process with a distinctly different way of working and expected outcomes from "design." It is unfortunate that this dialectic persists, but it is encouraging that more productive models are emerging, as discussed later in the topic "Choosing the Project and Structuring Exercises" (p. 437).

The second problem is the flip side of ecology as a social construct. Ecology does contain rules of thumb and information with reasonable probabilities, testable hypotheses, and observable phenomena. Currently, however, when ecological criteria become entangled in social, programmatic, personal, and institutional values, the science tends to be simplified or dismissed in favor of these other criteria.

In applying ecology, good intentions are not enough. Some tolerance might be allowed in users' behavioral and psychological reactions within a design. Indeed, some argue that such diversity and nonprescriptive possibilities should be encouraged. Plus, even problematic aesthetic and social negative effects are more easily remedied.

Yet, with ecology, incorrectly or inappropriately applied ecological knowledge can have severe and long-range repercussions to an ecosystem. Biologically positive ecological actions require specific, coordinated, and predictable actions—more specific, methodological, empirical analyses. And because monitoring is so critical to ecological questions, the methods and their application to a landscape must have clear criteria, specific procedures, and be testable. Consequently, ecologically concerned studios must be concerned with articulating decision criteria in a different way than usually applied in studios.

Originality is not always best, and it is a dangerous presumption when dealing with natural systems. Yet, its pervasiveness as an *implied* design criterion makes the demand for the explicit definition of criteria all the more important. Unfortunately, an often heard criterion is "One of my primary criteria was to preserve wildlife"; yet, it is so nonspecific that it cannot be falsified. A criterion such as this makes any design acceptable but does little in advancing ecologically positive landscape change. Too often, students (and professionals) operate on generalities and fail to pursue gaps in data. Rather than stopping and diligently finding the information, they try to "creatively" design their ways out of unknowns. Consequently, studio instructors should reinforce that ecological understandings demand a specificity to design criteria. Granted, sometimes finding an answer to an ecological question may consume a designer or a studio. Yet this is an equally important lesson. Students need to know when to acknowledge gaps in knowledge and how to operate on assumptions (which we all must do in all designs).

Designing requires that we operate in many modes. In part, design works intuitively, exploring generalities and shifting laterally between aspects. And though instructors do not often call attention to it, design also operates empirically, stochastically, methodologically, and mechanistically—all good things. Clearly, ecology operates more in the latter modes but is also value-laden, part of larger conceptual paradigms, and relative to context—making it much like any other design issue. The more that we can help students find the coincidences between creative activities, the more likely designers are to use the information.

All of this presents studio teaching with important learning opportunities, ecological and otherwise. First, studio is an excellent opportunity for demonstrating that ecological understandings must be prioritized to fit cultural expectations. In the face of development or management, it is rarely possible to preserve natural systems' functions in an undisturbed state. Designers and communities must make choices. This requires designers to be specific about decision criteria demands. And they must be much more precise about the relationships between the human-created and non-human-created environments.

Second, studio presents a forum for students to grapple with the concept that "good" does not equal "meeting the designer's decision criteria," that the decision criteria of a culture (or the designers' personal criteria) may not withstand the scrutiny of science. Students learn that not only do cultural values mold scientific paradigms but that science tests culture. Just as science and culture maintain a reciprocal relationship, so should students' designs.

Third, students' understandings that ecology is a social construction become finer tuned. By articulating decision criteria, they understand the intimate correlation between scientists' questions about ecological process and cultural or personal perceptions and values. Take, for example, the issue of fire in the Midwest, which has received major funding. There would be few, if any, prairie/savannah landscapes without human intervention. Some scientists argue that they are no less natural because of human intervention, a decision based in part on aesthetic values about the role of the prairies as an historic artifact and symbol. Other scientists argue that they are not worth maintaining because they are not natural, that is, left alone to operate without human intervention. By critiquing the science, students begin to learn how they must untangle the values motivating the decisions from observable, testable, replicable science: the hypotheses, the methodology, and the results.

Finally, and perhaps most important, students are forced to frame their own values in relation to others', situating the relationships between environmentally concerned and ecologically defensible actions. By demanding that they are rigorous about the ecological criteria, students must decipher the source of their values and understand that they are not absolutes but relative to a host of other values.

Assumed + relational aesthetics.

The phrase "ecological design" is commonly bantered about in discussions as a shorthand for a particular aesthetic expression (usually related to naturalistic forms and seemingly minimal interventions). These same discussions tend to cluster around particular project types such as ecological restoration efforts, ecotourism, greenways, river planning, nature interpretation centers, and the like.

While these might certainly be among the list of appropriate projects, the list could include *any* project, including projects where the ecology is quite engineered, such as a dam, a landfill, or a subdivision. Just as any other conceptual aspect of a project, ecological content may or may not ultimately be manifest in the physical expression of the design. The important matter is not that the project "looks" ecological but that it addresses the dynamics of the landscape of which the project is a part—of both the landscape's processes and its forms.

One of the studio critic's roles is to teach students to critically distinguish assumed values and aesthetics and to make significant and rich relationships between projects' many conceptual issues. It is also the critic's role to teach students to translate and transform those issues into meaningful built forms. If successful, ecological understandings and projects' aesthetics are released from their constriction to any particular physical expression, and the integration of ecological understandings becomes truly creative.

Dynamic processes + static forms.

As our ecological understandings increase, so does our realization that the drivers of a landscape are the processes—the biological, chemical, and physiological operations of which the physical elements are manifestations. Until recently, all but the most radical design has treated the landscape mainly as static, eschewing the importance of the underlying dynamics of the forms it manipulates. More and more we are addressing the dynamics and viewing how the physical forms change over time.[4] Yet the idea of process is not enough. More rigorous ecological knowledge necessitates our understanding the actual processes.

Perhaps all studios should consider the social, cultural, and economic dynamics that will necessarily change built landscapes over time. But the issue is imperative for designs that seriously consider ecology. Therefore, studios should address multiple temporal scales. Some instructors choose benchmarks in vegetation growth, such as 1 year, 5 years, 15 years, and 50 years, based on human lifetimes and expectations. Some opt for the benchmarks within the ecosystem type (Figure 16-1).

The benefits of attention to specific, scientific knowledge are showing themselves to be not only biological but also aesthetic. And the continued aesthetic content reveals itself as an exciting potential. Considering how these processes combine in complex interactions that evolve and change over time allows designers to contribute to people's physical health and their psychological experiences.

Stochastic + deterministic outcomes.

Many events and processes that change ecosystems are stochastic, that is, there is a probability, not a certainty, that a set of phenomena will occur. For example, species respond to habitat fragmentation stochastically because there is an increased chance they will be adversely affected by changing climates or biotic interactions. Landscape design solutions are typically deterministic, representing a singular outcome, and the most optimistic one at that, implying that events or processes occur with certainty.

It is neither ecologically acceptable nor realistic to assume that a design will work biologically. Sometimes these failures will destroy both the ecosystems and the artful intentions of the project. This does not mean that we should design for failure. Rather, we should acknowledge that uncertainty is

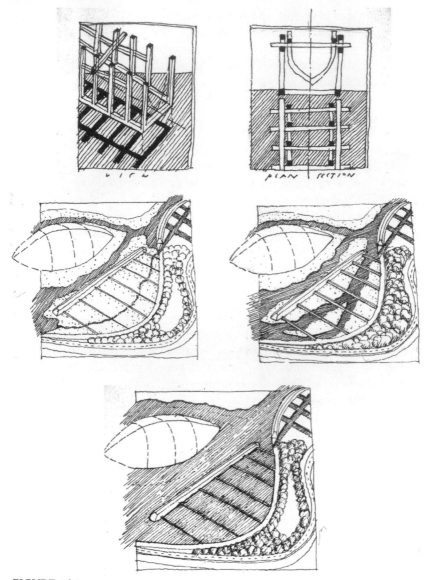

FIGURE 16-1.
"Civic Infrastructure" studio, exploring processes over time as a way of developing a design. Instead of approaching the project as an effort in producing a static, deterministic, and singular design, the designer accepted the geomorphical realities and used them to his advantage. (Instructor: Kathy Poole; designer: Keith McPeters.)

a critical component of designs and that we need to be responsible for what the possible outcomes might be, both optimistic and pessimistic.

Uncertainty will remain intrinsic to ecology. There is rarely a clear link between one cause and a particular effect. And the chance of a surprise event like a hurricane or pest invasion is always a possibility, even if it is of small probability. Consequently, we need to be aware of this uncertainty, even accepting that very large uncertainty might be the case. However, this should not be an excuse for ignorance, of not being responsible for the range of possible outcomes.

Designers cannot eliminate uncertainty, but instructors can teach students the difference between intrinsic ecological uncertainty and a designer's ecological uncertainty. Designers can also be more ecologically rigorous by assessing the potential risks, considering the *probable* outcomes of a proposal rather than assuming that all of its variables will necessarily behave the way the designers intend. To consider the ecological probabilities presents fascinating design opportunities. It asks designers to consider *alternative* outcomes, to explore multiple expressions, various iterations—an activity that is almost always productive and enriching for designs.[5]

Multiple scales + discrete sites.

Whether professional design commissions or academic exercises, projects are rarely concerned with larger scales or adjacent sites. Yet, ecological knowledge demonstrates that most sites are not autonomous. Intervening in them has consequences at larger and smaller scales than the built, visible physical changes. For example, knowing whether a wetland restoration will likely be important for waterfowl production within a region requires knowing the characteristics of the water and vegetation on-site *and* the condition of the surrounding upland *and* the pattern of wetlands in the surrounding region. On the other end of the scale, ecological evidence has shown "the tyranny of small decisions," that small spatial scale interventions may sometimes have tremendous repercussions on an entire system.[6]

Consequently, even if the site is very small, students should address the site's larger context. And if the site is a regional one, students should address something very small, such as a hedgerow, a subbasin, or a singular house lot.

Many instructors broach the issue through a hierarchy of scales, helping students to master understandings in telescoping fashion, large to small: region to landscape to site to detail. Increasing numbers of instructors are finding more success by having students "jump" between scales throughout the project: site to regional to detail to landscape, and so forth.

Focusing on multiple scales will not only increase the success of the project but may also be key to repairing what seems a split in landscape architecture, the rift between "site designers" and "regional planners." To realize that design *is* regional planning and regional planning *is* site design provides literal common ground on which to repair a damaging fragmentation of landscape design as a discipline. It is also critical for creating designs that are ecologically successful.

Progressive knowledge + rules of thumb.

Ecology as a discipline progresses. We learn more. We improve our understandings. Consequently, any individual's information (book or person) may be more *or* less up-to-date. Furthermore, our understanding of ecological processes may not always translate between ecosystems or may not translate to larger or smaller scales. Traditionally, landscape designers have been taught to rely on ecological "laws," depending on sc\ience to be a constant, monolithic body of knowledge.

Designers need to be critical about how they apply ecological knowledge and not be unthinking consumers of whatever ecology comes down the media pipeline—a skill that instructors must instill early in designers' careers. In the studio, instructors teach students to distinguish critically between different designs. And instructors teach students how to evaluate a particular aesthetic theory within its cultural context, as part of a particular place and time. The same applies to ecology. Students need to understand that ecological theories operate in particular places and times—and that some have now been proven otherwise or are now being questioned with different lenses.

Students also need to understand that a group of ecologists is much like a group of landscape architects, diverse, having different views that are equally valuable. For example, a population biologist may react differently to a design proposal than someone interested in biogeochemistry because the questions and approaches are different. If designers learn to pursue multifaceted and growing ecological knowledge, they will continually find new content and ways of viewing their designs—and new ways of interacting with their ecologist colleagues.

Designing a Well-Crafted Project

In exploring these issues, we begin by identifying key relationships between ecology and studio that warrant specific attention. We frame the subject to reflect the typical process employed by instructors in designing a well-crafted studio project: (1) identifying students' incoming ecological knowledge levels, and projecting desired outgoing understandings; (2) articulating processes and products to develop those understandings; (3) choosing a project that provides an appropriate medium for exploration, and structuring exercises that will lead students to desired goals; and (4) integrating ecolo-

Steps	Repertoire of Case Studies		
	I. Beginning	II. Intermediate	III. Advanced
Identifying Knowledge Levels			
Articulating Products			
Choosing a Project + Structuring Exercises			
Integrating Ecologists into Studio			

FIGURE 16-2.
Structure of case study repertoire.

gists into studio. For each step in the project-development process, we discuss its primary issues in relation to developing ecological knowledge. Accompanying each set of principles is a repertoire of case studies illustrating three tiered levels of ecological knowledge (Figure 16-2):

I. Beginning: Operating with rules of thumb.
II. Intermediate: Manipulating spatial and temporal aspects.
III. Advanced: Integrating complex processes.[7]

Rather than proposing "ideal" models for each studio level, our intention is to provide a repertoire of successful strategies and tactics that can help all studio instructors in various institutional and curricular situations. We conclude with broader conceptual reflections on how current impediments to ecological understandings might be minimized and how successful models might be enhanced.

Idenifying Levels of Ecological Understandings

If students are to use ecological understandings appropriately, then instructors must craft projects to ensure students' success. While this is true of all design projects, it is perhaps particularly true of those concerning natural systems. The ecology can be so difficult for students that they become overwhelmed, spending all of their time grappling with the natural systems issues and neglecting other important design issues. On the other hand, if instructors duplicate information at no higher level of sophistication, then students are neither challenged to use it nor convinced to pursue it as rigorously as other design values. Whichever the case, landscape architecture is clearly not following its own design and history models of developing an ecological "language."

In addressing this gap, we identify a general framework for assessing appropriate incoming and outgoing ecological knowledge levels when designing a studio. We have organized it by "ecological knowledge levels" of Beginning, Intermediate, and Advanced. For each knowledge level, we identify what students need to know coming into the studio and what they need to know by the end of the studio.

Beginning level.

For beginning-level studios, students must have a basic talking knowledge of ecological principles. This means that they may have very little specific knowledge of particular ecological relationships but understand general concepts such as water flow and nutrient cycling. Otherwise, students will lack the proper natural systems context in which to consider their designs.

Some students may acquire this knowledge through natural science courses taken before they enter design school. Otherwise, design faculty must ensure that the introduction of ecological knowledge be coincident with other design content. Three pathways are apparent:

1. Ecology or landscape ecology course taught within design school
2. Natural science courses taken after they enter design school
3. Related courses within design school that incorporate ecological understandings

By the end of the studio, students should have a better talking knowledge of ecological understandings and be able to operate by "rules of thumb."

- Apply general concepts such as watershed considerations, connectivity, or habitat preservation.
- Know when to garner additional expertise.
- Know who to call and what general questions to ask so that experts can give specific direction for a particular design proposal strategy.
- Understand and demonstrate the nonquantitative ecological effects of their proposals, for example, how groundwater infiltration will be changed or how the connectivity for mammals will be diminished or enhanced.
- Speculate on a design's evolution past its life in the designer's hands.
- Integrate ecological issues with a *limited* set of cultural, social, or political issues.

Students should be able to confidently execute specific skills:

- Go out in the field and "read" a site's major ecological patterns and processes.
- Map elements and general processes and systems (*not* multiple systems or dynamic interactions between systems).
- Construct simple scientific hypotheses and conduct simple field experiments.

- Critique case studies ecologically.
- Explore design in stepped spatial and temporal scales—going from larger/longer to smaller/shorter or vice versa, as opposed to jumping from large to small to large to median to large, and so forth.

Intermediate level.

For intermediate-level studios, students should have more specific knowledge of ecological principles and the natural systems of which their designs are parts—either from formal environmental sciences coursework, seminars, and previous studios.

By the completion of the intermediate-level studio, students should be able to apply information regarding spatial and temporal processes:

- Integrate a few ecological issues with other issues in a complex way.
- Demonstrate reasonable confidence levels of the effectiveness of design proposals.
- Employ simple to moderately sophisticated formal methods such as multiple-variable geographic information systems models.
- Understand relationships between ecosystem structure and function.

Specific skills should include the following:

- Map multiple systems or dynamic interactions between systems and processes.
- Identify case studies that should be consulted, or know where to look for appropriate examples.
- Access the primary ecological literature pertinent to design problems, or be able to keep abreast of important advancements in ecology through the scientific literature.
- Compare alternative hypotheses and make defensible predictions of design proposals' effects.
- Develop strategies for monitoring a design's effects.

Advanced level.

In advanced-level studios, instructors should develop students' abilities to understand spatial and temporal processes into abilities to manipulate them. For example, rather than planning for the inevitability of flooding or plant/animal population dynamics, students should be able to create designs that positively react to an ecological process. They should also be able to design the appropriate structure for a desired ecological function with a reasonable probability of success.

By the completion of the studio, students should demonstrate a significant advance in the synthesis of ecological aspects with other design issues.

Rather than integrating a few ecological and cultural issues, students should be able to synthesize many complex variables—ecological and otherwise.

- Assess the primary ecological issues with little assistance.
- Converse with ecologists on mathematical and methodological terms.
- Represent individual processes and systems dynamically.
- Create and defend stochastic models.
- Evaluate design proposals by more sophisticated formal models.
- Make professional-level recommendations—more than one—and articulate their evolution, attaching probabilities to various aspects' occurrences.

♦ ♦ ♦

Repertoire: Levels of Understanding

BEGINNING LEVEL: SOLVABLE WITH RULES OF THUMB

- California Polytechnic University, Pomona (Cal Poly Pomona) offers core courses in which students gain ecological knowledge. It also introduces ecological concepts in every studio. In some cases, studios and core courses work in tandem in their introduction of material.

- The University of Virginia introduces ecological knowledge as inseparable from cultural understandings. Therefore, it is blended into lecture courses plus seminars. This is partly by design and partly a function of having only one course devoted to scientific principles.

- Each University of Michigan studio not only includes but is also grounded on ecological concepts. As part of the School of Natural Resources, numerous ecology-oriented courses are available.

INTERMEDIATE LEVEL: APPLYING SPATIAL AND TEMPORAL PROCESSES

- At Cal Poly Pomona, undergraduate and graduate planting design studios emphasize ecology in the following way: Students are asked to prepare traditional cost estimates for a design; they then estimate water usage, biomass, oxygen generation, carbon storage, and energy content of their design and calculate the energy required to pump water, fertilize, and maintain it (Figure 16-3). They see that some designs result in more energy embodied and some result in more energy expended.

- In her Civic Hydrology seminar, Kathy Poole integrates the ecology and engineering of water with the study of its theory and history. Students are exposed to wetland processes and introduced to the technical skills that they will need in wetland construction and rehabilitation. Part of her intention is to offer the seminar as preparation for water-related studios or for a studio running coincident to the seminar.

- Virginia Tech's cross-listed course *Nature and American Values* introduces students to the evolving relationship between nature and American society, emphasizing the values that underlie forest, park, and wildlife management. The course, initiated by Bruce Hull of the department of forestry, includes issues of wilderness, sustainability, biodiversity, hunting, old growth, suburban sprawl, and environmental activism. Two earth science courses are required of all landscape architecture undergraduates, while graduates can elect several ecology-related courses. At Virginia Tech, the combination of studio projects, exercises/project work in landscape ecology, and land analysis theory courses provides opportunities to apply eco logical knowledge gained in outside courses such as *Nature and American Values* and other electives.

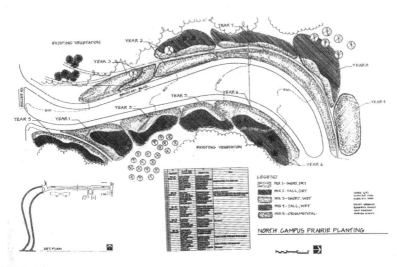

FIGURE 16-3.
Similar to the Cal Poly Pomona example noted, this University of Michigan studio exercise quantified ecological principles and linked them to design.

- At the University of Illinois, Doug Johnston's GIS/Modeling class uses ecological models to evaluate spatial design and planning decisions. Given Johnston's research, students use hydrological models most often, allowing them to learn about the hydrological processes as they learn how to use GIS modeling in their designs.

ADVANCED LEVEL: COMPLEX UNDERSTANDINGS OF ECOLOGICAL PROCESSES

- In many of his studios, particularly those relating to planting design, Robert Grese challenges University of Michigan students to prepare long-term management plans for design schemes. This forces students to confront issues relating to change, management, and monitoring necessary for designs based on ecological models. In one studio, he gave students a list of species found on local sites from which they could collect seeds and assigned them the problem of creating a prairie on a portion of the campus. Told that approximately one-half acre could be planted each of the first three years, the assumption was that in a few years, the planted sites could be used as sources for collecting seed to enlarge the planting. The assignment challenged students to investigate the autecology of each species and develop strategies for a phased planting and management plan for at least a fifteen-year period. Not only did they develop materials, labor budgets, and plant marketing strategies, but they also generated extremely imaginative strategies for making such a project work—both aesthetically and ecologically.

Articulating Processes + Products: Communicating Ecological Understandings

After assessing students' incoming and outgoing levels of ecological understandings, instructors often decide the form of the studio's products. We recognize that inherent within instructors' decisions products are decisions in which issues of process are inextricable: goals and objectives, evaluating ideas, showing implications, determining relevant information, group dynamics, encouraging creative design explorations, and so forth. For the sake of clarity, we defer process questions to a later section, "Choosing the Project and Structuring Exercises," and devote this section to the physical products themselves.

Products can be physical, such as drawings, diagrams, maps, models, narratives, graphs, and animations, or they can be intellectual or experiential understandings. Whatever the form, we identify five characteristics of products that are key with regard to ecological content.

Multiple outcomes.

All studios should perhaps address uncertainty and acknowledge the possibility for more than one possible solution and the change of any one solution over time. In fact, given the dynamic and evolutionary nature of natural systems in time, studios that directly address ecological content should highlight the limitations of a singular solution that presumes that it will maintain the forms given at the design's proposal or construction. Consequently, studios should communicate design proposals in less deterministic forms, exploring numerous potential future evolutionary forms.

Many instructors require that students project their designs into the future to whatever time frame is appropriate to the project. For example, Kathy Poole has required that students project a riverside scheme through 10-year, 50-year, and 100-year storms to speculate on how much the design might be altered by one of these events. Others take the technique further by requiring that each student's final design be represented as multiple possible outcomes rather than the more typical singular design "solution."

Multiple scales.

Most studios explore design projects at multiple scales. Yet, the importance of demanding these representations as final products is critical to studios that wish for rigorous treatment of ecological material. Not only should students present this material, but they should also be taught to corroborate it and demonstrate how their designs are derivative of their understandings.

While almost all of us require students to consider the larger context in the preliminary analyses or design explorations of a project, many instructors require students to portray the effects of their design proposals on landscape-scale systems as a postdesign exercise.

Narratives.

Not only should students be expected to recount their decision criteria in a graphic form, but they should also be able to critique their designs' effectiveness and account for their limitations in words. This is particularly important with ecological issues because design exercises always operate with gaps in data and unknown consequences—which are quite difficult to portray graphically.

Many instructors effectively use narrative exercises in all of their studios, ecologically concerned or not, to help students clarify their own ideas and help them articulate them to studio instructors or the public. We suggest that this is even more important for an ecologically concerned studio.

Null and status quo hypotheses.

Given the potential long-term ecological effects of a design's construction, it is important to know when to be restrained in design and when to do *nothing* and let nature take its course. Consequently, design products might include a "null hypothesis." It is similar to physicians' needing to understand a healthy body in order to diagnose an ill body. Carl Steinitz maintains that it is important that a null hypothesis exercise come early in the semester and that the students be required to reevaluate their proposals against it at strategic points in the semester.

Conversely, a status quo exercise can demonstrate what will happen if designs do nothing different from conventional, market-driven, and regionally consistent development. The technique is illustrated beautifully by Randall Arendt (1994, 1996, 1999) in his books on developing subdivisions and local planning ordinances. Kathy Poole finds the exercise an eye-opener on the consequences of conventional planning and an opportunity to gain critical insight into the repercussions of current law and development policies. It also gives students confidence in their design abilities early in the project. Like the null hypothesis approach, the status quo solution can be used as a measure against which to compare design proposals.

Quantification.

Differing from some other design content, ecology is as much a quantitative endeavor as a qualitative experience. Therefore, a studio's products should represent appropriate aspects of a design proposal quantitatively. To properly address ecology demands products such as graphs, tabulated formulas, calculations, and spatially referenced measurements, for example:

- percent of impervious cover, amount or rates of stormwater runoff and infiltration
- Index of Biotic Integrity (Karr et al. 1986)
- Floristic Quality Index (Swink and Wilhelm 1994 as cited in Skabelund and Borneman 1999)

Many instructors require students to assess the probability that their designs will fail or produce unexpected results. Students must then demonstrate how they will address these consequences.

At Cal Poly, students are required to estimate the water-use demands for their designs—not in general terms, but through formulas that demand precision and conclude with a discrete quantity (see Figure 16-4). In an ecosystem where water quantity and longevity are serious issues, planting design is not only a formal endeavor but a politically volatile and "engineered" one as well. The exercise helps students learn the direct relationship between aesthetics and water quantity.

However quantification is addressed, it is critical that instructors discuss with students the value of the exercise in relation to the other issues being addressed by the studio. And perhaps most important, the exercises must be linked not only to the assessment of the design but also to its aesthetics.

LA 342 -   **STUDIO WATER ESTIMATE EXERCISE:**
PART 2   **INSTRUCTIONS:**

Estimate the water use of our case study landscape for July and for the entire year using the following water use formula:

EWU = Estimated Water Use in Gallons.

$$EWU = \frac{(Et_O)\ (PF)\ (HA)\ (0.62)}{IE}$$

ET_O = Reference Evapotranspiratin (of Tall Fescue) In inches

PF = Plant Factor (water needs relative to turfgrass)
HA = Hydrozone area in square feet (planting area)
0.62 = Conversion factor to obtain gallons per square foot
IE = irrigation efficiency

CALCULATION VALUES:

ET_O = 8.5 inches in July

ET_O = 65 inches for the entire year

PF for warm season turf = 8
PF for cool season turf = 1.0
PF for ground cover = .3
PF for mixed native trees and shrubs = .25

SHOW CALCULATIONS:
Warm Season Turfgrass·   40,500 s.f.

July: Entire Year:

FIGURE 16-4.
California State Polytechnic Quantification Exercise.

Multiple products.

Designs require multiple graphic products to effectively communicate their intentions: diagrams, plans, sections, models, perspectives, and interpretive constructions. And specific aspects of a landscape require specific products (e.g., land-form necessitating a model, or the materiality of a bench requiring a section). Consequently, ecological intentions necessitate tools appropriate to the discipline. And ecologically concerned studios should expand the types of products students use to communicate their designs to themselves and others.

♦ ♦ ♦

Repertoire: Articulating
Processes and Products

BEGINNING LEVEL: SOLVABLE WITH
RULES OF THUMB

• In the undergraduate planting design studio at California Polytechnic University, Pomona, students design a demonstration garden for a water conservation district in the Southern California foothills. They focus on fitting plants to appropriate microclimates, demonstrating adapted Mediterranean plants, and designing for low maintenance. They are provided with rules-of-thumb ecological information. After students complete their designs, instructors ask them to examine their work as if it were thirty years later. They are surprised to see that most plants adapted to this climate die out within two to three years. And without fire to help regenerate the plants, they are left with a high-maintenance garden. Students resolve the issues, some by reintroducing fire, and some implementing water conservation. Most bring back much better designs the second time.

• At the University of Illinois, Doug Johnston consciously structured a studio to explore multiple scales simultaneously. The semester-long efforts were grouped into three scales: an element of a site, the site, and the region around the site. Each scale was addressed at an abstract and a concrete level. Three overall studio goals predominate: (1) Address an element of ecology as the subject of the design; (2) incorporate an understanding of ecological processes into the design; and (3) use pedagogical objectives in their implementation (to explain, demonstrate, or reveal a process that acts in the landscape).[8]

• At the University of Oregon, students in the third term of the Plants curriculum engage ecological concepts through active design and experimentation at the university's Urban Farm. Following some preliminary work in the preceding term, teams of students design experiments to be executed in 5- by 20-foot garden plots. A series of exercises guides the students through the basic concepts of experimentation, forming a hypothesis and a null hypothesis, and gives examples of the types of ecological parameters to investigate (fertility, pest control, companion planting, etc.). Students design a planting plan, develop an interpretive sign explaining the experiment, monitor the design's progress, and propose a design to succeed the existing one in the next growing cycle (with or without an experimental bent). Finally, they develop or find a recipe that uses ingredients grown on their plots. Rather than stressing the rigorous collection and analysis of data, the exercises emphasize understanding of ecological systems and concepts in ways that connect ecology to landscape architecture (and everyday life).

- The University of Minnesota has recently incorporated environmental review into its first-year studio. It accomplishes this in studio through field trips and in-class review of designed work. During the preliminary design phase, the class visits demonstration sites to consider how parking lot and road design can be configured to minimize stormwater runoff. They also consider how other design elements can be altered to change ecological and experiential outcomes. As a class exercise early in the term, students pose specific design criteria for both experiential and ecological issues they feel are most important to the site. Identification of these criteria is an outgrowth of a written analysis of site features. The class arrives at approximately twelve experiential and twelve ecological criteria it would use to evaluate their designs throughout the studio. Students use these criteria informally throughout the quarter to generate design ideas and to consider ways to accomplish specific goals as they refine their designs. Two weeks before the end of the term, students (in groups) give formal presentations of their designs in near final form. Students are also assigned to "assessment teams" for this review session. Each assessment team serves as the lead critics for one design team. The assessment team is responsible for querying the design team regarding their assumptions and predictions related to the criteria. Each member of the assessment team prepares a written review of the design and with their group prepares an oral presentation. The assessment teams make their presentations, characterizing the extent to which criteria have been achieved and identifying the most important needs for further design improvements. The last two weeks of the term are used to respond to the review.

INTERMEDIATE LEVEL: APPLYING SPATIAL AND TEMPORAL PROCESSES

- As a way of emphasizing the importance of postconstruction monitoring, students in graduate planning seminar are asked to select from the published Cal Poly Pomona 606 studio projects and follow the project's fate since publication. Students read the reports produced by previous classes, contact clients, and track the trajectory of the plan. Students find that some become shelf documents; some are partially implemented. Many clients report that the student's design process has been utilized while the plan itself has not. Others, especially those that have been guideline or alternative oriented, have had longer legs. A few have been largely implemented. And one inspired massive redesign of a Los Angeles area drainage channel for habitat and water quality improvement.

- Kathy Poole's work focuses on water, in part because more vividly than most, rivers are dynamic systems. In her graduate-level studio at the University of Virginia, Civic Infrastructure: Reframing the Anacostia River, Bladensburg, Maryland, she challenged students to make the dynamism part of their designs. The scientific principles of river processes and change were reinforced through field visits and a daylong river dynamics workshop by an ecologist. Students also considered dynamics historically by documenting history and projecting their own designs as part the river's evolution. What makes exercises like this distinct is not that designs are viewed as form upon which processes occur—as if the forms are baselines upon which to register processes. Rather, the reciprocating actions of human intention and natural processes are seen as equal and interacting partners in the continuous formation of landscapes. Change is not only acknowledged but also welcomed. Thus, the time projections do not occur after the design is proposed. Instead, the time projections *are* the design. There is no "baseline" design, and they welcome change. Plus, students are "designing" a management plan. There is no distinction between analysis and design, process and product.

ADVANCED LEVEL: COMPLEX UNDERSTANDINGS OF ECOLOGICAL PROCESSES

- At Harvard's Graduate School of Design, Carl Steinitz addresses the issue of varying products directly. He fills a semester with projects of varying lengths (from one hour to four weeks) and demands various products from each. Some are traditional qualitative descriptions in the form of drawings and models. Many of the products are directly related to the quantitative, scientific demands of ecology: a GIS model and analysis of the design; multiple-outcome scenarios; and a response to an "event" (such as doubling the program) that demands reevaluation of the ecological consequences. Most important, Steinitz never distinguishes between types of products; they are all means to a complete design. His studios, which he always co-teaches with ecologists (as well as involving others from many disciplines), produce what he calls "Alternative Futures," a collection of scenarios developed with different criteria, and expressing multiple values. And within each team's work, the students are required to forecast multiple physical manifestations of their theses.

- At the University of Illinois, Terry Harkness and Doug Johnston offered a project that involved a "restoration" of a series of water bodies. The notion of restoration was somewhat more complex, involving considerable public confusion regarding what is "natural" and multiple histories (wetland, Civil Conservation Corps, stormwater management, park). Students in this class did considerable work in formal analysis of the hydrological characteristics of the basins, modeling stream flow, stage height, frequency and duration, and the like to gain a clearer understanding of the nature of the ecology of the site and to reveal that nature through their designs.

- Doug Johnston notes that sometimes the ecological investigation is too demanding to address ecology's science and its formal design all within the course of a semester. In one of his advanced studios, he joined with an aquatic ecologist to teach a mix of landscape architecture and planning students to restore a lake within the city. The project included field surveys of water quality, inventory, habitat assessment, and the like. Students focused more on remediation and enhancement plans for water quality, wildlife, and plant species than for human activities beyond the traditional trail and boardwalk. In the subsequent semester, Johnston completed a public service project that looked explicitly at public accessibility and incorporated more formalized design concepts.

Choosing the Project and Structuring Exercises

A primary task in designing a well-crafted studio is choosing a project that provides an appropriate medium for exploration and that can accomplish desired products and objectives. The first aspect of choosing a project that specifically addresses ecology is to ensure that ecology content is not only present but also necessary to successfully address the project. Otherwise, it will almost necessarily be overlooked. Furthermore, because environmental review of proposed landscape change is increasingly common and often required, it is important for students to know how to minimize impacts to sensitive areas without limiting their creative responses to design.

For example, Lee Skabelund in his Landscape Planning & Management Studio for Floyd County, Virginia, encouraged students to seriously consider how much alteration an area can withstand without losing its integrity and to critically ask where are sensitive landscapes, and are there ecological thresholds for these places. To effectively address these questions, students had to go beyond what they themselves knew and search out experts in ecology and related fields through interviews and by reading pertinent scientific literature.[9] The project demanded that students carefully think about the implications of their site-scale plant selections and planting designs, the way they handled stormwater runoff, the relationships between site and surroundings, and how these factors would likely change twenty-five to fifty years in the future. And because the project included both county and site scales, students could not approach physical design as something separate from the scientific questions.

On the other hand, overprivileging ecological issues can have a negative effect, especially if the studio has "ecological design" in its title. Sometimes, students assume particular aesthetics and methodological approaches. And they can assume that other studios need not address ecological issues.[10] Nonetheless, since ecology has not been the traditional focus, Susan Galatowitsch's humorous analogy about "overprivileging" ecological information is pertinent. She contends that since planning and landscape programs have traditionally been far more attentive to designs centered on human needs, that a kind of "ecological affirmative action" at universities would be acceptable and effective in correcting years of imbalance.

Whatever instructors' positions, it is critical that they address four key aspects in crafting project exercises: knowing the site; knowing the scientific literature; knowing relevant precedents; and articulating degrees to which students can grapple with the six key issues specific to ecology within studio as outlined at the beginning of the chapter.

Knowing the site.

One of the most important ways to know a site is to address it directly—visit it. Steinitz says, "Go. Go often. And never go alone," that is, always take an ecologist (as well as professionals from other disciplines). Few would disagree with this strategy, but the critical matter is how to leverage field experiences in studio. To Steinitz's adage, we add two corollaries: (1) Map the site, and (2) consider using perennial sites.

All the ecological understandings in the world are worth little if students do not know how to translate those understandings into forms and processes that are readily applicable to design. While it is not the only medium, an important one for design studio is mapping, that is, accurately representing physical objects, processes, and phenomena in two- and three-dimensional forms. By ensuring that ecological understandings are translated into forms in which designers are facile, students are more likely to translate data into useful information and literally *situate* abstract concepts into physical spaces. We recommend no particular methods or strategies but identify two important aspects of mapping ecological content: empirical accuracy and multiplicity of products.

In recording and translating site qualities, mapping must be spatially and quantifiably correct. Impressionistic renderings and expressionistic mappings of *personal* interpretation are just that—personal expressions that are likely not to possess enough empirical rigor to be useful to the science of ecology.

These kinds of mappings may accomplish other objectives but will do little to reinforce scientific understandings. And when used alone, they may even negatively impact students' understandings of ecological content.

This is not to suggest that mappings should be "dry" or lack creativity. In beginning-level studios, Kathy Poole charges students with making maps that are scientifically rigorous and experientially meaningful. They must be empirically correct and evocative.[11] For example, when mapping hydrology, they locate springs, creeks, and groundwater depth, but they simultaneously map aesthetic impressions and consequences of "wetness," requiring them to exercise aesthetic judgment and establish relationships between design and analysis before there is any differentiation in students' minds (as in later discussions when quantifying a design proposal is necessary).[12] She repeats mapping exercises throughout the semester, maintaining that a key aspect is to have students map the same site over and over and over again, each time looking at a new element, process, and set of relationships. In essence, mapping is a way of constructing the site, both conceptually and physically (Figure 16-5).

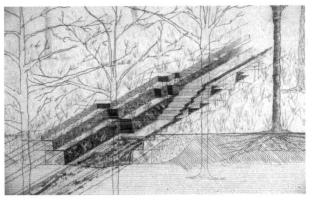

(a)

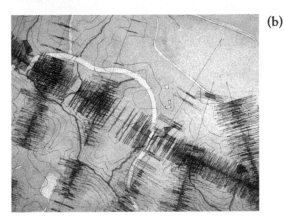

(b)

FIGURE 16-5.
Sites can be constructed through careful mapping, in these cases by mapping the relationships between soil and vegetation history. In (a), Laurel Haarlow inflects a "room" within a wood that is both an outgrowth of her mapping exercise and a creative addition to the mapping. In (b), Jen Parker and Carol Innes's quantitative graphing (in transects) of the relationships of soil to vegetation types not only gives a statistical description but also offers approaches for a design's possible spatial structure and texture.
(Instructor: Kathy Poole)

The second mapping aspect of note stresses repetition in a different way. No matter how many times a site is mapped, the site should be mapped in multiple modes. Traditional design media such as drawings, sections, and models should be complemented by modes typically related to scientific studies such as tables, graphs, and computer models. Susan Galatowitsch explores multiple modes of observation and site recording with the students at the University of Minnesota during their week-long field course at Itasca Biological Station. The field station is within one of the state's most significant historic and natural areas (and is very popular for recreation). They visit eight to ten sites during the week and do a variety of sketches, including sections and details to capture specific qualities of the site. Then they record the vegetation composition of the site they sketched using a releve technique.[13] The students compare sketches and vegetation data from several sites to consider questions related to experiential and ecological conditions. They usually realize that they can make more interesting and precise comparisons by using a combination of graphical and tabular data.

Using a perennial site.

Just as any project is potentially ecological, any site holds potential for building ecological knowledge. Nonetheless, we call special attention to what might be termed perennial sites as an extremely effective tool. By using a site year after year, instructors build a strong knowledge base, offering studios more data to process. Perennial sites also offer an important political opportunity for universities because as more sites are being consumed for development, long-term, high-volume data study sites allow landscape programs to work with other programs (on campus and off) to keep the sites in public ownership.

At Harvard's Graduate School of Design, upon entering the landscape architecture program, students live for a week at Harvard Forest, which is located in Petersham, Massachusetts, about an hour from the school. Not only does important social bonding occur, but students also become familiar with the ecology in which they will be studying. That initial understanding is built upon in a second-year core studio when they return to the site on a bus loaded with ecologists and other experts of various disciplines. Students continue to explore the ecology through their designs and through GIS site-location and design-evaluation exercises. They must also revisit the site to complete a required presentation exercise in which they must photograph a scene and show before and after views. By the end of their second year, students have explored the ecology of one place many times in numerous ways. Carl Steinitz is fortunate in that Harvard Forest is also a National Science Foundation–sponsored Long Term Ecological Research site, giving him access to an immense body of information in the forms of both data and people.

At the University of Michigan, the School of Natural Resources and Environment has formed a partnership with a foundation that runs a camp on the dunes of Lake Michigan as a perennial "tool" to teach students ecology. Students spend three days in September at the camp, studying ecological patterns and referring to analyses that have been prepared by a professional consultant. Working in teams, students explore specific habitats and prepare analyses of what they learn. Back at Ann Arbor, students investigate more general topics relating to the types of ecosystems found on the property. Students then develop preliminary design schemes for an environmental center on approximately 7–8 acres of the camp property. The perennial issue comes into play the following semester when students return to the project and the site to further explore other options for the environmental center.

Two drawbacks of perennial sites are noteworthy. First, they can become trampled by successive classes, destroying the very natural system that one is interested in studying or perhaps preserving. Second, limiting a studio level to one site limits a university's ability to actively involve a broad range of communities in the studio process, making perennial sites a particular liability for land grant universities that have strong outreach mandates. Nonetheless, we maintain that field experience is particularly important, given our collective sense of fieldwork being dropped from other courses.

Knowing relevant precedents.

Part of building a design repertoire is stocking it with the designs of others, case studies both historical and contemporary. The same is true with ecological issues. Admittedly, finding ecologically concerned case studies is problematic since the availability of published case studies and other precedents is still somewhat limited. Yet, as more of these projects are reported in the literature and the history of ecological restoration work is further developed, this situation is improving. In the meantime there are tactics within studios that prove effective.

Some instructors ask students to cite sources for the ecological design ideas in the same way that they cite precedents for design ideas. This teaches them to appreciate the value of building an "ecological case study repertoire," and will hopefully instantiate it as a standard practice in design development.

Kathy Poole challenges students to consider again, this time through ecological eyes, historical precedents that students already know from landscape history and theory courses. The natural systems of projects like the Villa Gamberaia or Stourhead or Jefferson's plan for the University of Virginia may have been addressed *theoretically*, but the ecology of the projects (inevitably) has never been analyzed. By having students apply ecological knowledge to familiar historical precedents, the ecological lessons are easier for them to absorb. And they learn to see that ecological issues are relevant to gardens on which great intervention has been exercised as well as to relatively natural or naturalized landscapes.

Knowing the scientific literature.

The need for students to be critical consumers of ecological information is really an issue of the larger landscape architecture curriculum. However, studio instructors must often address it more immediately—within that semester's studio. We offer four general strategies for increasing students' knowledge of scientific literature and helping them to distinguish between, as Chapter 5 notes, "Science, Nonscience, and Nonsense."

Scientific readings should always be included as part of exercise sequencing. Acquiescing to the argument that designers do not read or that they cannot understand scientific language sends negative messages about rigor in general and scientific rigor in particular. In fact, Joan Woodward argues that design instructors shoulder much of the responsibility because we remain complacent with the argument that we are keeping students too busy with graphic tasks for them to read. She acts constructively to remedy this by having students read excerpts from texts on the implications of evolution and other ecological concepts. Students must prepare reaction papers to each reading, find the link to their project, and participate in class discussions. Her hypothesis is supported in the students' work at Cal Poly: The reading never fails to make their site mappings better. Students also gain a window on their classmates' worlds, as occurred when one student rejected all the texts on evolution due to his fundamentalist Christian background. It is also a good lesson in managing conflict and framing strong, grounded arguments, which is what they will have to do in professional practice.

Susan Galatowitsch maintains that it is not critical that students understand the entire math or read all of the data of a particular article, just that they make connections. For example, she overheard third-year students relating to first-year students the quirky details of a paper on sustaining predator-prey dynamics in human-maintained landscapes. The students excitedly talked about how the author studied the effects of habitat fragmentation by creating a living landscape of oranges, grease, lint, and mites. Galatowitsch was delighted because students would clearly remember some of the key concepts. Of equal importance, the article helped students to understand the amazing things that are awaiting their discovery in the scientific literature.

Choosing key articles or chapters that are the most pertinent to the studio project helps students from becoming overwhelmed with the science. At Utah State University, Richard Toth's studios take the first two weeks of the semester for reading applicable ecological literature, requiring students to turn in summaries/reviews of each article or chapter they read. This makes students aware of sources of ecological information that they can turn to later in the semester as well as for future studio and professional work.[14]

Accompanying each reading with a discussion period is critical. Otherwise, students do not have the opportunity to clarify troublesome portions or make relationships between the text and the studio project. At Virginia Tech, Lee Skabelund and Patrick Miller have reserved an hour of studio each week for student-led discussions of ecological planning and design literature. They have found additional ways to connect literature and design: presenting findings from a published landscape planning study, requiring students to provide feedback on readings to each other, involving ecologists and other invited guests in discussions of readings and project work, and sprinkling relevant readings throughout the studio sequence.

Last, exercises that demand students seek out information from ecologists and scientific literature can activate their ecological learning. For example, in one studio Skabelund asked students to choose a species—black bear, for example—and decide how much space the species needed within a particular area. Instead of giving students the answers (which he sometimes does not know anyway), he gave students literature citations and specific people to contact. Students arrived at conclusions themselves and developed literature search skills.

Guiding students' explorations of key ecological concepts.

Finally, we return to the six key issues specific to ecology within studio that we outlined at the beginning of the chapter: (1) general social constructs + specific methods; (2) assumed + relational aesthetics; (3) dynamics processes + static forms; (4) stochastic + deterministic outcomes; (5) multiple scales + discrete sites; and (6) progressive knowledge + rules of thumb. While each studio should address each of these issues, specific studios must address three questions: the *degree* to which students can grapple with any given issue, the *relationship* of the issues, and the *responsibility* given to students.

1. *Degrees:* Healthy ecosystems tend to be extraordinarily complex—spatially, temporally, and in their processes and dynamics. Yet, talented and experienced scientists, let alone design students, understand any given ecosystem only in part. Therefore, studio instructors should carefully define a set of ecosystem aspects and the levels of understanding and effort required by students to successfully complete the project. For example, beginning-level students best approach scale incrementally and hierarchically, either step-by-step from regional to area to site, or vice versa. And intermediate students can predict probable failures conceptually but lack the tools to quantify the probabilities.

2. *Relationships:* The relationships in any design are infinitely complex in

terms of the possible synthesis that a designer might find. However, not all designs involve infinite conflict between cultural models and ecological health. Nor do all designs need necessarily make relationships between all of ecosystems' aspects to be successful. Consequently, studio instructors should know beforehand which relationships must be addressed for students to arrive at a successful solution. For example, must space and time be addressed simultaneously, or can they be approached separately or in parallel? Or how much uncertainty is inherent when applying a particular program to a particular ecosystem?

3. *Responsibilities:* Studio instructors play a variety of roles. Using a professional practice analogy, these roles range from managing principles to consultant to client. The more advanced the studio, the greater responsibility the student should shoulder for defining the issues (ecological and otherwise), for identifying the relationships between issues, for choosing applicable models, and for structuring and timing pathways for action and decision making. Plus, instructors should help students learn when and where their responsibility ends. For example, students should be able to decide when design alone can solve the problem, when they need a scientist as consultant, and when there is no definitive "solution," only designs that involve choices in values and specific ecological consequences, none of which may maintain all of the ecosystem's complexity, diversity, or other measure of ecological health.

♦ ♦ ♦

Repertoire: Choosing the Project and Structuring Exercises

BEGINNING LEVEL: SOLVABLE WITH RULES OF THUMB

KEY CHARACTERISTICS

- Instructors act as managing principals, and the students are interns.
- Project approaches scale and time incrementally, in a step-by-step fashion.
- Scale and time are explored separately, not simultaneously.
- Project involves conflict between ecological values and cultural constructs but is also not consumed by it.

- Program is given to students.
- Limited ecological issues with many possible solutions exist.
- Hypotheses are evaluated by students' own decision criteria, but instructors provide students with models upon which they base their decision criteria.
- Designers without consultants can solve the problem.
- Uncertainty is minimized.

- The site for Carl Steinitz's second-year graduate landscape planning and design studio is located in nearby Petersham because for beginning-level studios he believes that easy accessibility and multiple visits to the site are critical. Within a project area, students choose a site of a definable scale (about 1 square kilometer). The project is very loosely defined for the students, for example, a new community. Students are given some program, but ultimately, the students must define their own projects. However, teams of students are assigned different ideologies that demand that they not default to their own values and attitudes; for example, students must design for a religion such as Buddhism or Mormonism one semester, or in another semester for a particular social group such as Alzheimer's patients or Chinese immigrants. This demands that students be rigorous about articulating the decision criteria upon which the project will be judged. Students are required to address ecological issues offered by an ecologist who is also a primary studio instructor. Steinitz issues students a list of ecological standards (decision criteria) that students' teams must further articulate according to the cultural construct in which they're working, which reinforces the lesson that ecology is a social construct and not a monolithic scientific body of knowledge. Students test their own decision criteria with geographic information systems (GIS) models that they devise themselves, once at the beginning of the project and again between midsemester and semester's end. Therefore, and importantly, the quantifiable and methodological models evaluate the formal and experiential designs.[15]

- The Design Center for Regenerative Studies was the vision of John Lyle, who was a leader in bringing ecology to the forefront as a design concern. Continuing his legacy, Cal Poly Pomona employs two kinds of long-term study tactics, both of place and time. First, in almost every year of both graduate and undergraduate classes, at least one project is assigned that utilizes the on-campus, 16-acre Center for Regenerative Studies as the design project site. Some students are residents at the Center and have daily knowledge of the design site. Others visit on a frequent basis and have worked on projects at other stages in their design education, giving a sense of continuity and access to follow-up views of their work. Projects range from planting design exercises for creating demonstration gardens to developing plans for wildlife habitat enhancement, to planning physical campus connections, to restoring *Stipa pulchra* grasslands. Frequent, easy, required, and desired contact with the Center creates an important frame of design reference, enhancing student understanding of landscape dynamics of one place over time during their tenure at the university.

- Cal Poly also designates three weeks of the school year, one week each quarter, for field trips. It is each faculty member's prerogative to cancel classes and participate in a weeklong field experience. Faculty and students typically visit urban and remote sites in California, Nevada, and Arizona and conduct field documentation exercises, meet with practitioners in these locations, and work on on-site design projects. Most combine construction, design, and planting design emphases during the trips. The benefits are that students gain intense field, design, and social experience. The department also sends a clear message about the value of field experience.

INTERMEDIATE LEVEL: APPLYING SPATIAL AND TEMPORAL PROCESSES

KEY CHARACTERISTICS

- Instructors act as the project landscape architects, and the students are the project designers. Instructors are articulating much of the general direction, shouldering the administrative duties, maintaining overriding power, and suggesting personnel working strategies. Yet, the primary decision-making responsibilities rest on the students.
- Project involves conflict.
- Students define the project.
- Multiple issues with many possible solutions.
- Hypotheses evaluated by formal models.
- Real project with profound consequences.
- Physical design alone *may* solve the problem.
- Major amount of uncertainty.
- Successful proposal *must* address multiple temporal and spatial scales.

- Four students in Cal Poly's 606 Studio utilized their knowledge of spatial and temporal processes to address complex issues in a six-month study, "The Shaping of Owens Lake" (Figure 16-6). The project was commissioned by an air pollution control district in an effort to coordinate multiple management plans and scientific studies for the nation's worst dust pollution source. Water has been diverted from Owens Lake to Los Angeles since 1913, leaving a dry lakebed and hazardous dust source threatening the health of forty thousand people in the vicinity. Utilizing scores of independently produced scientific studies, students systematically analyzed physical and social issues, focusing on the regional landscape structure, function, and change, and then narrowing to the specific lake environment. Through this analysis, it became clear to them that certain landscape processes produced nonemissive environments within the lakebed. They set out to maximize these processes and resulting landscape structures, and to minimize emissive processes. They developed a "toolbox" of both wet and dry remediation techniques to be applied in certain characteristic times and environments throughout the lakebed. An understanding of temporal scale was critical, and they couched their recommendations in the understanding that the lake was part of a gradual warming and drying climatic condition since the Pleistocene (but had been speeded up by Los Angeles's water diversions). Therefore, they favored the dry remediation techniques, such as creating sand dunes and using sand fences, since these conditions would become more prevalent in the future. A key step in understanding temporal processes was identifying ecological markers that would alert managers over time as to the success or failure of remediations: for example, the expansion of saltgrass patches and growth of spring mounds. Students then derived three scenarios based upon various combinations of remediations from their toolbox, demonstrating how tasks within these scenarios could be phased to meet project objectives and be redesigned as needed if ecological markers indicated the necessity. They then forayed beyond the project boundaries and time frame by considering long-term and large-scale implications of the lake becoming stabilized.[16]

• Anne Whiston Spirn and Kathy Poole commonly outline their projects as an infrastructure, a fundamental necessity for human habitation. Because stormwater, a common infrastructure choice for both, is both natural and engineered, students are required to address the ecology. Since stormwater is a sequential activity (rain> runoff> collection> conveyance> detention> release> etc.), decision criteria and evaluative models stress process. In fact, the project is framed so that it cannot be solved without designing a process. And because the infrastructure is the primary concern of the studio, it is more difficult for them to treat it as "background" structure upon which they apply aesthetic expression. Rather, they are encouraged to understand the dynamics as a way of finding expressive content. Also, students imagine how the processes evolve over time by imagining what the future processes—and resultant forms—might be. And in these projections students must provide multiple projections rather than a singular "solution" (Figure 16-7).

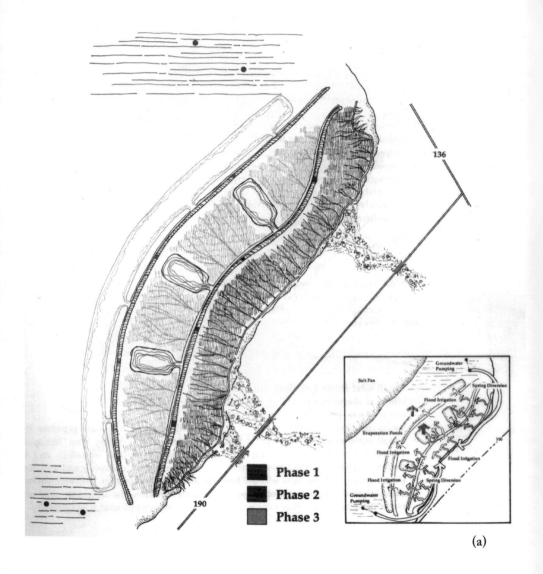

Phase 1
Phase 2
Phase 3

(a)

FIGURE 16-6.
In Cal Poly's 606 Studio, "The Shaping of Owens Lake," students apply their in-depth studies of spatial and temporal processes. In (a), students proposed an initial design with a set of aggressive interventions to accelerate the establishment of saltgrass along the lake margin. In (b), students projected the consequences of their design fifty years beyond its initial construction, speculating on the ecological function restoration as well as the rebuilding of public trust.

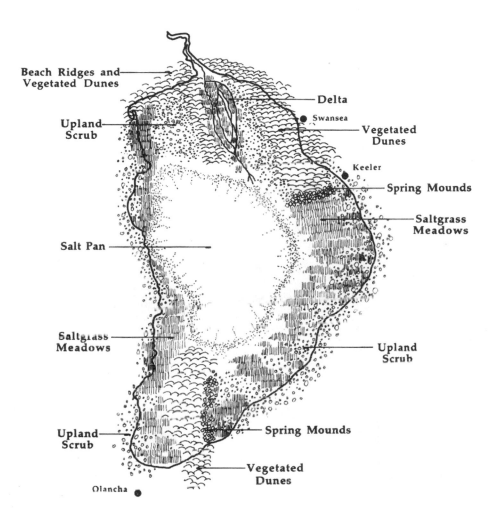

Beach Ridges and
Vegetated Dunes

Delta

Upland
Scrub

Swansea

Vegetated
Dunes

Keeler

Spring Mounds

Saltgrass
Meadows

Salt Pan

Saltgrass
Meadows

Upland
Scrub

Upland
Scrub

Spring Mounds

Vegetated
Dunes

Olancha

(b)

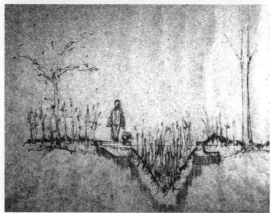

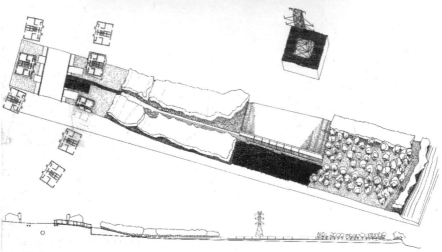

FIGURE 16-7.
Ecology as infrastructure. Rodrigo Abela used the process of drainage from top of bluff to river to determine more than its engineering. Water channels become the skeletal structure upon which the design is based.
(Instructors: Robin Dripps, Kathy Poole, and Daniel Bluestone.)

ADVANCED LEVEL: COMPLEX UNDERSTANDINGS OF ECOLOGICAL PROCESSES

KEY CHARACTERISTICS

- Instructor is a co-consultant with the students, giving them broad authority in guiding the content, methods, and products of the studio. Like the primary investigator, the instructor facilitates the process, maintaining a "big-picture" view and making relationships between the many components.
- Designers should not be able to solve the problem on their own; they should need to engage consultants.
- Designs must be evaluated by sophisticated models that students are capable of defining.
- No ideal solution is apparent; valuing of various aspects and opinions is necessary.

- "Biodiversity and Landscape Planning: Alternative Futures for the Region of Camp Pendleton, California" explored how urban growth and change in the rapidly developing area located between San Diego and Los Angeles might influence the biodiversity of the area. A team of investigators from the Harvard University Graduate School of Design (GSD), Utah State University, the National Biological Service, the USDA Forest Service, The Nature Conservancy, and The Biodiversity Research Consortium conducted the research (Figure 16-8). The intent of the research was to examine the connections among urban, suburban, and rural development and the consequent stresses on native habitats and biodiversity in a region that is one of the most biologically diverse in the continental United States. The major product of the research was a computer-based GIS and a set of models that evaluate the complex dynamic processes of the very large study area and the possible impacts on biodiversity resulting from changes in land use. A 1995 graduate-level studio at the GSD applied the research framework in the design and comparison of the implications of six alternative regional conservation/development strategies for the study region. The 1996 studio had the challenging task of proposing a "best alternative design."

  The studio's products included a range of media and intentions, spanning qualitative and spatial proposals, strategic development guidelines, and quantitative predictive models: diagrams demonstrating the performance of various aspects (best to worse); prioritization diagrams of various lands according to a particular value, not discrete designs; quantitative evaluations of species numbers according to proposed changes; modeling of various forms of physical buildout; and supportive diagrams of the

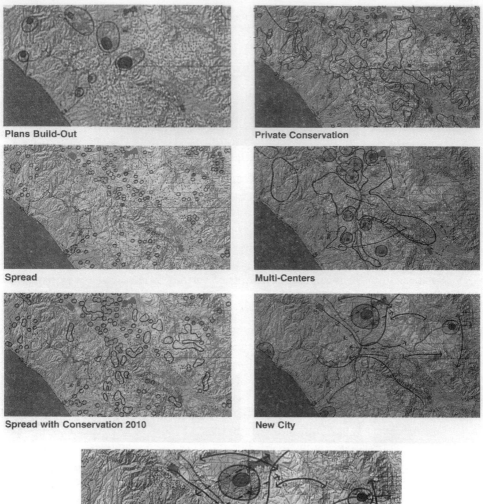

Plans Build-Out

Private Conservation

Spread

Multi-Centers

Spread with Conservation 2010

New City

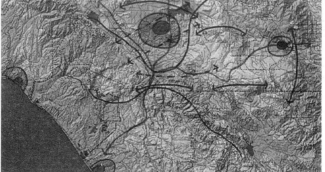

FIGURE 16-8.

"Concept Diagrams," Camp Pendleton Studio. Teams of students developed five "alternative futures" to the predominant regional trend of "spread" low-density rural residential and clustered single-family residential development. In addition to traditional design development of particular forms and spaces, part of the studio's work was to compare the economic, social, and environmental implications of the designs. (Instructor: Carl Steinitz.)

quantitative consequences of development (like hydrographs). All of these products accompanied traditional drawings showing design alternatives. Typically considered "analysis" items, the more quantitative products were actually post-design-proposal evaluations, tests of design criteria that students were required to produce early in the design's development.

Integrating Ecologists into Studios

Landscape design studios have a long tradition of inviting those of "other" disciplines into studio, such as architects, planners, and the clients involved in a particular studio project. And with the increasing attention given to natural systems, ecologists are being invited into studio and included as instructors. This is a positive and well-intentioned act; however, neither designers nor ecologists are overwhelmingly cheering the results.

Much discussion in this subject sounds remarkably like "How can designers strategize ways to include ecology without the ecologists?" or "How can designers make it so ecologists are not necessary?" A more constructive approach would be to ask how *each* of us could make our own expertise effective in various collaborative relationships. In our own analysis, we have identified four general models and summarize the benefits and limitations of each.[17]

Ecologists, whether designated to the studio or visiting, need to be included in the evaluation of studio projects, that is, giving grades or offering critiques. Until students are held responsible and accountable for the ecological information, they will regard it as peripheral. Yet, a key obstacle for involving ecologists in studio is the format of traditional studio reviews. They are familiar contexts for designers but new to ecologists, who are often a bit uncomfortable as part of a jury that inevitably focuses on compositional, cultural, or experiential issues. Consequently, instructors need to devise alternative review configurations. Some studio instructors structure presentations and evaluations that focus on ecological issues at strategic points before the final review. Others have "roving" reviews where various reviewers speak one-on-one with students.[18] Some find that ecologists are most effective when they operate like the studio instructor, giving desk critiques.

Most ecologists come to studio to teach, not to be a convenience. Often considered "walking encyclopedias" or substitutes for field guidebooks, they are viewed as dispensaries for ecological information. Rather than expecting ecologists to drill down information to an "answer," instructors should craft studio exercises in ways that allow ecologists to be critics in similar ways that design critics have always operated. Like designers, ecologists are different people with different agendas and skills who are there to help students articulate their own observations of a place and its issues. In this context, the ecologist is released from being a "factual fountain" or a "scientist," one of those beings foreign to design that students erroneously assume will not like their designs because scientists "expect certain things," or "do not care about aesthetics," or "do not understand our designs." Rather, the ecologist becomes a consulting partner in making a design better—on many levels.

Designers should not attempt to take on the role of ecologist (anymore than we should attempt to take on the role of sociologist, historian, or geologist). Showing relationships between ecology (and all other disciplines) and design is what designers are best at doing. And it is an important and formidable task that deserves the majority of our attention. Similarly, instructors should afford ecologists the same opportunity to stay within their own disciplines, to exercise their expertise. To be rigorous about their knowledge will only advance the application of ecology in design. Integrating ecology and design is less about blurring disciplines than finding ways of allowing the values of each to flourish—and creating opportunities for the disciplines to discover what those values are.

◆ ◆ ◆

Repertoire: Integrating Ecologists into Studio

DESIGNERS + ECOLOGIST AS CO-INSTRUCTORS

BENEFITS

- Students are more likely to see the ecologist's role and critique on equal terms with the design instructor.
- Ecology is part of forming the project, daily discussions, and exercises and, therefore, more likely to permeate content and methods.
- When ecologists are creative partners in forming projects, their teaching commitment is likely to be greater.

LIMITATIONS + IMPLEMENTATION OBSTACLES

- Ecology can become privileged content, diminishing the ultimate goal of integrating it.
- Institutional barriers make full or joint appointments almost a necessity. Low program enrollments make this difficult; one way to address this is to have an ecologist appointed to multiple studios simultaneously.

CASE STUDIES

- At Rutgers University, ecologist Jean Marie Hartman holds a full-time appointment within the Department of Landscape Architecture with association with the graduate program in Ecology and Evolution. She maintains a full scientific research platform with funding from major science institutions (usually in excess of a half million dollars). Trained with a Master of Science in Landscape Architecture and a Ph.D. in Plant Ecology, Hartman enjoys that one of her primary teaching responsibilities is teaching design studio, usually without landscape architect colleagues. This enables her to integrate ecological content in studio without the label of being a "scientist."

- At the University of Minnesota, ecologist Susan Galatowitsch has worked with a 37 percent appointment to landscape architecture (the rest is in horticulture). One of her most effective roles as a visiting ecologist occurred in a situation where the payoff was delayed. Students spent a semester with her in the fall field ecology course. By the spring studio, students were willing to trust her as a visiting critic to studio.

- At the Graduate School of Design, Carl Steinitz Petersham studio has an ecologist (also of the Department of Landscape Architecture faculty) assigned to the studio full-time (Figure 16-9). Steinitz coordinates the studio, but the ecologist has a major role.

FIGURE 16-9.
Harvard's Planning and Design of Landscapes studio always has an ecologist as co-instructor with Steinitz. (Designers: Nancy Parmentier, Cynthia Jensen, and Caroline Pendleton-Parker.)

ECOLOGIST BROUGHT IN AT STRATEGIC POINTS (IN-HOUSE OR VISITING FACULTY NOT APPOINTED TO STUDIO)

BENEFITS

- Students receive focused attention on particular ecological issues.
- Awareness of ecological content is maintained.
- Ecology is less likely to become privileged.
- Designers are challenged with demonstrating their value to scientists and improving interdisciplinary relationships.

LIMITATIONS + IMPLEMENTATION OBSTACLES

- It is a major time commitment on both faculty necessitating either (1) reciprocal visits to each other's classes, or (2) long-term collegial relationships between designer and ecologist (provided there are multiple points of contact).
- It is difficult for the complexity of most content ecological issues to be addressed in a "surgical strike" mode.
- Even though visiting ecologists may equip students with knowledge, students most often cannot readily apply it, necessitating more extensive strategic measures on instructors' parts (addressed in detail below).

CASE STUDIES

- In 1996, Robert Grese involved students in the development of a plan for a "prairie heritage" park for Brownstown Township, Michigan, a unique remnant of lake-plain prairie (less than 1 percent out of 160,000 acres remained) and one threatened by development. Working with Nature Conservancy ecologists who were diligently working to preserve the site, the students prepared designs that helped local people learn to appreciate the prairie habitat as a valuable resource and build support for the site's protection.
- In 1998 at Virginia Tech, Lee Skabelund invited a soil scientist/geomorphologist familiar with wetland systems to review student inventory and analysis work prepared for his landscape planning and management studio. In another studio, Skabelund invited an entomologist to join a studio discussion about conserving biological diversity, causing students to consider the concept in a much deeper way.

ECOLOGIST AS CLIENT

BENEFITS

- Students benefit from hearing ecologists discuss ecology in design projects from the perspective of ecologists, a rare and insightful position.
- An ecologist is much more likely to demand more scientific rigor of students.
- The studio receives "free" professional advice.

LIMITATIONS + IMPLEMENTATION OBSTACLES

- Projects tend to be infrequent in some locales.
- Matching of client and educational objectives may be difficult.

CASE STUDY

- Students in Cal Poly 606 Studio, a culmination studio, write proposals and pursue potential clients, often sparking the creation of a funded project where none existed previously (ninety-eight projects completed over twenty three years). Most clients are designers, administrators, and concerned citizens; several have been sponsored by ecologists.

STUDIO PROJECTS THAT ARE COINCIDENT WITH OR OUTGROWTHS OF RESEARCH

BENEFITS

- Greater scientific rigor demanded of students' work.
- Often students are exposed to quantifiable modes of thinking.
- Students must really exercise aesthetic creativity to avoid producing a "pretty science experiment."
- Once research is funded, the studio greatly benefits and may itself be funded.

LIMITATIONS + IMPLEMENTATION OBSTACLES

- Though there is a tremendous amount of information available, reformatting that information can demand extreme amounts of preparation on the part of the instructor.
- Having ecologist research collaborators does not equal having studio collaborators; therefore, all the same institutional obstacles exist.

CASE STUDY

• Kathy Poole is undertaking a long-term research project on Boston's Back Bay Fens with faculty members in environmental sciences and civil engineering. The project will construct a history of the Back Bay from its relatively unmanipulated seventeenth-century form to its present-day form, examining it as an evolving ecological, social, and institutional space. The effort also includes a digitized visual simulation of the Fens in its present state and a geographics information system scripting program that enacts ecological dynamics in the landscape according to proposed physical and biophysical changes. Design studios use the research as a way of exploring ecological dynamics. Conversely, the studios will test the adequacy and effectiveness of the ecological data and the project's modeling capabilities.

Removing Impediments + Enhancing Successes

Our repertoire of topical case studies offers specific strategies and tactics for integrating ecology within design studios. We hope that we conveyed our subject with optimism, showing that there are numerous positive and effective efforts in landscape architecture programs all across the country. We have a firm foundation upon which to build.

We close by synthesizing the ideas explored throughout the chapter, pointing to how we can further this project. Some of our observations are issues contained within studio and can be addressed by any studio instructor. Most suggest that there are larger curricular, if not institutional, issues that need to be addressed so that our efforts will be advanced in ways we have not yet fully imagined.

Ecology Is Not Privileged Content but It Is Central

Ecology needs to be treated as a necessary design element, not as a peripheral or secondary issue. It needs to be introduced at the beginning of the curriculum because this is where we state, as a department or program, what our values are. Our compartmentalization of it in "ecological design studios" and "landscape ecology" courses is valuable in terms of focusing on issues. However, when these courses are stand-alones and the content is not reinforced throughout the curriculum, ecology's importance is diminished.

The relationship of ecology to other courses could be strengthened if it were clearly articulated and discussed among faculty and with students. To critically reconsider one content's relationship among various courses helps each of us think and work with clearer and broadened perspectives.

Collaborative Teaching

We should focus on increasing collaborative teaching opportunities in studio. In some programs, this will mean joint appointments to departments. In others, it will mean being creative about academic lines and Faculty Teaching Equivalents (FTEs) so that ecologists can be a part of studio *and* continue to teach within their disciplines. It is critical that scientists remain scientists, still doing excellent, international-level research and operating within scientific circles.

We should also make sure that we train ecologists to work with and understand designers. Concerned ecologists and designers are already encouraging collegial, research, and teaching partnerships. Yet, it will take institutional changes to propel it forward. Models such as the University of Michigan's School of Natural Resources and Environment are rare. Here, the Landscape Program is strongly integrated with the other disciplines that are housed within the school. Ecology classes are taught by ecologists, and many landscape classes benefit from fellow students whose main studies are in resource ecology, conservation biology, environmental education, and resource policy (who sometimes make up three-quarters of the total students in a class).

If we cannot change the institution, we can sometimes buy our way into the collaborative, interdisciplinary situations that will benefit ecological content in studios. Team teaching is not monetarily efficient for the institution. And it is certainly not time efficient for instructors, often demanding twice as much to coordinate well. Yet, it is extraordinarily efficient in terms of student learning. It is also a very effective and efficient way of teaching junior faculty. Therefore, we might pursue grants for buying teaching release time. We can also be creative about leveraging options (money and otherwise) to "buy" ecologists for studios, a technique used by over half of the authors. For example, Joan Woodward has "bought" an ecologist to teach an environmental planning course that she has coordinated with a concurrent design studio. And Doug Johnston has agreed to teach University of Illinois planning students in exchange for having the planning department fund the ecologist to participate.

Producing Students with Realistic Expectations and Skills

Being a designer who sincerely cares about ecology does not make an ecologist. The same is true of designers who actually know a significant amount about ecology. It would be sheer folly to send the message that students' knowledge levels—whether working, talking, or quite sophisticated—were sufficient for successfully solving even moderately sophisticated ecological problems. Therefore, it is important to teach students how to make decisions given certain data levels.

We suggest two additional strategies for ensuring that students operate within their own levels of understanding. First, teaching failures—instances where people thought they had enough ecological knowledge but did not and ended up with disasters—are as important as teaching successes.

Another effective strategy is to knowingly require students to produce something that they do not know or do not know how to figure out. It is not a matter of designing for their failure. Quite the contrary, it is an opportunity for them to acknowledge that they lack data, knowledge, or methodological tools. Knowing the need to understand precedes the act of educating themselves or seeking expertise. This is particularly true of students who are most interested in formal aesthetics or who have no predilection toward ecology. If students have a compelling need for more ecological information for the sake of the design, then they are more likely to understand the value of ecology.

While we do not want to produce overconfident students, we do want to instill in students a sense of responsibility regarding ecological content. Landscape architects should see themselves as competent and responsible professionals as they apply ecological concepts to built landscapes. Currently, our training of students is giving them reason to abdicate responsibility or be timid with ecology. Generally speaking, they do not know enough, and we are not demanding that they be responsible for either knowing what to do or knowing when they should seek outside help. Their level of confidence, competence, and responsibility should be as equivalent for ecology as it is for aesthetics, construction, planting design, and history.

Linking History and Ecology

Finally, we need to amplify the relationship of history to ecology. In the teaching of landscape history at most programs, we give little attention to key figures and places in either ecology or ecologically concerned design. Just as students have a working knowledge of key figures in garden, park, and urban planning history such as Le Notre, Olmsted, and Halprin, they should have a working knowledge of key figures in ecology. This is not to suggest that the pertinent history resides within a few giants; rather, it is about four to six important people per topic.

Designers would also benefit from being familiar with thinkers important to environmentalism in a larger sense such as Emerson and Muir, not to mention scientists who contextualize their ecological knowledge within larger philosophical frames such as Aldo Leopold and E. O. Wilson. Susan Galatowitsch, an ecologist, addresses the issue in her Landscape Ecology class, teaching students ecological concepts (e.g., island biogeography) through the scientists (Robert MacArthur and E. O. Wilson). Her position is that showing science is the product of human creativity is more interesting for students and that design students respond better to personalities than to abstract scientific concepts.

Contemporary ecological history is as important as past ecological history. While BLA and MLA program students cannot acquire a thorough repertoire of key contemporary scientists, they can be offered a talking knowledge of key scientists and laboratories that they should follow throughout their careers. Just as we expose them to a relatively small handful of key contemporary designers, we would do well to give them guidance on who the key ecologists are.

Almost all studio instructors utilize case studies in their studio teaching. Most often, we use them to discuss formal, programmatic, cultural, and (sometimes) political relationships. In other words, we discuss environmental relationships but rarely include much ecological discussion. We could easily expand our critiques to include scientific information. Robert Grese, a Jens Jensen scholar, overlays his discussion of Jensen's interpretation of a region's native landscape with more empirical discussion of how those designs have changed over time. For example, Jensen's work at Lincoln Memorial Garden in Springfield, Illinois, was intended to represent the native landscape of Illinois that Abraham Lincoln experienced as a youth. Critical study of the site reveals that some of the plantings now closely resemble native ecosystems while other portions of the site have failed miserably. This is perhaps because of lack of attention to soil characteristics and Jensen's lack of understanding of certain ecosystem processes such as fire and threats by invasive exotic species. As students begin to understand the 60+ year history of a place like Lincoln Memorial Garden, they not only expand their critical skills but also build an ecological repertoire of case studies, an important part of their design tool box.

Final Reflections

While the discussion of history may at first glance seem peripheral to studio teaching, it is critical to teaching ecology in studio well. This is particularly important given the current, popular ecological design press stressing the return of landscapes to self-maintaining, "natural" states.

Designers of all expertise levels are gravitating toward low-technology strategies and "natural" aesthetics, eschewing the possibility of more intense, higher-technology solutions. This is unfortunate given that greater invention is sometimes necessary to remedy an ecological problem or to pressure something of ecological value. It is also unfortunate given that such a view tends to stifle creative imagination for solutions that through their high intervention allow people's connections to the natural world to be poignant and richly expressive.

When ecological history is taught well, students learn the history of science and its place within the history of design and its relationships within larger cultural frameworks. This enables students to discover the range of ecological conceptual frames and the extraordinary array of ecological expressions that are possible. A more rigorous understanding of ecological processes coupled with a more complete understanding of ecological history would not only allow designers to approach their work with less bias but also equip them to be more ecologically critical about their work.[19]

Ultimately, it is this ability to be critical about the relationships of ecology and design that is the most important matter. Professionals who can apply ecological knowledge in a way that illuminates new relationships between culture and nature will add immeasurably to the human experience. And designers who can work effectively (if not always easily) with ecologists and other professionals to maintain or reestablish healthy ecological systems will bolster our hope that the relationship between humans and the rest of nature will continue to be stronger and more profound.

As studio instructors, it is our role to foster environments where students can learn to build healthy ecosystems and further the human experience—to do so across both space and time, with both confidence and reservation, and with both judicious evaluation and resonating beauty.

Citations

Ahern, J. 1999. Integration of landscape ecology and landscape design: an evolutionary process. Pages 119–123 in Issues in landscape ecology. J. A. Weins and M. R. Moss, editors. International Association for Landscape Ecology: Fifth World Congress. Pioneer Press, Greeley, Colorado, USA.

Arendt, R. 1994. Rural by design: maintaining small town. Planners Press, American Planning Association, Chicago, Illinois, USA.

———. 1996. Conservation design for subdivisions: a practical guide to creating open space networks. Island Press, Washington, D.C., USA.

———. 1999. Growing greener: putting conservation into local plans and ordinances. Island Press, Washington, D.C., USA.

Cronon, W. 1991. Nature's metropolis: Chicago and the Great West. W. W. Norton. New York, New York, USA.

Dramstad, W. E., James D. Olson, and R. T. T. Forman. 1996. Landscape ecology principles in landscape architecture and land-use planning. Harvard University Graduate School of Design, Island Press, and the American Society of Landscape Architects. Washington, D.C., USA.

Evernden, N. 1992. Social creation of nature. Johns Hopkins University Press, Baltimore, Maryland, USA.

Karr, J., K. Fausch, P. L. Angermeier, P. R. Yant, and I. J. Schlosser. 1986. Assessing biological integrity in running waters: a method and its rationale. Illinois Natural History Survey, Special Publication #5, Champaign, Illinois, USA.

Lewis, P. 1996. Tomorrow by design: A regional design process for sustainability. John Wiley and Sons, New York, New York, USA.

Lyle, J. T. 1985. Design for human ecosystems. Van Nostrand Reinhold, New York, New York, USA.

Marsh, W. 1998. Landscape planning: environmental applications, 3rd edition. John Wiley and Sons, New York, New York, USA.

Nassauer, J., ed. 1997. Placing nature: Culture and landscape ecology. Island Press, Washington, D.C., USA.

Skabelund, L. R., and D. Borneman. 1999. "Ecological restoration and park development as interpretive and public education tools in the urban-suburban matrix: Furstenberg Nature Park, Ann Arbor, Michigan." Pages 218-225 in On the frontiers of conservation: proceedings of the 10th conference on research and resource management in parks and on public lands. D. Harmon, ed. The George Wright Society, Hancock, Michigan, USA.

Steiner, F. 1991. The living landscape: an ecological approach to landscape planning. McGraw-Hill, New York, New York, USA.

Swink, F. S., and G. S. Wilhelm. 1994. Plants of the Chicago region. Academy of Science, Indianapolis, Indiana, USA.

Van der Ryn, S., and S. Cowan. 1996 Ecological design. Island Press, Washington, D.C., USA.

Worster, D. 1977. Nature's economy: a history of ecological ideas. Cambridge University Press. Cambridge, UK.

Appendix 16A

Representative Curricula Highlighting Scientific Ecological Content

A key question in our discussions was how each landscape architecture program's curricula widely differed, necessitating that we approach our task as developing a repertoire for varying situations. No program is "ideal"; rather, each represents certain values. Nonetheless, the issue of curricula structure is an important one, especially with regard to our discussion. As a supplement to this discussion, we offer three curricula outlines, *broadly* representative of landscape programs across the country, current as of January 2000. The overriding criteria for categorizing curricula is the amount of ecological content consciously included within the core curriculum.

We acknowledge that ecology can be incorporated in courses in many ways. In the outline we continue to utilize our opening definition of ecology as concerning scientific issues. For efficiency we highlight only courses where ecology constitutes a significant portion of the course.

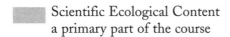 Scientific Ecological Content   Scientific Ecological Content
a primary part of the course   a secondary part of the course

Design Emphasis: University of Virginia, MLA first professional degree

In its focus on design as a synthetic activity and product, the department emphasizes the theoretical aspects of ecology as a cultural construct. Weavings of these theoretical explorations of ecology and scientific ecological contentrate the core of several seminars offered by landscape faculty. The University of Virginia operates on the semester system.

Summer Session

Landscape Design Studio   Landscape Representation   Computers

First Term	*Second Term*	*Third Term*	*Fourth Term*	*Fifth Term*	*Sixth Term*
Landscape Design Studio	Landscape Design Studio	Landscape Design Studio	Landscape Design Studio	Landscape Design Studio	Option Studio
Plants + Environment	Plants + Environment	Site Engineering	Material Detail + Expression	Material Detail + Expression	Road Design
Landscape Representation	Landscape Representation	Theories of Modern Architecture	Planted Form	Elective	Professional Practice
Landform + Grading	Applied Ecology	Elective	Architectural History Elective	Elective	Elective
Quest for Order*	History of Landscape Architecture				

* A multidisciplinary history-theory course, co-taught by faculty from each of the school's four departments: architectural history, architecture, environmental planning, and landscape architecture.

Increased Ecological Emphasis: Harvard University, MLA first professional degree

Around 1982 Harvard made a commitment to integrating ecology within its curriculum by creating two positions for ecologists (or scientists doing work applicable to ecology). Roughly six natural systems distributional electives are offered by the department and others in the University. Eight units of electives must be in natural systems to be selected from a list of approved courses issued by the department each year; one of these courses must be taken during the third term. Harvard operates on the semester system.

Pre-Semester

Site Ecology and Plant Communities

First Term	Second Term	Third Term	Fourth Term	Fifth Term	Sixth Term
Landscape Design Studio	Landscape Design Studio	Planning + Design of Landscapes Studio	Planning + Design of Landscapes Studio	Option Studio	Option Studio
Landscape Representation	Landscape Representation	Fundamentals of CAD	Methods of Landscape Planning	9 units distributional electives	10–14 units distributional electives
History of Landscape Architecture	History of Landscape Architecture	Landscape Architectural Theory			
Fundamentals of Landscape Technology	Fundamentals of Landscape Technology	Natural Systems Elective	Landscape Technology	Natural Systems Elective	
Plants as Design Material	Plants in Design	Site Planning	Plants in Design		

Primary Ecological Emphasis: California Polytechnic University, Pomona MLA first professional degree

The program is committed to involving ecology in all aspects of the program. CalPoly operates on the quarter system.

Summer Term	First Term	Second Term	Third Term	Fourth Term
Landscape Design Studio	Landscape Design Studio	Landscape Construction + Design		Methods + Applications for Landscape Architecture
Design Graphics	Landscape Construction Design	Seminar on Theory + Literature	Plant Ecology + Planting Design	Landscape Planting
Landscape Awareness	Seminar on the Profession		Seminar on Professional Directions	Ecology

Fifth Term	Sixth Term	Seventh Term	Eighth Term	Ninth Term
Design Research	Environmental Analysis	Landscape Technology	Ecosystematic Landscape Design	Ecosystematic Landscape Design
Landscape Design and Natural Processes		Graduate Seminar	Graduate Seminar	Master's Degree Project
Landscape Ecology				

Appendix 16B

Public Service Studios

While all studios are potentially "ecological," public service studios present some particularly attractive advantages for strengthening students' ecological understandings. For this discussion, we define public service broadly. Private citizens, corporate interests, public agencies, or studio instructors who see a public need can initiate the projects. They can be under contract, foundation-funded, or pro bono. What is common to these projects is that they involve real physical sites and are concerned primarily with issues that affect the general public.

One of the strongest advantages is that they involve real sites with real people, politics, physiologies, biologies, and chemistries. Unlike history, theory, and aesthetic considerations that can be abstract and are not often subject to the same empirical judgment as scientific method, ecology is based in biophysical processes. Faced, on the one hand, with visible problems such as erosion, exotic plant invasion, or evidenced poor water quality or, on the other hand, with healthy-functioning landscapes with balanced grassland-to-elk populations, high biodiversity, or rare species, it is difficult to skirt or merely conceptualize the ecological issues. Criteria must be articulated, and successes must be demonstrated.

Hypothetical projects can often allow students to remain emotionally neutral because students' designs will never actually do any harm or good because they will not be built. Public service projects not only have real consequences that affect people's lives but also have constituents that are vested in the potential outcomes. Consequently, students must take their work seriously, and the projects usually demand that students reconcile conflicting intentions, only some of which concern ecology, reinforcing the importance of decision criteria and necessity for framing ecology socially.

Since the public is usually involved, these studios are excellent mechanisms for students to explore how to represent complex systems in relatively straightforward, understandable terms. In the process of making the ecology, their intentions, and their designs "imagable," students become more articulate about the project's ecology and its relationship to design.

Finally, public service projects often demand follow-up that hypothetical projects do not. This presents compelling opportunities for instilling in students the importance of monitoring the ecological effects of the designs.

Our advocacy of public service projects is not meant to dismiss their challenges (all of which extend beyond ecological project content):

- No matter how casual the client-institution relationships, it is worth taking the extra step to get at least of letter of understanding or agreement to make sure the products and scope are clear (Iowa State's financial arrangements, in essence, necessitate such agreements as all of their public service studios are funded—$1,000 to $100,000).[20]
- Seriously address the issue of intellectual property rights (at Iowa State University, the department owns them and makes copies available for interested parties).
- Seriously address the issue of faculty and students continuing the project's work as a professional (Iowa State prohibits faculty from doing future paid work and is moving toward the same prohibition for students).
- Devise a strategy for the inevitable follow-up of the studio—in terms of either documentation or monitoring (Carl Steinitz always builds in funding for someone to compile a document, post Web sites, etc.).
- Be clear about the neutrality of the studio in the crafting of the project and any contract. Tim Keller and Carl Steinitz, both of whom have long-standing experience with public service studios, advocate that the studios offer no "solutions." Instead, they suggest presenting "alternative futures."
- If contracted to do professional-level work, have professionals in studio and legal insurance/assurance.

Clearly, the content of projects differs for different levels of students' ecological expertise. Beginning-level students may only be capable of generating ideas to be taken up by professionals. Students with more proficiency model and project the ecological consequences of their designs. And some schools, like California Polytechnic University, Pomona, have contracts stating that they are doing professional-level work, which fosters the necessity of ecologists within the studio. Whatever the level of ecological understanding, choosing a public service project often reinforces students' ecological learning.

Notes

1. Jack Ahern discusses the potential for "reciprocal integration" of landscape ecology and design in his short paper in *Issues in Landscape Ecology*, Weins and

Moses, editors (1999), which builds upon earlier works by Nassauer, editor (1997), Lyle (1985), and others.

2. Ecology as a social construct is well represented by Chapter 2 in this volume. See also Evernden 1992, Worster 1977, and Cronon 1991.

3. There are many exciting examples from outside of our working group. We have limited ourselves to personal cases or examples of which we have personal experience. Also we include some examples from nonstudio courses. While we acknowledge the differences between studio and courses of other types, in our discussions we have found it counterproductive to make distinctions. Instead, we have taken a pragmatic approach: If it works, consider using it or adapting it.

4. For a discussion of the differences between processes, dynamics, and creativity—and their relationship to design—see Kathy Poole, "Creative Infrastructure: Dynamic Intersections of Urban Ecology and Civic Life," Proceedings 2000 of the Annual Conference of the American Collegiate Schools of Architecture, Los Angeles, 11–15 March 2000. For an in-depth discussion of the relationships between ideas of ecology to design, see *Landscape Journal*, Special Issue: Nature, Form, and Meaning 7(2): Fall 1988, guest-edited by Anne Whiston Spirn.

5. For example, in some studios, Poole will initiate a "surprise" exercise. Since her work focuses on water, typically she "floods" the site. Students do not decide how to redesign to make their designs protected from such an event. Instead, they accept the flood as a real event after which they must decide how to deal with the effects. They can decide to rebuild differently. They can decide to move their site. They can decide to change their construction technique. They can devise a way to allow the flood's memory to resonate. They can decide how to make the site stay flooded. They can decide how to make the site flood more often. In this way, the process of flooding becomes part of the process of design.

6. As noted ecologist Eugene Odum coined in an article of the same name, "Environmental Degradation and the Tyranny of Small Decisions."

7. While we focus our efforts on ecological content, we realize that the process and content of our approach are applicable to the structuring of all design studies, regardless of the level of ecological emphasis.

8. It is worth noting that Johnston's approach is quite different from some other "regional" studios, for example, Steinitz. Rather, it is a very focused and definitively not a comprehensive approach to teaching studio.

9. As discussed later in "Integrating Ecologists within Studio," research venues are also excellent teaching and learning opportunities. In his work on the proposed New River Parkway Environmental Assessment and New River Parkway Land Management System (I-64 to Hinton, West Virginia), Lee Skabelund has involved many disciplines with ties to ecology (state, federal, and local agencies in additional to several subconsultants). In the course of the project, more than forty Virginia Tech students have been involved, including students from landscape architecture, planning, environmental design, and architecture. The project is supported by a multi-million-dollar funding base.

10. In some curricula there are good reasons for having studios that focus on ecological issues. When other faculty are not interested or prepared to explore ecological content, it may not receive appropriate concentration within the studio

for students to learn enough to be ecologically competent and confident practitioners.

11. The exercise is also effective as a "time out" to design as late as three-quarters through the semester. The point is to have students re-view the site through a mapping exercise that has them look at the site anew or reevaluate a forgotten relationship.

12. We note four sources for poetic representations of quantitative information. Edward R. Tufte has written three books on the subject (*The Visual Display of Quantitative Information* 1983, *Envisioning Information* 1990, and *Visual Explanations* 1997). See also James Corner and Alex S. Maclean, *Taking Measure across the American Landscape* (New Haven: Yale University Press, 1986).

13. The releve technique entails recording all of the species found in a large plot (15 by 20 meters for forests) and annotating this list with the vertical strata in which the species was found and the aerial cover the species achieved for each strata.

14. Lee Skabelund tried this idea in his landscape architecture studio, teaming up with an urban affairs and planning studio that was addressing the same study area as his students. Skabelund added a discussion session at the beginning of the second week of the semester to initiate dialogue between the planning and landscape architecture students. This combined studio discussion broadened the issues that students addressed but did not allow for as much depth in some of the issues most relevant to landscape architecture students. A deeper understanding of ecology, applied to design and planning, was possible as students read from Marsh 1998; Van der Ryn and Cowan 1996; Dramstad, Olson, and Forman 1996; Lewis 1996; Steiner 1991; and other texts and articles.

15. It is important to note that in addition to the studio, Steinitz teaches "Theories + Methods" as a prerequisite to the Landscape Planning and Design Studio, ensuring that students can complete the quantifiable and methodological models asked of them.

16. As described in *Landscape Architecture* magazine, "Implementation of the plan will follow a sequence like that of natural succession. As vegetative cover is established and air pollution is reduced, it is hoped that a diversity of natural life and human uses will return to the basin." The project won an ASLA merit award for planning and urban design in 1995. "Winners Circle," no author, November 1995, p. 60.

17. We identified one other model: Studio Project Undertaking Ecological Research through Design Interventions. The advantages are attractive: Greater scientific rigor is demanded of students' work, students are exposed to quantifiable modes of thinking and rigorous methodologies, and students are exposed to the importance of monitoring. The disadvantages, however, are formidable and why we do not include the option within the body of the paper: It is difficult to design a project that is not postoccupancy, and it is not a reciprocal activity because the scientist must maintain creative control. The lynchpin is that the data come in over years, not weeks, so there is no way to construct designs and monitor and alter them. Perhaps the real opportunity for the model is for design and ecology faculty to collaborate and implement the model, not as a studio function but as a living case study from which students can learn.

18. Individual discussions between critic and student are a standard, proven review technique for any kind of studio project, ecologically concerned or not, at the University of Oregon.

19. Recent publications offer considerable encouragement. *Eco-Revelatory Design: Nature Constructed/Nature Revealed*, Special Issue 1998, catalogues a set of commissioned landscape architectural designs currently touring the nation. George F. Thompson's and Frederick R. Steiner's 1997 book (in which they were editors), *Ecological Design and Planning*, contains a series of essays that frames ecology by viewing it in "Retrospect," in "Prospect," and in "Respect." And William Cronon edited a collection of essays in *Uncommon Ground: Toward Reinventing Nature*, 1995.

20. Harvard's public service contract not only protects Harvard from lawsuits but protects the university from a particular political or personal view by stipulating that the studio will advocate neutrality in its presentation of numerous possible outcomes.

CHAPTER 17

From Theory to Practice: Educational Outcomes in the World of Professional Practice

René Senos, Carolyn A. Adams,
Dean Apostol, and Jurgen Hess

In this chapter we consider the core competencies required for ecological design, and propose tactics for creating a dynamic learning cycle between education and practice. We explore how academic programs can best prepare students for a socially and environmentally responsible practice by asking several key questions: How can we integrate science and design in education and practice? What are the roles of ethics and markets? What skills do practitioners need to perform ecological design? What are the opportunities to apply ecological knowledge in practice? How can links between education and practice be reinforced to ensure that learning is reciprocal and relevant?

In addressing these questions we evaluate the strengths of landscape architects and planners that apply to ecological design and planning, as well as professional traits that may run counter to an ecological imperative. We incorporate a practitioner's perspective on building an ecological design career and recommend strategies to foster a vibrant practice based on ecological values and principles. Finally, we consider how practitioners and educators can foster lifelong learning that offers practitioners updated tools and critical pathways for inquiry.

Opportunities for Growth: Linking Science and Landscape Planning

Many educators and practitioners in this volume and elsewhere agree that ethical design and planning practice entails an ecological perspective. This imperative encompasses urban centers and greenspaces, suburban and rural development, forestry, agriculture, and watershed management. It requires simultaneously mending ecosystems and social systems. Furthermore, an inclusive design approach that sees natural and cultural processes as intertwined may help us move beyond language—and practice—that divides the world into "natural" and "human" (Spirn 1984; Chapters 2 and 3). The connections between ecology and design should be clear. Both focus on the dynamics of the landscape, its constituents and processes. Landscape architecture's rich tradition of designing human landscapes offers applied ecologists an informed perspective for including people as integral components of ecosystems. In turn, designers and planners can learn from ecologists how to incorporate the complex dynamics of ecological communities into design.

Human societies are faced with mounting problems that go to the core of the relationship between culture and nature. In the Pacific Northwest, treasured salmon may be headed for extinction. We look in alarm at what we have done to rivers and streams and ask hydrologists and fish biologists to offer solutions. Scientists provide study after study to identify the problems: overfishing, clear-cutting, farming, dams, subdivisions, and urban development. The very building of our culture has created changes to rivers and streams that salmon cannot survive. Scientists and engineers offer technical fixes for some of the site-specific problems they uncover: plant a riparian zone there, replace this culvert, reconnect that old side channel, and so forth. But the problem is not merely an accumulation of small insults. As biologist Jim Lichatowich (1999) observes, restoring salmon requires that we humans rewrite our story by building a new salmon culture.

How can we chart this new course? How will science contribute to our understanding, and what role will planners and designers assume in crafting solutions? Landscape architecture often is described as an applied art. As such, it has to do with creative skills brought to practical problems. Landscape architects rely on intuition at least as much as they do on empiricism or reason. This can be a strength, but it also can pose a challenge when it comes to interacting with scientists. Scientists want to know what data support that intuition. They want to see the rigor in our research.

How can the natural sciences be harnessed to the problems of landscape planning and design? Different uses of ecology offer potential intersections with landscape theory and practice. Ecology is a *science* of understanding interconnected relationships. It has been thought of as a *philosophy*, a synthetic approach to characterizing the world (Karr and Chu 1999). Finally, although it is not typically viewed in this manner, it may be used to derive an *aesthetic* that draws on an appreciation of underlying processes (Nassauer 1997). These ways of characterizing ecology suggest entry points for imaginatively reconstructing design curriculum and practice. For example, we can draw lessons from ecology about productivity and energy cycling, and apply them to community design. Creative reimagining of human dwelling in the world may represent the next step in design evolution, a maturation of our relationship with nature. Rather than attempt to control nature or imitate natural forms aesthetically, nature's example of coexistence may inspire us to change our relationship to the environment and to each other. The designer's challenge is to bring these lessons into the world of practice.

New opportunities for ecological design and planning are emerging, and landscape architects need to respond to this demand to stay relevant (Chapter 5). For example, the Environmental Protection Agency (EPA) is encouraging communities nationwide to promote locally created projects that address environmental problems through sustainable development strategies funded by the EPA's Sustainable Development Challenge Grant (SDCG). EPA Administrator Carol Browner is a strong advocate for the involvement of the landscape architects in these projects, urging design leaders to increase their involvement by working on projects within the program's scope, and galvanizing local communities to apply for grants (LAND October 1999).

ASLA has tremendous leadership potential, demonstrated by the selection of the American Society of Landscape Architects (ASLA) as a key partner in the White House Livable Communities Program, an executive program to reduce urban sprawl and create healthy communities (LAND April 2000). The ASLA is a critical educator and role model for both the practitioner and the public. This is demonstrated by the professional standards ASLA sets, the kinds of projects it awards, and even the types of landscapes the organization publicly recognizes. Recently, ASLA designated the Columbia Gorge Scenic Area as a Medallion Landscape, a positive step in the direction of bringing attention to a culturally and ecologically significant bioregion (Figure 17-1). The landscape design and planning profession must demonstrate these and many more examples of strong initiative and leadership to make a watershed impact on our communities.

FIGURE 17-1.
Multnomah Falls, Columbia Gorge Scenic Area. Design educators and practitioners can bridge the leadership gap in ecological design with greater emphasis on ethics and stronger connections between curriculum and practice. The American Society of Landscape Architects recently nominated this area as a Medallion Landscape. (Photo by Jeff Lanza.)

Reimagining Olmsted

We see a troubling disparity between standard practice and the demand for ecological design. Despite growing interest in an interdisciplinary approach, the relationship between ecological science and the design profession is curiously tenuous. Certainly there are ample opportunities for ecological practice. The listing of salmon runs under the federal Endangered Species Act in the Pacific Northwest is an unprecedented action that impacts high-density urban centers such as Portland and Seattle. Salmon defy human-made boundaries in their anadromous course between ocean and spawning stream, and for the first time an endangered fish species has landed in urban dwellers' backyards. This problem represents a unique opportunity to move beyond "single-species" approaches to whole community design. A few landscape architects are taking the lead, as Steve Moddemeyer at Seattle Public Utilities has done by promoting community-based efforts to salmon restoration (Seattle Public Utilities 2000). Why aren't more landscape designers and planners at the forefront of this issue?

One reason we generally do not engage these projects may be that they fall outside the standard realm of practice, especially private practice, and it is easier to stay with the norm than to set trends. A recent survey by the ASLA of its members determined that the top three clients of the profession are developers, homeowners, and architects, respectively. Far and away the biggest project types are residential projects at 27 percent, while other project types such as waterfronts, wetlands, parks, historic, stormwater, and agricultural project types each comprise less than 1 percent of landscape architecture work reported (ALSA 1998). While not all landscape architects are members of ASLA and some types of practice such as public practice may be underrepresented, these figures nonetheless reveal the prevailing work emphasis of the profession. The statistics suggest what many of us have observed, that our profession has relatively few models of ecological practice to reference. Many designers are familiar with prototypical modern or postmodern landscape design projects, but how many can name models of sustainable design?

Nonetheless, students enter design programs with a fervent desire to become environmental stewards, a desire that is encouraged and supported in their education. When they enter the workforce, many fledgling practitioners find that private-sector practice is market-driven, and that most paying projects focus on traditional design problems such as public use, aesthetics, circulation, and site engineering. Political, social, and economic pressures carve out the "regulatory landscape," which shapes how practice is actually carried out (Chapter 7). These pressures may or may not be addressed within education programs even when those programs promote ecological design. Design programs need to address economics and politics, since costs and policy drive private and public practice.

These issues reflect the underlying question of "who leads" landscape architecture with regard to ecological practice. Educators try to anticipate emerging trends and offer leadership while at the same time serving the needs of the profession. But because practice is oriented to market and client, environmental ethics introduced in school may or may not be reflected in practice. There are too few "leading lights" of ecological design and planning. Many more practitioners would like to incorporate ecological design principles in their firm, but are at a loss for the right resources. At the same time, there are other designers who may never question what organisms can actually inhabit their proposed landscapes or what impacts their designs have on ecological processes. Most of us fall somewhere in between.

How can we locate ecology in the heart of landscape architecture and planning? All too frequently ecology is an afterthought to designs, an appendage rather than the organizing framework. University studios and design firms must both commit to constructing solutions that animate our landscapes. Designers and planners could be true visionaries in changing the form and function of our cities, farmlands, forests, neighborhoods, urban plazas, streams, backyards, and yes, golf courses. If Frederick Law Olmsted were designing landscapes today, what would he be doing? Reimagine Olmsted, and you might end up with another "emerald necklace," a restorative greenway that connects the cultural and ecological pulses of the community.

The veil between practice and education must be made more permeable so that the values and skills of students match the realities of practice, and practice is elevated to a higher level by fresh faces and ideals. Socially and environmentally conscientious students are graduating to a profession that offers few opportunities for ecological practice. New practitioners are often faced with the unexpected choice of either suspending values or detouring to another type of practice. Multidisciplinary or engineering firms are actively recruiting design and planning graduate students, and offering higher salaries. The number of landscape architects entering engineering firms is 9 percent, up from 4 percent in 1988 (1998). This shift is a boon for practitioners' livelihoods, and begins to establish the kind of cross-disciplinary dialogue required to solve complex problems, but it may mean that some practitioners most capable of developing and marketing ecological design in traditional design firms are being lost to other professions.

Stronger partnerships with universities could benefit design professionals by providing technical support for sustainable design and planning. Information and expertise need to flow freely between universities, research labs, and design offices so that practitioners can respond to the market with ecological design solutions that are based on "best science," and address complicated social, economic, or policy issues. This is an important point, because the profession's credibility rests on sound methodology, and may explain why some landscape designers and planners are reticent to venture into the fast-changing world of ecology. Without adequate skills and resources, designers risk not only their professional integrity but also violating the precautionary principle and causing harm to the environment (Chapters 8 and 12).

If education and practice become more committed to ecological design, there may be ways to increase designers' and planners' influence on environmental design and policy. One approach may be to adopt a basic code of ecological ethics that spans the various types of practice. Designers and planners can be key educators of the public as part of developing markets for ecological design. They can change what they recommend to clients while still responding to client's concerns with sensitivity. Finally, they can become activists in promoting environmental and social equity, speaking up for the well-being of all landscape constituents. Now that we better understand the profound interconnections embodied in Earth's distressed ecosystems, it is time to ask the question, "Who is the real client?" ASLA's published objection to current standards for plant genetic testing is a positive step in this direction (LAND September 2000). As leaders in the public realm, landscape designers and planners can be of great service to the immediate client and the client environment.

The Nature of Practice: Where Does Ecology Fit?

The diverse ways in which practitioners interpret and apply design theory principles reflect cultural perceptions of human relationship to nature, and affect how natural and built elements of landscapes are interwoven (Spirn 1984; Chapter 2). On one hand, the breadth of the discipline allows designers and planners to work in many realms, including private firms, public agencies, universities, and nonprofit organizations, with numerous opportunities for influencing and participating in landscape decisions. On the other hand, this diversification leads to questions of identity within the profession and confusion in the public mind—who or what is a landscape architect?

Combining these disparate aspects under a unified body of theory and practice is a daunting challenge. Tracing the roots and branches of practice, however, may clarify the design profession's identity and integrity, and help determine desirable directions. The American Society of Landscape Architecture has attempted to "map the territory" of the profession in order to guide practitioners in market development (ASLA 1997). This information can help educators be more responsive to the needs of the profession; better prepare students for practice; and assist practitioners' choice of a career path. Defining the field promotes greater public awareness of the role of landscape architects in designing healthy landscapes. In turn, the public can require greater accountability from the profession.

It is crucial for the design profession to perform such self-appraisal in terms of ecological accountability. Sandra Steingraber, Catherine Howett, and others urge us to examine our environmental design ethics. Is an ecological perspective truly "optional" in view of the current state of the world or when we consider basic human health issues such as clean air, soil, water, and food (Steingraber 1997; Chapter 8; Howett 1998)? A fundamental question confronting both educators and practitioners is whether designers and planners should strive to make an ecological framework the central organizing principle for their work (Chapter 12). Just as design schools are under pressure to reprogram curricula according to ecological imperatives, so too are design and planning firms compelled to change and adapt their structures. This pressure comes from many sides: environmental and health activists, scientists, public agencies, concerned citizens, and students, all seeking relevant and thoughtful design approaches to degraded landscapes.

Practitioners have a wide array of choices both inside and outside the circle of standard practice to incorporate ecology. Most landscape architects and physical planners continue to work in traditional settings in private offices, public agencies, and university departments. A recent survey of ASLA members indicates that 78 percent are in private practice, 17 percent public, and 5 percent academic, findings consistent with previous studies (ASLA 1998). Increasingly, practitioners discover opportunities within and outside of these settings to focus on environmental issues. New specialties are evolving in response to complex problems and more sophisticated problem-solving technologies. Today's landscape architect may focus on spatial analyses, watershed planning, natural resource management, ecological restoration, or community participatory design. Powerful opportunities for cross-pollination occur when researchers and educators step into practice, or when practitioners teach in a studio setting.

As social and environmental problems increase in complexity, new tools are being developed and refined to study landscapes at different resolutions and in different ways. Technologies such as remote sensing and geographical information systems offer sophisticated analytic tools (Chapter 9). New types of research are informing planning for large landscapes (Chapters 15 and 10; Hulse et al. 2000). These advances are redefining the design profession, and blurring the boundaries between landscape architects, cartographers, and planners. It is increasingly common for landscape design students to venture into other professions and assume roles not directly related to traditional design. Students may take positions that deal with land management, watershed coordinators, conservation planners, or policy administrators.

Given the expanding nature of practice, clearly no single design and planning program can be all things to all students; rather, schools should identify their philosophical and programmatic strengths, and clearly set those forth (Chapters 16 and 14). Nor will all the branches of practice grow identically. However, we believe there are core values and skills that need to be an intrinsic part of any student's learning, and, we hope, a part of every practitioner's work, for designers to provide meaningful solutions to environmental problems.

Skills for an Ecological Practice

What are the core competencies that enable students and practitioners to responsibly translate cultural and ecological knowledge into physical form? How do landscape designers link knowledge to values? As one Shire Conference participant asked, "How do you learn to think from the heart?" Ask any student, professional, or professor, and you will hear a wide range of answers. However, several overriding themes consistently rise to the surface.

Problem Definition

If we consider the skill base that most landscape architects have to draw upon through academic study and professional practice, we find that landscape designers and planners offer the following competencies. First and foremost, a designer's training and practice involves practical problem-solving skills

TABLE 17-1. Problem-solving skills in ecological design

- Proficiency in multiscalar synthetic thinking.
- Understand and ensure ecological content in design programs.
- Know policy and regulations that are based on ecological health.
- Ability to ask the right questions.
- Use critical thinking to assemble the right colleagues to consult and work with.
- Reinterpret complex ecological issues for inclusion in design programs, policy briefs, or community presentations.
- Recast simple design problems in terms of complex ecological contexts.

that may be applied to ecological design and planning (Table 17-1). The problems designers and planners engage range from fairly simple and straightforward (planting plans for suburban branch banks) to very complex (new towns, national parks, management plans for wildlife refuges). As learners, landscape designers become acquainted with natural science education, including botany, geology, ecology, soils, and hydrology. This familiarity provides a basis for understanding what science-based colleagues are talking about, and what information their data sets or maps might be conveying. Through exposure to urban and rural planning, designers comprehend the complex interactions between politics, economics, transportation, and other social issues that drive landscape projects. This bird's-eyeview enables landscape architects and planners to view ecological problems within a social context, whereas natural science colleagues may have a much harder time integrating issues across landscape and social scales.

Design theory and research historically have directed a great deal of inquiry into the problem of how one defines problems (Steinitz 1995; Chapter 10). This discussion is even more vital in design education today. Designers and planners must be able to define environmental problems clearly, accurately, and comprehensively to prescribe solutions that are compatible with the social and physical landscape. This entails comprehending the economic, political, cultural, and ecological forces at work in any given landscape system prior to promoting specific change.

As a first step, designers and planners develop proficiency in multiscalar synthetic thinking and visualization. They must consider how a project fits within both smaller and larger contexts just as a camera moves between wide-angle and telephoto view, alternatively bringing into perspective an entire watershed or a storm drain within that watershed. While the focus of a problem may be directed on a single element, the designer keeps sight of the entire picture. Conversely, when attention is on the big picture, the designer remains aware of all the constituent parts (Chapter 3). Good environmental designers are cognizant of the patterns and processes that link landscapes and cultures over temporal and spatial scales.

While it is essential for designers to be able to think across gradients from fine-scale to broad-scale design, it is equally important to translate issues across gradients of complexity. Practitioners must take complex ecological issues and interpret them in comprehensible form for inclusion in policy briefs, design plans, or community presentations. Conversely, practitioners must also recast "simple" design problems in terms of complex ecological contexts. For example, the cumulative effects of individual homeowners' disconnected downspouts on native landscapes need to be factored into watershed planning.

How does a designer gain the experience and knowledge necessary for ecological fluency? A competent practitioner must be committed to understanding and ensuring ecological content in all design. While a grasp of key ecological principles is a fundamental requirement, perhaps a more significant quality is that of the "dedicated seeker" who actively questions and seeks updated information, tools, methods, and expertise. This approach entails a "transdisciplinary" approach, gathering the right experts for collaboration to unify diverse perspectives (Chapter 15). It also requires keeping abreast of policy and regulations that impact land-use choices. Finally, a reflexive ability to ask the crucial questions may be the most important skill a designer or planner can cultivate.

Analysis and Synthesis

Design students develop strong skills in analysis and synthesis in the course of addressing environmental problems that range from backyards to bioregions. These are best illustrated in studios that analyze overlays of various resources, while working toward a synthesis that harmonizes human uses with the land (McHarg 1992). At its best, this process allows designers to marry the craft of design (form giving) with ecology (dynamic relationships between organisms and their physical and biological environments). Our confidence in taking marker to tracing paper, pen to Mylar, or mouse to digitizer board is unmatched by other natural resource professionals. Designers typically have just a few short weeks to pin a solution up on the wall, and we learn to get to it.

Synthesis describes a process for how designers frame and answer questions. Designers and planners are adept at collecting, deciphering, and analyzing complex data that facilitate deep understanding of a site toward to a thoughtful design solution. But landscape architects and planners do more than just collect information. We gather input from many sources, distill the essentials, and critically make the synthesis spatially explicit. Unlike single-focus approaches to environmental issues, a design approach considers multiple facets of a particular problem. Designers communicate this view to other professionals, officials, and citizen groups, and incorporate their feedback. This process theoretically enables us to determine a solution that best fits the needs of a place or the people who use that place.

Landscape architects can use their skills as synthesizers and visual interpreters to develop design solutions that bridge scientific data with complex social and ecological processes. To do this, design practitioners must work with natural resource scientists to understand site-specific and watershed-specific processes. Interdisciplinary collaboration is essential, and we need to push beyond the limits of our knowledge by gathering the right team of experts together. The complexity of landscape issues often requires a much more comprehensive analysis than the standard overlay model of soils, climate, vegetation, and the like. We need to know when to bring in the hydrologists, forest engineers, wetlands biologists, economists, and our other colleagues to develop a full picture of a landscape's patterns and processes. In a way, this approach revisits the McHarg model of landscape architecture: McHarg's interdisciplinary studios at the University of Pennsylvania were legendary, and he hired more natural scientists than designers on the faculty.

Communication and Facilitation

Designers and planners are required to be excellent communicators visually, verbally, and in writing. With drawing hand connected to the brain, we offer tremendous creativity in communicating ideas and promoting appropriate action on the ground. Designers are adept at translating complex issues into images easily grasped by nonprofessional audiences. Large-scale ecological problems are inherently public problems that require processes and products that can be understood by nontechnicians. Effective visual communication allows us to show spatial form and relationships in ecosystems and to illustrate how alternative solutions will modify these places.

Yet we have to be careful to use graphics tools honestly and wisely. We can become mesmerized by pretty pictures and forget to dig deeper into content. If we can fool ourselves, perhaps we can also fool our colleagues and the public. One of the authors recently attended a seminar at the University of British Columbia, where the main topic was the use of "visualization tools" to help forest managers develop plans and communicate with the public (UBC Forestry Conference February 1999). This discussion explored how difficult it is to portray complex ecological concepts with two-dimensional pictures, no matter how advanced the technology. In fact, this problem goes deeper, back into our distant past as a profession that reshaped entire landscapes in order to achieve "picturesque" qualities for clients with too much disposable wealth. We must exercise graphic skills responsibly to get past the surface of the picture and into the depth of landscape function and form.

As capable writers and public speakers, designers can effectively relay complex ecological design ideas. Practitioners require good training in listening, facilitation, mediation, and conflict resolution to bridge communication and value differences across divergent stakeholder groups. Yet we must overcome discomfort with public involvement and develop ways to engage people in co-creating solutions that are meaningful to them. Increased recognition that environmental and social problems are intrinsically linked is prompting more designers to shift from *ego*centric to *eco*centric problem-solving strategies. Innovative landscape architects and planners solicit public input at the onset of a project, so that design solutions informed by technical information rise out of the community itself, and are generally more representative of public values and concerns (Hester 1990; Apostol et al. 1998; Johnson and Campbell 1999; Hulse et al. 2000).

Solutions and Implementation

Landscape architects can adopt a prominent role as leaders in ecological and social problem solving by harnessing relevant professional skills to the task. Landscape designers and planners work in public forums; are effective 'translators" of spatial form; synthesize complex data and information; mediate conflicting interests; and are able to visualize the impact of various human choices on the landscape. Our work is oriented toward finding practical solutions, and doing so requires an ability to think "out-of-the-box" and adapt to change through innovation, creativity, inspiration, and integration.

Practitioners must be literate in basic scientific, political, and economic concepts to engage in dialogue with scientists and develop appropriate design proposals (Chapter 13). It is not enough to install a project and walk away, and this notion points to another fundamental tension between standard practice and ecological practice. Ecological interventions are typically long-term ventures. Natural and social changes require time, patience, and a tolerance for uncertainty. Plans must be fluid, dynamic, and adaptive, and use a time frame that extends beyond human life spans. Designers and planners must understand the fundamentals of developing monitoring plans, as well as protocols for adjusting strategies based on what works in the landscape.

Ecological Design Values

Where does a commitment to ecological design come from? In a social and political climate deeply divided on environmental issues, the will to adhere to ecological principles regardless of the regulatory bottom line may arise foremost from ethics and values. In landscape architecture and planning, these ethics and values need to be stated explicitly by individual practitioners, by academic programs' mission statements, and by the relevant professional societies. We believe that programs have an obligation to model core values to their students, and to offer classes in ethics and professional practice. We also believe that the design profession should cast its definition of health, safety, and welfare broadly to include practices that impact social and ecological systems. Ideally, these values are activated in the professional community by practitioners' voices and built works. To this end, we argue that a fundamental principle for all design is to *practice an ecological ethic that promotes social and environmental health for the whole land community, human and nonhuman.* The following values are derivatives of this first basic principle:

- *Do no harm.* A basic tenet borrowed from the medical profession as well as many spiritual traditions (Sanskrit word for "non-harming": *avihimsa*), "do no harm" essentially means that a designer's decisions do not detract from whole community health. Problems are defined not only according to cultural, aesthetic, or functional parameters, but also in terms of ecological health. Another version of this tenet is the precautionary principle described by Steingraber (Chapter 8) and Hill et al. (Chapter 12), which shifts requirements for the weight of evidence toward protecting humans and the environment when there is uncertainty about outcomes. Design programs need to teach students to ask the vital questions, Am I doing harm here? and Do I know enough to be relatively confident that I am not causing greater damage?

- *Create wellness.* This credo extends the "do no harm" approach toward the deliberate cultivation of community health. "Do no harm" or the precautionary principle set minimal standards of behavior, while the "create wellness" principle requires the practitioner to actively promote a healthy circumstance in the social and physical environment. Design programs need to teach students to inquire, How can I facilitate whole community health?

- *Consistently question whether one's intervention will, in fact, achieve no harm or move ecological systems toward wellness.* Continual scrutiny, analysis, selfappraisal, and adaptive decision making are necessary to ensure that one's interventions are moving a system toward a healthy state. Design programs need to teach students to constantly question and clarify design choices. Practitioners need to test their designs against the feedback of scientists, citizens, policy makers, and the state of an ecosystem.

- *Reconcile ethics and values with pressures and constraints of private practice.* Practitioners actively balance client's demands and pressures with personal ethics, and are ethically bound to inform clients of the likely ecological impacts of particular actions. At the same time that the role of the designer or planner as a public educator and leader could be elevated, educational programs need to help practitioners develop a market that is ecologically and economically viable.

A Practitioner's View: Carol Franklin

Carol Franklin is cofounder of Andropogon Associates, a prominent landscape architecture firm dedicated to ecologically sound design, and an adjunct faculty at the University of Pennsylvania, Department of Landscape Architecture. Franklin has dedicated a lifetime practice to integrating natural, cultural, and historic concerns in an overall social-ecological design approach. This integrative approach is exemplified in Andropogon's long involvement at Morris Arboretum (Figure 17-2). We asked Franklin her view of what is important in a designer's education and how to connect curriculum with practice.

Franklin says that students of the landscape must understand the patterns and processes of a place. Skilled landscape architects are able to read the landscape and answer the question, What is the problem here? They comprehend the key ecological and cultural processes that determine landscape structure and function over time, and how these might be altered through design interventions. Designers are landscape and cultural interpreters, telling the story of a place, its past, its present, and what it could be. Collaboration with other experts and the community is critical. Designers need to know the limits of their strengths and weaknesses, and when to bring in the right consultants. Franklin points to Allen Savory's holistic management model (Savory and Butterfield 1999), which demonstrates effective dialogue and partnership.

In addition to a broad knowledge base and a fervent commitment to "right thinking," students and professionals who wish to succeed in an ecological practice require fundamental skills. Franklin has found that excellent organizational skills, writing, and technical skills such as grading and knowledge of hydrological processes are particularly important. These practical skills must then be applied to addressing the synergistic effects of multiple impacts in complex, interactive, dynamic landscape systems. The ability to model systems is both quantitatively and qualitatively necessary.

Landscape architects offer creativity, passion, and tenaciousness to solving problems in new ways, as Franklin and her Andropogon partner Leslie Sauer have done in building their successful practice. She believes that enriching people's experience, not depriving them, is the key to changing human behavior toward the environment. Designers need to create places that facilitate a sensual relationship to nature at the same time they heal wounded

FIGURE 17-2.
Morris Arboretum, University of Pennsylvania. Carol Franklin and Leslie
Sauer of Andropogon Associates, Inc., have spent twenty years reinterpreting
the meaning of the arboretum as a public educational experience of cultural
landscapes. In the famous Fernery, young children eagerly fill out an arboretum challenge quiz and search for the fern that looks like a lion. (Photo by
René Senos.)

landscapes. Their work must become both meaningful and wildly imaginative, a true marriage of science and art. Frank Lloyd Wright's Falling Water
(a house catapulted over a stream) is iconic of the potential partnership of
humans and nature. Yet Falling Water is located at the nexus of the highest
concentration of acid rain in the United States, and its stream and surrounding forest are subject to the pressures of pollution, sport hunting, sewage
dumping, and other grievances common to our landscapes. It is not enough
to cast iconic relationships; landscape designers must address our day-to-day
affairs with nature.

A design or planning solution needs to grow out of the place itself; or, as
Franklin poignantly states, a designer's role is to "dismantle the dysfunctional
and save the sacred." A personal value system or value-centered practice is
essential to keep practitioners on the path to this kind of practice. Franklin's
advice to students: "Follow your bliss, and build a world around what you
believe is right. A firm commitment to making the earth whole is what sustains you, and keeping an eye on the legacy you will leave. Respect complex-

ity, and simple, elegant expressions emerge." The ideal Franklin aspires to, and that the work of Andropogon has clearly achieved, is that all great work should be blindingly obvious.

Strategies for Lifelong Learning

Lifelong learning in the design and planning fields is vital to ecologically and socially responsible practice. As the costs of misinformed decisions in terms of environmental accountability grow steeper, practitioners have an ethical obligation to stay current in the field. How can an expanding body of knowledge be funneled into practice at different stages of career growth? Design schools can sow the seeds of ethics and skills, some of which bear fruit immediately, while others lie dormant until years later. One such seed is to inspire people to learn how to learn. Students and practitioners develop new knowledge and skills one project at a time. Knowledge, however, changes constantly, especially with respect to scientific inquiry and public policy, and designers and planners must efficiently and thoughtfully "consume research" and apply it to their proposals.

If educators and practitioners agree that the core of practice includes environmental ethics, skills, and knowledge, what strategies can be used to facilitate their development while recognizing diverse perspectives? As discussed by several authors in this volume, one such strategy is to enhance exchange between the worlds of education and practice. We suggest the following concepts to encourage reciprocal and ongoing learning:

Develop common and innovative approaches for education and practice.

- Designers and educators develop and employ common ecological diagnostic tools, as well as incorporate the latest scientific tools for assessing landscape health.
- Designers and educators employ "gamuts" or ecological checklists, a strategy modeled on the medical profession's system of always asking specific questions in the same order.
- Promote inquiry and experimentation in design studios toward devising creative solutions to pressing classes of social and ecological problems.

Develop support systems that link practice and research.

- Facilitate the mindful consumption and application of research by students and practitioners as a key component of design processes.
- Focus research in ways that are directly relevant to landscape design and planning.
- Direct practice-oriented and ecologically based problems from professional practice to schools and institutions as a basis for research,

including theses problems.

- Establish collaboration between practitioners and researchers to measure ecological condition baselines and monitor change through design.

Build bridges to joint learning between education and practice.

- Provide externships for students and faculty to participate in practice.
- Buy the time of master professionals to participate in educational curricula.
- Involve practitioners and applied ecologists in "vertical studios," including conservation biologists and restoration ecologists.
- Design CD-Rom and Internet-based learning to assist practitioners to keep up with relevant ecological research.

Conclusions

We must be clear that tying ecological theory to private and public landscape practice is the single most important educational outcome that underpins this chapter and, yes, the entire book. It is a useless endeavor for academics to speculate theory only to engage in fascinating but frivolous dialogue with other academics. Landscape planners and designers are *practitioners*—that is, we deal in actions and applications. In the past we have relied on tenets largely founded in principles of aesthetics, cultural interpretations, and perceptions about human behavior and needs. This has served us fairly well in the past, but is insufficient for advancing into the twenty-first century as the leading landscape design profession—the one that considers the local and regional biota whether the site under study is 1 acre or 1 million acres. Academics must engage directly and constantly with practitioners to tie practice to ecological theory and help determine how this theory will become a central part of our profession's foundational thinking. This is, of course, a two-way street. It is equally important that public and private practitioners engage directly and frequently with academics. There must be a passionate and intentional cross-fertilization. It is not enough to train tomorrow's practitioners; academicians must rely on today's practitioners to enrich curricula and use it as a conduit for emerging theory to make its way into practice.

So what would a leadership role in ecological design mean for the landscape professionals? It might mean that we have to remake our profession in fundamental ways. We may have to begin by learning how to design parking lots and gardens as functional and vital parts of regional ecosystems. It will certainly mean hiring non–landscape designers into our firms or sharing already limited fees with ecologists and other scientists so that we can learn from them and they from us. It will require asking fundamental questions about maintaining or enhancing ecological patterns and processes on *all* projects, including seemingly innocuous projects that we continue to reward as "good design." It also means that landscape practitioners interact with ecologically based professional organizations and read peer-reviewed journal articles discussing new scientific findings or theories. We simply must commit to better science grounding and understanding.

Will something called ecological design evolve? Who will become the ecological designers? Not surprisingly, "ecological design" as a marketing tool and identified area of expertise already exists, and landscape professions are only minor players. Enter the phrase into any Web browser and you will get about 300,000 hits. The entries tend to focus on ecologically benign building materials for architectural use, but the vocabulary is increasingly expanding into the language of engineers, interior architects, and so on. The term tends to be used fairly narrowly in its scope and certainly does not meet the expectations of David Orr when he stated that ecological design would be the right scale, be simple, use resources efficiently, and be resilient and durable (Orr 1994). More important, he said it would solve more than one problem at a time. People who understand harmony, patterns, and systems produce good ecological design. Orr went on to call for the development of the "ecological design arts" (p. 102). To some degree this is already occurring in certain curricula, but so far, the essential components of synthesis and collaboration as they relate to conceptual thinking and design remain primarily mastered by the landscape professions.

So how do we capitalize on our strengths and, at the same time, build our capacity to give meaning to scientific analysis or findings? Designers and planners must increase collaboration with colleagues in forestry, planning, wetlands biology, and policy administration. Designers can and should be able to provide interpretation, context, and scale. We must be able to take ecological findings and translate them into landscape form and function. This approach differs from our historical methods of using personal inspiration to give form.

We must also train ourselves to be effective bridges between science and human communities. This entails building skills in active listening, facilitation, and mediation. Designers need to think of synthesis as a skill that can be taught and learned. We should embellish the skill of facilitation with graphical interpretations that are easily understood. We also need to bolster our ability to bring together multiple interests in participatory community design so that our design proposals give expression to both social and ecological values.

Last, we must reenvision the design profession within an ecological context. This means the core of our work should be ecologically literate, no matter what the scale or complexity of the problem. We should not easily accept designs or plans that degrade ecosystems or social systems. We must begin to hold ourselves to a higher standard, and ask whether our designs "reveal the processes which sustain life" (Hough 1995, p. 30). This accountability should be reflected in our professional ethics, publications, juried awards, and, especially, in the social and physical form of built projects. It needs to be reflected by a diversity of voices and ideas rooted in the common value of culturally and ecologically vibrant communities.

Citations

American Society of Landscape Architects. 1997. Map of the territory: survey data on the size, scope, and direction of landscape architecture practice. American Society of Landscape Architects, Washington, D.C., USA.

————. 1998. National salary survey of landscape architects. American Society of Landscape Architects, Washington, D.C., USA.

Apostol, D., M. Sinclair, and B. R. Johnson. 1998. Design your own watershed: top-down meets bottom up in the Applegate Valley. Pages 106–108 in Proceedings of ASLA Annual Meeting, October 1998. American Society of Landscape Architects, Portland, Oregon, USA.

Hester, R. T. 1990. Community design primer. Ridge Times Press, Mendocino, California, USA.

Howett, C. 1998. Ecological values in twentieth-century landscape design: a history and hermeneutics. Landscape Journal, Special Issue: Eco-revelatory design: nature constructed/nature revealed. pp. 80–98.

Hulse, D., J. Eilers, K. Freemark, D. White, and C. Hummon. 2000. Planning alternative future landscapes in Oregon: evaluating effects on water quality and biodiversity. Landscape Journal 19: 1–19.

Johnson, B. R., and R. Campbell. 1999. Ecology and participation in landscape-based planning within the Pacific Northwest. Policy Studies Journal 27: 503–529.

Karr, J. R., and E. W. Chu. 1999. Restoring life in running waters: Better biological monitoring. Island Press, Washington, D.C., USA.

Lichatowich, J. 1999. Salmon without rivers: a history of the Pacific salmon crisis. Island Press, Washington, D.C., USA.

McHarg, I. 1992. Design with nature. John Wiley and Sons, New York. Originally published 1969 by the Natural History Press, Garden City, New York, New York, USA.

Nassauer, J. I. 1997. Placing nature: culture and landscape ecology. Island Press, Washington, D.C., USA.

Orr, D. 1994. Earth in mind: on education, environment, and the human prospect. Island Press, Washington, D.C., USA.

Savory, A., and J. Butterfield. 1999. Holistic management: a new framework for decision making. 2nd edition. Island Press, Washington, D.C., USA.

Scach, J. C. 2000. Genetically modified plants need closer scrutiny. Landscape Architecture News Digest (LAND) 42: 3–5.

Seattle Public Utilities. <http://www.ci.seattle.wa.us/util>. October 29, 2000.

Spirn, A. W. 1984. The granite garden: urban nature and human design. Basic Books, Inc., New York, New York, USA.

Steingraber, S. 1997. Living downstream: an ecologist looks at cancer and the environment. Addison-Wesley Publishing, Reading, Massachusetts, USA.

Steinitz, C. Fall 1995. Design is a verb; design is a noun. Landscape Journal 14: 188–200.

Welsh, B. 1999. The ties that bind: ASLA leaders meet with federal officials to discuss partnerships with agencies. Landscape Architecture News Digest (LAND) 7: 1–2.

———. April 2000. ASLA is chosen as key partner on White House livable communities program. Landscape Architecture New Digest (LAND): 42: 1–2.

CHAPTER 18

Conclusions: Frameworks for Learning

Kristina Hill and Bart R. Johnson

Our intention in this book was to offer students, practitioners, and theorists a contemporary perspective on the potential role of ecological thinking in design education and, to a lesser degree, on the role of design thinking in environmental education. We purposely brought a range of perspectives together in a "quilt" of ideas and teaching strategies, without expecting or asking for consensus. Our hope is that the multiple voices included may provoke a diversity of thought, dialogue, and debate within and across the disciplines of design and ecology.

In this final chapter we reflect on the perspectives contained in this volume of work, marking out areas of common ground while also surveying the issues that may be points of contention. As we see them, these areas of common ground and points of possible disagreement form a critical foundation for future dialogue among designers and ecologists. New ideas may emerge from each of the areas and points noted here that will allow us to map the issues this book has addressed, placing them in a new framework. We look forward to those future dialogues, even as we retrace the lines of thinking that our authors used to frame today's issues.

Common Ground, Creative Tension

Most of the authors who were invited to participate in the 1998 Shire Conference and contribute to this book were teaching ecology-related courses in university programs at the time. For that reason, it is not surprising to find that our authors have many ideas, perceptions, and values in common. They do not represent a cross section of opinion in the design disciplines, because our intention was to articulate the ideas and approaches that are currently being used to bring ecological knowledge into design education. These shared ideas and approaches can also help to establish a sense of intellectual community for those educators who believe that ecological knowledge could or should inform design theory and practice more it does today. The Shire Conference brought those teachers together, and we hope that this book will both enlarge that group and provide it with a foundation for future work.

In an important way, the goals for teaching and learning that are articulated in this book are also a signal that times have changed. In the past, there may have been advocates of ecological knowledge who implied that a knowledge of human cultures was a less important part of design theory and practice than a knowledge of natural systems. Our authors did not express that position. If we look at what they've written, we find broad agreement that a dialectical, reciprocal approach is needed to understand the relationships between cultural and natural processes. In the next few pages, we trace that dialectic over the sections of this book.

Theories of Nature in Ecology and Design

In the first section of this book, Spirn, Pulliam and Johnson, and Forman agreed that our ideas about nature are shifting. In some cases these changes are happening slowly, so that it can be difficult to track them and note their significance. Spirn noted that the opportunity to create professional and academic dialogues around this process of shifting ideas often founders, and that it does so particularly when we abandon a spirit of open inquiry and instead advance dogmatic or polemical arguments. She recommends that we avoid seeking singular definitions of nature as an authority for our work. But Forman calls for more explicit use of theory in design, from many fields—ecology, sociology, economics, and others—so that we can explicitly test our ideas and learn from our experiments. This concept of replicable testing as a source of authority is fundamental to the scientific method, and Pulliam and Johnson point out that it has lead to theoretical shifts in ecology. Can explicit testing of theories and hypotheses be adopted as part of our design methods, or will shifts in design theory be driven more by intellectual trends and ethical positions than by the addition of conditional knowledge that results from testing?

The idea of design as experiment dates back at least to the 1920s (Jaggard 1921), but consistent methods for evaluating the performance of designed spaces did not evolve in the decades that followed. Individual designers still conduct tests of new materials and technologies, but the results are often not widely disseminated, or they are not accepted by practicing professionals (Ferguson 2000). Designs for landscapes often involve complex programs, and this complexity makes it difficult to isolate individual interactions among the many elements of a design, the behavior of its users over time, or its construction process. It is likely that if designs are treated as experiments, then the tests of designs will either need to be very specific, focused on the performance of single elements or materials, or will have to take a more qualitative approach using case studies. In any case, it will not be a simple matter to build theory in design with rigorous, replicable tests.

It may be especially difficult to avoid the trap of arguing from authority if we cannot argue from the evidence produced by replicable tests. Perhaps we need to reconsider whether we might argue from multiple rather than singular authorities, drawing on the dialogic model established in cultural studies and feminist theory (see, for example, Spretnak 1997, Meyer 1997, and Bakhtin 1981). If we take this approach, our authors have pointed out that we would still need to maintain an open spirit of inquiry (Chapter 2) and identify at least a modest number of testable questions related to the performance of designed spaces (Chapter 4).

Perspectives on Theory and Practice

In this section of the book, our authors identified several "hot spots" where we need to pay special attention to the relationships between theory and practice. Jim Karr stressed the importance of sorting out the various uses of the term "ecology," which is sometimes intended to refer to a science, sometimes to a life philosophy or ethic, and sometimes to goal-oriented studies that guide the management of a biological commodity. In several places in this book, our authors seem to disagree about whether ecology is indeed only a science, or whether the same term should be used to refer both to a science and to a life philosophy.

This is a significant difference because we may not understand each other's educational goals if we do not clarify our use of the term "ecology" in design education. As Adams (Chapter 5) pointed out, we should help students sort out the difference between knowledge produced using the scientific method and ideologies that sometimes pass as knowledge because they promote particular truisms. For example, most of our authors might agree that ethically, humans must seek a new balance between satisfying their needs and allowing other species to meet their own needs. But current theories of ecology do not support the idea that the dynamic processes that shape the relationships among species display an *intrinsic* balance (Chapter 3). We need to clarify the root concepts in our arguments by clearly defining the version of ecology that we think should play a stronger role in future design curricula. Perhaps the answer for some of us will be that we need to include ecology as both a life philosophy and a science, but that answer must be clear. This clarity is critical to our ability to communicate with colleagues in the life sciences, since they will typically expect that our use of the term "ecology" refers to the science, not the philosophy.

Health is a concept that has similar problems of definition (and opportunities for redefinition). Steingraber and Hill both discuss human health in the context of ecological relationships, while Karr uses the term "health" to refer to an ecological condition that can be measured in the environment. As our ideas on these subjects develop, will we be able to think about our own health, in all its dimensions, as dependent on environmental conditions that include the health of other species? Will the science of ecology come to consider human interactions with the environment using the same theories that are used to explain the interactions of other species with their biological and physical environments? Human health and environmental health must eventually be seen as inseparable issues, if we are successful at bringing our studies of humans and our studies of the nonhuman world into an inclusive science of ecology.

The advantages of redefining infrastructure were also raised in this section, both by Wenk (Chapter 7) and by Hill (Chapter 8). Hill referred to a "restorative" infrastructure that supports human health, while Wenk noted opportunities for new aesthetic and ecological approaches to the design of infrastructure that provides water to human communities and treats polluted water. In addition, the concept of an ecological infrastructure of habitat networks that support biodiversity has been presented by others, including Forman (1990). Similarly, Meyer (1997) re-framed the significance of Olmsted's designs for park systems as a kind of infrastructure that serves both cultural and ecological needs. As we consider the design challenges this implies for current practice, described by Adams (Chapter 5) and Senos et al. (Chapter 17), we may well find that the opportunity to redefine infrastructure allows us to build a bridge from our discipline's past to its future. As Wenk notes, contemporary designers may underestimate the opportunities inherent in this perspective because they are typically focused on identifying design opportunities at the site scale.

Education for Practice

The authors of this section of the book were in agreement that design is a cultural practice, and cannot be simply an outgrowth of science or the scientific method. Knowledge of scientific evidence and theories is necessary to construct a desirable relationship between cultural and natural processes, but that knowledge is not a sufficient condition for good design. Steinitz (Chapter 10) was very clear in his opinion that ecological knowledge plays an important and distinct role in design, yet that role is neither unique nor necessarily dominant. Other impacts and processes may be equally important or more so. Although Steinitz does not offer a method for determining which issues are the most important, Hough (Chapter 11) suggests that an ethical framework based on maintaining the integrity of "place" may be the key. All three authors—Steinitz, Hough, and Nassauer—explicitly identified some type of cultural framework as a prerequisite for applying ecological knowledge in design.

Each author addressed the questions of how ecological knowledge enters into the design process and into design education. For Nassauer (Chapter 9), this can be accomplished by establishing a relationship of mutual respect and consultation between ecologists and designers. Steinitz added a set of hierarchical principles that could guide educators in setting expectations for the necessary depth of student skills at different levels in the curriculum, and a theoretical framework that provides a place for knowledge about natural processes and the impacts of change. Both Nassauer and Hough emphasized the importance of grounding ecological knowledge in the holistic qualities of specific places, including their historic processes, uses, and meanings as well as contemporary observations of their qualities. All of these authors appeared to agree that an understanding of ecological relationships and human impacts could be taught by introducing the skills of observation and an ethic of accountability early in the curriculum.

Clearly, some kind of philosophical or ethical framework is necessary in order to determine whether the impacts humans have on other forms of life are acceptable, and to weigh those impacts against others. Although Hough clearly stresses the need to teach ethics, the essays in this section leave the specifications for an ethical framework to the instructor and, ultimately, to the designer.

Prescriptions for Change

The group-authored essays that comprise this section of the book identified shortcomings in the current state of design education, and at the same time provided enough successful examples of alternative approaches to suggest that those shortcomings are already being addressed. The essays also provided different answers to the question of how design and ecology are related. Several seemed to take the position that designers and ecologists are mutually exclusive identities (Tamminga et al. Chapter 14, Ahern et al. Chapter 15, Poole et al. Chapter 16, Senos et al. Chapter 17). This could be interpreted to mean that students should probably not be encouraged to identify with both of those disciplinary titles. In contrast, two groups left the possibility open that designers might also have advanced education in ecology, and even lead or participate in traditional ecological research, while people who are trained as ecologists might practice design (Hill et al. Chapter 12 and John-

son et al. Chapter 13). Can one person belong to both disciplines successfully? While some of the authors seemed skeptical that this can be done, we
wonder what the implications of an affirmative answer would be for the relationship between design and science. Perhaps Steinitz' model of advanced
education in design, where studies in ecological science and studies of design
theories and methods are both acceptable topics, offers an insight here. Successfully mixing the identities of ecologist and designer in one person may
require education in both fields at least at the master's level, and perhaps success is even more likely at the Ph.D. level.

The group-authored chapters consistently identified a need for explicit
models of design impacts, and for equally explicit decision criteria to be used
when making choices in the design process. This curriculum emphasis
provides an alternative model to a design education that accepts implicit
decision-making and intuitive processes as sufficient for good design. Will
this be the way of the future? What role will design students and faculty play
in developing these impact models and decision criteria? How will faculty
resolve the conflicts created within the design disciplines when intuitive
processes are associated with the fine arts, while explicit decision making is
associated with engineering and science? Will we simply say that both methods of reasoning should be employed, or might we say that intuitive processes
are not sufficient by themselves to produce "good" designs? The latter statement seems to be in keeping with the positions our authors have taken, but
it is likely to produce tension within the discipline of design. Perhaps this is
the tension that has lead to debates about the relative importance of science
and art in landscape architecture. The question of whether design should be
informed by theories of art or theories of science has been rejected by our
authors, perhaps because it is less meaningful to them than the question of
whether it is ever appropriate to practice design using the theories of art
alone.

On Schools of Thought in Design

Throughout our work on this book, we hoped that our efforts might lead to
a modest redefinition of the relationship between designers' conceptions of
"humans" and "nature." Perhaps that ambition requires insights beyond the
current state of our reasoning. But after reviewing the chapters contained in
this book, there may be differences in perspective that we can identify. Identifying and naming those differences could be the keys to future dialogues
among designers and ecologists, or among designers with different values.
Similarly, if we could name and map the different choices we make in designing courses and curricula, we might be able to compare and reevaluate those
choices from a new perspective. These different choices might be thought of

TABLE 18-1. A matrix of differences among "schools of thought"

	Option 1	*Option 2*	*Option 3*
I. Defining "ecology"	Science	Life philosophy	Both
II. Defining the identities of designers and ecologists	Singular	Multiple	Transdisciplinary
III. Balancing cultural values and an awareness of ecological consequences	Clients' values frame issues	Professional ethics, including an awareness of consequences, frame issues	Designer's personal values frame issues
IV. Bridging theory and practice	Theory leads	Practice leads	Dialogue informs

as constituting different "schools of thought" about design education, particularly with regard to the role of ecology.

We wondered whether the writings assembled here suggest a typology of ideas that could allow us to distinguish such schools of thought (Table 18-1). Each of the items shown with Roman numerals is an area in which we think educators within the design disciplines might disagree. The "Options" columns suggest alternative emphases that a particular person or a particular educational program might choose. Granted, the options are caricatures. But if we consider their broader implications, we might find new ways to characterize the pedagogical and philosophical positions we take as individuals and as departments.

For example, in defining ecology an individual or a program might determine that this word refers to both a life philosophy and a science (Option I-3). This person or program might also determine that designers should be identified as designers only (and not wear more than one disciplinary "hat" in collaborations) (Option II-1). Finally, the same person or program might decide that, in his or her approach, the designer's personal values should be treated as the primary "frame" for determining whether a particular design is appropriate (Option III-3). This hypothetical individual or school could be said to adhere to a different school of thought about ecology in design than a person or educational program that makes a different set of choices among the options listed in Table 18-1. Although we have suggested that these choices might be conscious, it is just as likely that they may be unconscious selections that are reinforced through subtle pressures and cultures within a program or a school. Identifying schools of thought could empower both individuals and programs to make conscious choices about where they stand.

We consider it likely that individuals do make these kinds of choices, sometimes contingent on, or influenced by, the institutional or professional contexts in which they find themselves. It may be within the scope of some educational programs to try to find common ground on these issues. It may even be that groups of like-minded people or departments will distinguish themselves by forming a more complete educational philosophy using some of the options listed above, as designers of a different era did by forming the Bauhaus school in Germany, or the Ecole de Beaux Arts in Paris. Will there be a "Seattle School" of landscape architecture, for example, or a Pacific Northwest "block"? A new Penn or Harvard model? An "Urban School," with an emphasis on infrastructure design? We look forward to future dialogues in which ideas about design and applied ecology might unfold in our educational programs, leading to new insights and new relationships. It might be that the graduates of new "hybrid" educational programs could begin to heal fragmented landscapes by healing the fragmented disciplines that exist in university settings (Chapters 14 and 12).

In Closing

When Olmsted wrote that landscape architecture was an art informed by the sciences, he may or may not have anticipated an era in which some designers would feel they had to choose sides between art and science, between "rational analysis" and "intuitive inspiration." We would like to think that the chapters in this book might help to eliminate this apparent dichotomy, either through promoting well-informed and mutually respectful collaboration among designers and ecologists, or by offering opportunities to completely reframe the issues. This is not something we see as a purely academic dilemma.

In the eyes of many of the people who wrote this book, we live in a critical time. A time when humanity's choices, perhaps over the next several decades, have the potential to preserve or eliminate much of Earth's cultural and biological diversity. In the first chapter of this book, we proposed something we called "landscape realism." Whether that particular term catches hold is irrelevant. It is a marker that points to a particular ethical and pragmatic starting point for inquiry in the design fields. We have proposed landscape realism as a philosophy that requires an honest appraisal of human relationships with other forms of life. This does not require the elevation of other species' needs above human needs. But it does require a careful mapping of the relationships among humans and among species, and emphasizes the importance of choosing to act ethically within those relationships. Perhaps the greatest challenge of bringing ecology and design into a closer working relationship is that it forces us to confront the question of whether humans can successfully share their current living environments with the other forms of life that evolved in those places. We look forward to future discussions with our colleagues in design and in the natural sciences on this question, and to the development and testing of new prototypes for built space and managed places that may allow us to give form to the diverse and dynamic garden that human-designed landscapes could become.

Finally, we would like to thank the many authors who contributed to this book for sharing their insights and talents with us in this project. We hope that other practitioners, students, and faculty will find these authors' ideas as engaging as we ourselves have found them and that through a reexamination of some of our assumptions, the polarization of art and science within the design fields might end in the early years of the twenty-first century. A revitalized era of creative collaboration between ecology and design could be the result.

Citations

Bakhtin, M. 1981. The dialogic imagination: four essays. University of Texas Press, Austin, Texas, USA.

Ferguson, B. 2000. Research in landscape architecture. Panel discussion, American Society of Landscape Architects, Annual Exposition, St. Louis, Missouri, USA.

Forman, R. T. T 1990. Ecologically sustainable landscapes: the role of spatial configuration. In I. Zonneveld and R. Forman, editors. Changing landscapes: An ecological perspective. Springer: New York, New York, USA.

Jaggard, W. R. 1921. Experimental cottages, a report on the work of the Department at Amesbury, Wiltshire. Dept. of Scientific and Industrial Research, London, H.M. Stationery Office, UK.

Meyer, E. 1997. The expanded field of landscape architecture. Pages 45–79 in G. Thompson and F. Steiner, editors. Ecological design and planning. John Wiley and Sons, New York, New York, USA.

Spretnak, C. 1997. The resurgence of the real: body, nature, and place in a hypermodern world. Addison-Wesley, Reading, Massachusetts, USA.

Notes on Primary Authors

Carolyn A. Adams, Director, Watershed Science Institute, North Carolina State University

Carolyn A. Adams is the Director of the USDA Natural Resources Conservation Service's Watershed Science Institute with headquarters in North Carolina. Adams has a B.L.A. from Mississippi State University, an M.S. from Texas Tech University, and was a Loeb Fellow in Advanced Environmental Studies at Harvard University. Her twenty-five-plus years in landscape architecture include work with a major southeastern engineering consulting firm in Chattanooga, Tennessee. She served nine years with the American Society of Landscape Architect's Roster of Visiting Evaluators and the Landscape Architecture Accreditation Board.

Jack Ahern, University of Massachusetts

Jack Ahern is a Professor of Landscape Architecture and Head of the Department of Landscape Architecture and Regional Planning at the University of Massachusetts. He holds a B.S. in Environmental Design from the University of Massachusetts and an M.L.A. from the University of Pennsylvania. He co-edited *Greenways: The Beginning of an International Movement* (1995) with Julius Fabos, and *Environmental Challenges in an Expanding Urban World and the Role of Emerging Information Technologies* (1997) with João Reis Machado. He is a fellow of the ASLA, and is past chair of the U.S. Division of the International Association for Landscape Ecology.

Richard T. T. Forman, Graduate School of Design, Harvard University

Dr. Forman is Professor of Advanced Environmental Studies in Landscape Ecology at Harvard University. He received a B.S. from Haverford College, and a Ph.D. from the University of Pennsylvania. He is an ecologist best known for his groundbreaking approaches in the development of landscape ecology, and recent work in developing road system ecology. He is a Fellow of the American Association for the Advancement of Science, and was named Distinguished Landscape Ecologist by the International Association of Landscape Ecology (USA). He is a past vice president of the Ecological Society of America. His major books include *Landscape Ecology* (1986) with Michel Godron, and *Land Mosaics* (1995).

Michael Hough, Faculty of Environmental Studies, York University

Michael Hough is a principal and founding partner in the landscape architecture firm of Hough Woodland Naylor Dance Leinster Limited in Toronto, and is a Professor at the Faculty of Environmental Studies at York University in Toronto. He holds a Diploma in Architecture from Edinburgh College of Art, Scotland, and an M.L.A. from the University of Pennsylvania. He was named a Distinguished Practitioner by the International Society for Landscape Ecology, and received the ASLA's Bradford Williams Medal for journalistic excellence 1989. His books include *Cities and Natural Processes* (1995) and *Out of Place* (1990). He is a past president of the Canadian Society of Landscape Architects.

Kristina Hill, Department of Landscape Architecture, University of Washington

Kristina Hill is an Associate Professor of Landscape Architecture and Director of the Graduate Program in Landscape Architecture at the University of Washington. She holds a B.S. in Geology from Tufts University, an M.L.A. and a Ph.D. in Landscape Architecture with a minor in Ecology from Harvard University. She has taught urban design at the Massachusetts Institute of Technology and was a Fulbright Scholar at Stockholm University in Sweden. Hill specializes in understanding the landscape ecology of urban areas, and incorporates ecological processes in her urban design work. She has served on the National Awards jury in Urban Design for the ASLA.

Bart R. Johnson, Department of Landscape Architecture, University of Oregon

Bart Johnson is an Associate Professor of Landscape Architecture at the University of Oregon. He holds a B.S. in Agronomy from Cornell University, and an M.L.A. in Landscape Architecture and a Ph.D. in Ecology from the University of Georgia. His primary focus in teaching, research, and practice is the integration of ecology with landscape design, planning, and management. Current projects include participatory landscape planning and biodiversity conservation in the Siskiyou Bioregion of southern Oregon and northern California, revitalizing urban ecosystems, restoration of Willamette Valley oak savanna in and around urban areas, and investigations of rock outcrops as evolutionary cradles and museums.

James R. Karr, University of Washington

James R. Karr is a Professor of Aquatic Science and Zoology and Adjunct Professor of Civil Engineering, Environmental Health, and Public Affairs at the University of Washington, Seattle. He holds a B.S. in Fish and Wildlife Biology from Iowa State University and an M.S. and a Ph.D. in Zoology from the University of Illinois, Urbana-Champaign. Karr is a Fellow in the American Association for the Advancement of Science and the American Ornithologists' Union. He has served on the editorial boards of *BioScience, Conservation Biology, Ecological Applications, Ecological Monographs, Ecology, Ecosystem Health, Freshwater Biology, Tropical Ecology, and Ecological Indicators.* He is author, with E. W. Chu, of *Restoring Life in Running Waters* (1999).

Joan Iverson Nassauer, School of Natural Resources and Environment, University of Michigan

Joan Iverson Nassauer, FASLA, is Professor of Landscape Architecture in the graduate program of the School of Natural Resources and Environment at the University of Michigan. She holds a B.L.A. from the University of Minnesota and an M.L.A. from Iowa State University. Her work has focused on understanding what local people value about their landscapes and using their values in ecological design and planning, watershed management, and urban landscape ecology. She is both editor and an author in the book *Placing Nature: Culture and Landscape Ecology* (1997).

David W. Orr, Chair, Environmental Studies Program, Oberlin College

David Orr is Professor of Environmental Studies at Oberlin College. He has a B.A. from Westminster College, an M.A. from Michigan State University, and a Ph.D. from the University of Pennsylvania. He is the author of many books and articles on the need to bring ecological understanding into education, including *Ecological Literacy: Education and the Transition to a Postmodern World* (1992) and *Earth in Mind: On Education, Environment, and the Human Prospect* (1994).

Kathy Poole, Principal, POOLE DESIGN, Landscape Architecture & Urban Design & Ecologically-Healthy Infrastructure, Charlottesville, VA

Kathy Poole is a practicing landscape architect who continues to guest teach, having taught for more than seven years at the University of Virginia and the University of Oregon. She holds a B.A. in Architecture from Clemson University and an M.L.A. from Harvard University. Her professional practice and her primary research interest concern what she calls Civic Hydrology, the possibilities for ecology, particularly water, to build better cities and to reinvigorate civic life. Multidisciplinary in nature, her work is a blend of technical, theoretical, and historical content. Her principal Civic Hydrology research venues include the mapping of what she terms "wetlands" and a historical urban design study of Boston's Back Bay Fens.

H. Ronald Pulliam, Institute of Ecology, University of Georgia

H. Ronald Pulliam is Regent's Professor of Ecology at the University of Georgia. He has a B.S. in Zoology from the University of Georgia and a Ph.D. in Zoology from Duke University. His major research contributions have been in the areas of biological diversity, community ecology, behavioral ecology, source-sink theory, and population dynamics in heterogeneous landscapes. Professor Pulliam has also served as president of the Ecological Society of America, Science Advisor to the United States Secretary of Interior, and director of the National Biological Service.

René Senos, Jones and Jones Architects and Landscape Architects, Seattle, Washington

René Senos is a landscape designer and planner with Jones and Jones Architects and Landscape Architects, an international practice committed to the integration of built environments and the conservation of natural and cultural resources. She holds an M.L.A. and a B.L.A. from the University of Oregon, and a B.S. in Sociology from the University of Delaware. Senos has worked extensively as a social advocate for families and children. Her current research focuses on the broader form of cultural-ecological restoration as a vehicle for healing communities and reconnecting people with place.

Anne Whiston Spirn, Department of Urban Studies and Planning, Massachusetts Institute of Technology

Professor Spirn is a landscape architect and author of *The Granite Garden: Urban Nature and Human Design* (1984) and *The Language of Landscape* (1998). She holds an A.B. from Radcliffe College and an M.L.A. from the University of Pennsylvania. Spirn practiced with Wallace McHarg Roberts and Todd before becoming a faculty member at Harvard's Graduate School of Design, where she served as Director of the Landscape Architecture Program, and is a past Chair of the Landscape Architecture Program at the University of Pennsylvania. She is now a Professor of Urban Design and Planning at MIT.

Sandra Steingraber, Somerville, Massachusetts

Dr. Steingraber holds a Ph.D. in Biology from the University of Michigan. She has worked as an ecologist, a teacher, and a poet, and recently wrote the acclaimed book *Living Downstream: An Ecologist Looks at Cancer and the*

Environment (1997). She has held fellowships at Radcliffe College, the University of Illinois, and Northeastern University. She was appointed to the National Action Plan on Breast Cancer, administered by the U.S. Department of Health and Human Services.

Carl Steinitz, Graduate School of Design, Harvard University

Carl Steinitz is the Alexander and Victoria Wiley Professor of Landscape Architecture and Planning at Harvard's Graduate School of Design. He received a Ph.D. in city and regional planning from MIT and holds an M.Arch. from MIT and a B.Arch. from Cornell University. Steinitz has devoted much of his career to improving methods by which planners and designers analyze information about large land areas and make decisions about conservation and development. The Council of Educators in Landscape Architecture (CELA) has named him an Outstanding Educator, and he received the Outstanding Practitioner Award from the International Society of Landscape Ecology (USA).

Ken Tamminga, Department of Landscape Architecture, Penn State University

Ken Tamminga is Associate Professor of Landscape Architecture at Penn State University. He holds a B.L.A. from the University of Guelph and a Master of Urban and Regional Planning from Queen's University, Canada. He specializes in interdisciplinary approaches to ecological restoration and design. He is a member of Penn State's InterCollege Graduate Ecology program and is Faculty Fellow with the newly established Center for Watershed Stewardship. Prior to joining Penn State, Ken was a practitioner in Toronto, where he managed waterfront, wildland, and greenway projects throughout the province.

William E. Wenk, Wenk Associates, Denver, Colorado

William Wenk, FASLA, is founder and president of Wenk Associates, Inc., a Denver-based landscape architecture firm. Their work creates urban landscapes that reinvent the city's infrastructure while making civic space. Wenk holds a B.S.L.A. and an M.L.A. from Michigan State University and the University of Oregon, respectively. He is Adjunct Associate Professor of Landscape Architecture at the University of Colorado in Denver, and is a Fellow of the American Society of Landscape Architects.

List of Contributors

CAROLYN A. ADAMS, Director, USDA-NRCS Watershed Science Institute, Raleigh, North Carolina.

JACK AHERN, Associate Professor, Department of Landscape Architecture and Regional Planning, University of Massachusetts, Amherst, Massachusetts.

DEAN APOSTOL, Landscape Architect, Portland, Oregon.

JULIA BADENHOPE, Associate Professor, Department of Landscape Architecture, Iowa State University, Ames, Iowa.

JON BURLEY, Assistant Professor, Landscape Architecture Program, Department of Geography, Michigan State University, East Lansing, Michigan.

DONNA ERICKSON, Associate Professor and Chair, Landscape Architecture Program, School of Natural Resources and Environment, University of Michigan, Ann Arbor, Michigan.

ED FIFE, FCSLA, Associate Professor, Program in Landscape Architecture, Faculty of Architecture, Landscape and Design, University of Toronto, Toronto, Ontario, Canada.

RICHARD T. T. FORMAN, Professor, Graduate School of Design, Harvard University, Cambridge, Massachusetts.

ROBERT FRANCE, Assistant Professor, Graduate School of Design, Harvard University, Cambridge, Massachusetts.

KATHRYN FREEMARK, Research Ecologist, National Wildlife Research Centre, Environment Canada, Ottawa, Ontario, Canada.

SUSAN GALATOWITSCH, Associate Professor, Departments of Horticultural Science and Landscape Architecture, University of Minnesota, St. Paul, Minnesota.

ROBERT E. GRESE, Director, Nichols Arboretum, and Associate Professor, School of Natural Resources and Environment, University of Michigan, Ann Arbor, Michigan.

JOHN HARRINGTON, Professor and Chair, Department of Landscape Architecture, University of Wisconsin-Madison, Madison, Wisconsin.

JURGEN HESS, Director, Columbia Gorge National Scenic Area, USDA Forest Service, Hood River, Oregon.

KRISTINA E. HILL, Associate Professor, Department of Landscape Architecture, University of Washington, Seattle, Washington.

MARK HOSTETLER, Assistant Professor, Extension Wildlife Specialist, Department of Wildlife Ecology and Conservation, University of Florida, Gainesville, Florida.

MICHAEL HOUGH, Principal, Hough, Woodland, Naylor, Dance, Leinster, Ltd., Toronto, Ontario, Canada.

DAVID HULSE, Professor, Department of Landscape Architecture, University of Oregon, Eugene, Oregon.

MARYCAROL ROSSITER HUNTER, Adjunct Faculty, Institute of Ecology, University of Georgia, Athens, Georgia.

BART R. JOHNSON, Assistant Professor, Department of Landscape Architecture, University of Oregon, Eugene, Oregon.

DOUG JOHNSTON, Associate Professor, Department of Landscape Architecture, University of Illinois at Urbana-Champaign, Champaign, Illinois.

GRANT JONES, Principal, Jones and Jones Architects and Landscape Architects, Seattle, Washington.

JAMES R. KARR, Professor, Department of Zoology, University of Washington, Seattle, Washington.

J. TIMOTHY KELLER, FASLA, Professor and Chair, Department of Landscape Architecture, College of Design, Iowa State University, Ames, Iowa.

MIRANDA MAUPIN, Senior Planner, Seattle Public Utilities, Seattle, Washington.

ROBERT Z. MELNICK, FASLA, Dean, School of Architecture and Allied Arts, University of Oregon, Eugene, Oregon.

APRIL MILLS, Graduate Student, Department of Landscape Architecture, University of Washington, Seattle, Washington.

LOUISE MOZINGO, Associate Professor, School of Landscape Architecture and Environmental Planning, University of California-Berkeley, Berkeley, California.

JOAN IVERSON NASSAUER, Professor, School of Natural Resources and Environment, University of Michigan, Ann Arbor, Michigan.

FORSTER NDUBISI, Professor of Landscape Architecture and Regional Planning, Director, Interdisciplinary Design Institute, Washington State University, Spokane, Washington.

DAVID W. ORR, Professor and Chair, Department of Environmental Studies, Oberlin College, Oberlin, Ohio.

KATHY POOLE, Principal, POOLE DESIGN, Landscape Architecture & Urban Design & Ecologically-Healthy Infrastructure.

H. RONALD PULLIAM, Professor, Institute of Ecology, University of Georgia, Athens, Georgia.

DAVID RICHEY, MLA Candidate, Department of Landscape Architecture, University of Oregon, Eugene, Oregon.

BARBARA RYDER, Planner, Facilities Operations, Washington State University, Pullman, Washington.

SALLY SCHAUMAN, Adjunct Professor, Department of Women's Studies, Duke University, Durham, North Carolina.

STEPHAN SCHMIDT, Graduate Student, Department of Landscape Architecture, University of Washington, Seattle, Washington.

RENÉ SENOS, Landscape Designer and Jones and Jones Architects and Landscape Architects Planner, Seattle, Washington.

JANET SILBERNAGEL, Assistant Professor, Department of Landscape Architecture, University of Wisconsin, Madison, Wisconsin.

LEE R. SKABELUND, Assistant Research Professor of Landscape Architecture, College of Architecture and Urban Studies, Virginia Polytechnic Institute & State University, Blacksburg, Virginia.

ANNE WHISTON SPIRN, Professor, Department of Urban Design and Planning, Massachusetts Institute of Technology, Cambridge, Massachusetts.

SANDRA STEINGRABER, Somerville, Massachusetts.

CARL STEINITZ, Professor, Graduate School of Design, Harvard University, Cambridge, Massachusetts.

KEN TAMMINGA, Associate Research Professor, Department of Landscape Architecture, Penn State University, University Park, Pennsylvania.

REBECCA TAYLOR, Planner, URS Greiner, Seattle, Washington.

WOOD TURNER, Graduate Student, Department of Landscape Architecture, University of Washington, Seattle, Washington.

WILLIAM E. WENK, Wenk Associates, Denver, Colorado.

DENIS WHITE, Geographer, National Health and Environmental Effects Research Laboratory, U.S. Environmental Protection Agency, Corvallis, Oregon.

JOAN WOODWARD, Professor, Department of Landscape Architecture, California State Polytechnic University, Pomona, California.

Index

Island Press Board of Directors

Chair, HENRY REATH
President, Collector's Reprints, Inc.

Vice-Chair, VICTOR M. SHER
Environmental Lawyer

Secretary, DANE NICHOLS
Chairman, The Natural Step, U.S.

Treasurer, DRUMMOND PIKE
President, The Tides Foundation

WILLIAM M. BACKER
Backer, Spielvogel, Bates (ret.)

ROBERT E. BAENSCH
Professor of Publishing, New York University

MABEL H. CABOT

DAVID C. COLE
Sunnyside Farms, LLC

CATHERINE M. CONOVER

GENE E. LIKENS
Director, The Institute of Ecosystem Studies

CAROLYN PEACHEY
Campbell Peachey & Associates

WILL ROGERS
Trust for Public Lands

CHARLES C. SAVITT
President, Center for Resource Economics/Island Press

SUSAN E. SECHLER
Director of Global Programs, The Rockefeller Foundation

PETER R. STEIN
Managing Partner, The Lyme Timber Company

RICHARD TRUDELL
Executive Director, American Indian Resources Institute

WREN WIRTH
President, The Winslow Foundation